HAZLITT: A LIFE

William Hazlitt, *c.*1802, self-portrait (prior to restoration).
Maidstone Museums and Art Galleries.

HAZLITT: A LIFE

From Winterslow
to Frith Street

STANLEY JONES

CLARENDON PRESS · OXFORD
1989

Oxford University Press, Walton Street, Oxford OX2 6DP
Oxford New York Toronto
Delhi Bombay Calcutta Madras Karachi
Petaling Jaya Singapore Hong Kong Tokyo
Nairobi Dar es Salaam Cape Town
Melbourne Auckland
and associated companies in
Berlin Ibadan

Oxford is a trade mark of Oxford University Press

Published in the United States
by Oxford University Press, New York

British Library Cataloguing in Publication Data
Jones, Stanley, 1916-
Hazlitt : a life : from Winterslow to Frith Street.
1. Essays in English. Hazlitt, William. Biographies
I. Title
824'.7
ISBN 0-19-812840-1

Library of Congress Cataloging-in-Publication Data
Jones, Stanley, 1916-
Hazlitt: a life from Winterslow to Frith Street / Stanley Jones.
p. cm.
Includes bibliographies and index.
1. Hazlitt, William, 1778-1830—Biography. 2. Authors,
English—19th century—Biography. I. Title.
PR4773.J66 1989
824'.7—dc19
[B]
ISBN 0-19-812840-1

Typeset by Burns & Smith, Derby

Printed in Great Britain by
Bookcraft Ltd, Bath, Avon

To Dorothea,
Eileen, Denise,
and Catriona

Preface

I hate to fill a book with things
That all the world knows.

Quoted by Hazlitt

THIS volume embodies the results of researches extending over a quarter of a century and originates in an enthusiasm dating from an even remoter moment in the past when the author was Junior School Fives Champion (proud title) in a year when a prescribed English text happened to be a famous essay by one William Hazlitt.

There are many biographies of Hazlitt, and I should not have entered the field if it had not been that, despite their excellence, so many obscure areas still persisted in his life. I have tried to fill some of these gaps, and if in so doing I have wanted space to devote to other important figures and events in his life, let it be my excuse that my predecessors have already done them honour.

For reasons of economy also, I have not thought it necessary, in a work already heavily burdened with notes, to provide evidence for well-known facts that are easily verifiable, if need be, through the indexes of the biographies. Nor have I quoted page-numbers where dates of letters and of diary-entries, given in the text, are a sufficient guide. On the other hand, difficulty of access has led me in the case of one privately printed book, Le Gallienne, to provide alternative references, where the contents are the same, in Bonner.

S.J

Glasgow

Acknowledgements

I GRATEFULLY acknowledge financial assistance from the following: from the Carnegie Trust for Scotland, the award of a Carnegie Fellowship at an early stage; from the British Academy, a very timely grant-in-aid; from the Leverhulme Trust, the award of a Leverhulme Emeritus Fellowship to enable me to complete the work. I am also indebted to the University of Glasgow for a period of study-leave.

Over so many years of wide-ranging investigations I have incurred many debts of gratitude. For particular kindnesses I wish to thank the following: Dr Robert Bertholf of the University of Buffalo for enabling me to examine all the Hazlitt papers and first editions in his curatorship; Mr L. C. Dorman, Lincoln's Inn, who put me in touch with the descendants of the Walkers; Mr John Fieldwalker, great-grandson of Sarah Walker's brother, and Commander Simon McCaskill, RN, a descendant of Isabella Hazlitt's sister, both of whom courteously welcomed a complete stranger; Professor J. M. Halliday, who went to much trouble to examine legal records on my behalf; Dr Elizabeth McGrath of the Warburg Institute, and Mr Clive Wainwright of the Victoria and Albert Museum, who firmly corrected some of my misconceptions at an important juncture; John Murray, Esq., CBE, for generously allowing me access to the Murray Archives; Mr Mark Roberts, Librarian, British Institute, Florence, for showing me a rare edition of Scott; Professor Charles E. Robinson, University of Delaware, for kindly granting me the use of material from his article on new Hazlitt letters;[1] Dr Alvise Zorzi, Rome, for a necessary, though abortive, investigation; Mrs M. Kingswell, Home Office, for many valuable suggestions; Mrs Olive Classe, Dr Angus Kennedy, and Mr Noel Peacock, my colleagues at the University of Glasgow, for practical help at various stages; Mr Henry Heaney, Glasgow University Librarian, for essential microfilms; and Miss Dorothy Black, Miss Sheila Craik, and Mrs Jean Robertson, of that Library, for almost daily assistance.

I was kindly received at the Archivio di Stato and the Biblioteca Nazionale in Venice, the Archivio di Stato at Turin, the Archives cantonales vaudoises at Lausanne, the Archives d'Etat at Geneva, and the Archives municipales at Boulogne, as well as at the Record Offices of Buckinghamshire, Cheshire, Hampshire, Middlesex, Shropshire,

---

[1] *The Keats – Shelley Review*, 2 (August 1987).

Wiltshire, and of the City of Edinburgh, the Borough of Holborn, the Borough of Westminster, the former Greater London Council, the Diocese of Salisbury; at the Public Record Office, and at Register House, Edinburgh. I am indebted also to the staffs of the Archives nationales, Paris, and of the following libraries: Bibliothèque nationale, Paris; Bibliothèque de l'Arsenal, Paris; British Library; Guildhall Library; Dr Williams's Library; Victoria and Albert Museum; Reading Public Library; Cambridge University Library; Trinity College Library, Cambridge; Shropshire County Library; Liverpool Public Library; Manchester Central Library; National Library of Scotland; Edinburgh University Library; Edinburgh Public Library; Mitchell Library, Glasgow; Aberdeen University Library; Inverness Public Library.

For permission to quote unpublished material I am obliged to Lord Abinger; Mr Michael Foot, MP; Mrs Hestor Marsden-Smedley; The Poetry/Rare Books Collection, University Libraries, SUNY, at Buffalo; Reading Public Library; Durham County Library; and the Watt Monument Library, Greenock.

I wish also to record my gratitude for the constant help and support of my wife. My greatest debt, however, is to Mr Michael Foot, for his boundless generosity and unflagging encouragement.

Contents

Abbreviations and Short Titles used in the Notes

References to the main work, *The Complete Works of William Hazlitt*, ed. P. P. Howe, 21 vols. (1930–4), are given as volume number, followed by page number, in arabic, without title or name of author.

Add. MSS	Additional Manuscripts, British Library.
Baker	Herschel Baker, *William Hazlitt* (Cambridge, Mass., and London, 1962).
Bewick	*Life and Letters of William Bewick*, ed. Thomas Landseer, 2 vols. (1871).
BL	The British Library.
Blackwood's	*Blackwood's Edinburgh Magazine*.
Bonner	*The Journals of Sarah and William Hazlitt, 1822–1831*, ed. Willard Hallam Bonner, University of Buffalo Studies, 24 (1959).
Broughton	Lord Broughton, *Recollections of a Long Life*, ed. Lady Dorchester, 6 vols. (1909, 1911).
Buffalo	Hazlitt papers in The Poetry/Rare Books Collection, University Libraries, SUNY, at Buffalo.
Carlyle	Thomas Carlyle, *Reminiscences*, ed. C. E. Norton (1932 edn.).
Carlyle, *Letters*	*The Collected Letters of Thomas and Jane Welsh Carlyle*, ed. C. R. Sanders and others (Durham, NC, 1970–).
Champneys	Basil Champneys, *Memoir and Correspondence of Coventry Patmore*, 2 vols. (1900).
Chorley	*The Letters of Mary Russell Mitford*, 2nd ser., ed. Henry Chorley, 2 vols. (1872).
Cockburn	Lord Cockburn, *Life of Francis Jeffrey*, 2 vols. (Edinburgh, 1852).
Constable	Thomas Constable, *Archibald Constable and his Literary Correspondents*, 3 vols. (Edinburgh, 1873).
CUL	Cambridge University Library.
Cunningham	Allan Cunningham, *The Life of Sir David Wilkie*, 3 vols. (1843).
DNB	*The Dictionary of National Biography*, L. Stephen, 63 vols. (1885–1900).
EA	*Etudes anglaises*.
ER	*The Edinburgh Review*.

EUL	Edinburgh University Library.
Farington	*The Diary of Joseph Farington, R.A.*, ed. Kenneth Garlick and others (New Haven and London, 1978–84).
Four Generations	W. C. Hazlitt, *Four Generations of a Literary Family*, 2 vols. (1897).
GM	*The Gentleman's Magazine.*
Godwin Diary	Manuscript diary of William Godwin in the possession of Lord Abinger.
Gordon	Mrs Gordon, '*Christopher North*': *A Memoir of John Wilson*, 2 vols. (Edinburgh, 1862).
Griggs	*The Collected Letters of Samuel Taylor Coleridge*, ed. Earl Leslie Griggs, 6 vols. (Oxford, 1956–71).
Haydon	*The Autobiography of Benjamin Robert Haydon*, ed. Edmund Blunden (Oxford, 1927).
Haydon, *Corr.*	*Benjamin Robert Haydon*: *Correspondence and Table Talk*, ed. Frederic W. Haydon, 2 vols. (1876).
Hazlitts, The	[W. C. Hazlitt], *The Hazlitts: An Account of their Origin and Descent*, 2 vols. (Edinburgh, 1911).
Houck	James A. Houck, *William Hazlitt : A Reference Guide* (Boston, Mass., 1977).
Howe	P. P. Howe, *The Life of William Hazlitt* (rev. edn., 1947).
Hunt	*The Autobiography of Leigh Hunt*, ed. J. E. Morpurgo (1948).
Hunt, *Corr.*	*The Correspondence of Leigh Hunt*, ed. T. Hunt, 2 vols. (1862).
Hunt, *Poems*	*The Poetical Works of Leigh Hunt*, ed. H. S. Milford (Oxford, 1923).
Keats	*The Letters of John Keats*, ed. Hyder E. Rollins, 2 vols. (Cambridge, 1958).
Keats Circle	*The Keats Circle*, ed. Hyder E. Rollins, 2 vols. (Cambridge, Mass., 1948).
Keynes	Geoffrey Keynes, *Bibliography of William Hazlitt* (2nd rev. edn., Godalming, 1981).
KSJ	*The Keats–Shelley Journal.*
Landré	Louis Landré, *Leigh Hunt*, 2 vols. (Paris, 1935).
Le Gallienne	*William Hazlitt, Liber Amoris*, ed. Richard Le Gallienne (privately printed, 1894).
L'Estrange	A. G. L'Estrange, *The Life of Mary Russell Mitford*, 3 vols. (1870).
Letters	*The Letters of William Hazlitt*, ed. Herschel M. Sikes, Willard H. Bonner, Gerald Lahey (1979).
LH	W. C. Hazlitt, *Lamb and Hazlitt* (1900).

Literary Remains	*The Literary Remains of the Late William Hazlitt*, ed. W. Hazlitt, 2 vols. (1836).
LM	*The London Magazine.*
Lucas	*The Letters of Charles and Mary Lamb*, ed. E. V. Lucas, 3 vols. (1935)
Lucas, *Life*	E. V. Lucas, *The Life of Charles Lamb* (4th rev. edn., 1907).
Maclean	Catherine M. Maclean, *Born Under Saturn* (1943).
Marrs	*The Letters of Charles and Mary Anne Lamb*, ed. Edwin W. Marrs, jun. (Ithaca and London, 1975–).
MC	*Morning Chronicle.*
Memoirs	W. C. Hazlitt, *Memoirs of William Hazlitt*, 2 vols. (1867).
Mitford, Letters	Holograph letters of Mary Russell Mitford, 6 vols., in Reading Public Library.
Moore	*Memoirs, Journal, and Correspondence of Thomas Moore*, ed. John Russell, 8 vols. (1853–6).
Moorman	Mary Moorman, *William Wordsworth*, 2 vols. (Oxford, 1957, 1965).
Morley	*Henry Crabb Robinson on Books and their Writers*, ed. Edith J. Morley, 3 vols. (1938). [consecutive pagination]
Moyne	*The Journal of Margaret Hazlitt*, ed. Ernest J. Moyne (Lawrence, Kan., 1967).
Murray Papers	Archives of the publishing house of John Murray.
NCBEL	*The New Cambridge Bibliography of English Literature*, ed. G. Watson, 5 vols. (Cambridge, 1974–7).
NLS	National Library of Scotland.
NMM	*New Monthly Magazine.*
NNHL	Stanley Jones, 'Nine New Hazlitt Letters and Some Others', *EA* 19 (1966), 263–77.
NQ	*Notes and Queries.*
Oliphant	Mrs Oliphant, *William Blackwood and his Sons*, 3 vols. (3rd rev. edn., 1897).
Parl. Deb.	*The Parliamentary Debates from the year 1803 to the present time . . . under the superintendence of T. C. Hansard*, 41 vols. (1812–20).
Patmore	Peter George Patmore, *My Friends and Acquaintance*, 3 vols. (1854).
Pinney Papers	Papers of the Pinney family deposited in Bristol University Library.
PMLA	*Publications of the Modern Language Association of America.*
Pope	*The Diary of Benjamin Robert Haydon*, ed. Willard B. Pope, 5 vols. (Cambridge, Mass., 1960, 1963).

PRO   Public Record Office, London.

Procter   [Bryan Waller Procter], 'My Recollections of the late
                        William Hazlitt', *New Monthly Magazine*, 29 (1830),
                        469–82.

Procter, *Fragment*   B. W. Procter, *An Autobiographical Fragment*, ed.
                        Coventry Patmore (1877).

Procter, *Lamb*   B. W. Procter, *Charles Lamb* (1866).

QR   *Quarterly Review*.

RES   *The Review of English Studies*.

RHE   Register House, Edinburgh.

Robinson Diary   Unpublished passages in the manuscript diary of
                        Henry Crabb Robinson, Dr Williams's Library,
                        London.

Robinson Papers   Unpublished papers of H. C. Robinson, Dr Williams's
                        Library.

Sadler   *Diary, Reminiscences, and Correspondence of Henry Crabb
                        Robinson*, ed. Thomas Sadler, 3 vols. (1869).

Scott   *The Letters of Sir Walter Scott*, ed. H. J. C. Grierson,
                        11 vols. (1932–7).

Smiles   Samuel Smiles, *A Publisher and his Friends*: *Memoir and
                        Correspondence of the Late John Murray*, 2 vols. (1891).

Southey, *Corr.*   *The Life and Correspondence of the Late Robert Southey*, ed.
                        Charles Cuthbert Southey, 6 vols. (1849–50).

Southey, *New Letters* *New Letters of Robert Southey*, ed. Kenneth Curry, 2 vols.
                        (New York and London, 1965).

Southey, *Selections*   *Selections from the Letters of Robert Southey*, ed. J. W.
                        Warter, 4 vols. (1856).

Talfourd, *Letters*   *The Letters of Charles Lamb*, ed. Thomas Noon Talfourd,
                        2 vols. (1837).

Talfourd, *Memorials*   T. N. Talfourd, *Final Memorials of Charles Lamb*, 2 vols.
                        (1848).

Ticknor   *Life, Letters, and Journals of George Ticknor*, ed. G. S.
                        Hillard, 2 vols. (Boston, Mass., 1876).

TLS   *The Times Literary Supplement*.

V. & A.   Victoria and Albert Museum, London.

Wardle   Ralph M. Wardle, *Hazlitt* (Lincoln, Neb., 1971).

Wordsworth   *The Letters of William and Dorothy Wordsworth*, ed. Ernest
                        de Selincourt, 2nd rev. edn. by Mary Moorman and
                        Alan G. Hill (Oxford, 1967–88).

Place of publication is indicated, above and in the footnotes, only when it is
other than London. In page references to any publications here named the
number alone is given, without the abbreviation 'p.'

Chronology, from 1808, of the Life of William Hazlitt (1778–1830)

1808   1 May, marries Sarah Stoddart (b. 1774). Nov., leaves London for Winterslow. [Peninsular War begins; Convention of Cintra.]

1809   Writes *A New and Improved Grammar of the English Tongue* (pub. Nov., supplanted 1810 by Godwin's *Outlines of English Grammar*), and a life of Thomas Holcroft (not pub. until 1816). First child dies. Lambs visit Winterslow. [Corunna; Talavera; Walcheren expedition.]

1810   Lambs' second visit. [Bank failures, including Salisbury Bank.]

1811   Sarah's widowed mother dies. Birth of son. Commissions for portraits in London. Attends Coleridge's lectures. [Economic depression; Luddism.]

1812   Lectures on philosophy. Meets Haydon at Northcote's. [Regency effective; Castlereagh Foreign Secretary; the Retreat from Moscow.]

1813   Parliamentary reporter on *Morning Chronicle*. Settles in York St. Parents leave Wem for Chertsey. Controversies with Mudford, Sterling, and Stoddart. [Henry Bridgwater marries Isabella Shaw (b. 1790). Southey Poet Laureate; Wordsworth Stamp-distributor. Oct., Battle of Leipzig.]

1814   Drama and art critic on *Chronicle*. Clashes with Perry, moves to *Champion* and *Examiner* (visits Leigh Hunt in jail). [Kean's London début; *The Excursion*; *Waverley*. Napoleon abdicates; peace celebrations.]

1815   Recommended by Lady Mackintosh to *Edinburgh Review*. John Scott's perfidy drives him from *Champion*. Wordsworth tries to set Lamb, Haydon, Hunt, and Robinson against him. The Montagus invite him to Yorkshire; at Bedford Square meets John Fearn (and J. R. Bell?). May, regular dramatic critic of *Examiner*. Contracts with Constable for *Round Table*. [Wordsworth, *Poems*. Corn Bill riots; the Hundred Days; Waterloo.]

1816   Reviews Schlegel in the *Edinburgh*. May, visits parents at Bath. Attacks the Government, Wordsworth, Coleridge, Southey, Stoddart, and the *Catalogue raisonné*. [*Alastor; Christabel*. Stoddart founds the *Correspondent* with French ultra-royalists; Robert Owen active; the Spa-Fields riot.]

1817   *The Round Table; Characters of Shakespear's Plays*. Dramatic critic of *The Times*. New friends: Keats, Reynolds, Clarke, Talfourd, Procter, Bewick, Patmore, Ogilvie, Henderson. [*Wat Tyler; Biographia Literaria*; Keats, *Poems*. Habeas Corpus suspended; Derbyshire Rising; trials of William Hone.]

1818 Lectures on poetry (M. R. Mitford attends). *A View of the English Stage*; *Lectures on the English Poets*; second edn. of *Characters of Shakespear's Plays*. Attacked by Tory journals (first, *New Monthly*; then *Quarterly*). Begins to write for Constable's *Edinburgh Magazine*; attacked by *Blackwood's*. [Lamb, *Works*; *Endymion*; *Heart of Midlothian*.]

1819 *A Letter to William Gifford, Esq.*; *Lectures on the English Comic Writers*; *Political Essays*. Parts from his wife. [Peterloo massacre; Castlereagh's Six Acts.]

1820 *Lectures on the Age of Elizabeth*. Contributes to *London Magazine*. Visits parents in Devon. Father dies. Meets Sarah Walker (b. 1800). [Haydon's *Christ's Entry into Jerusalem* exhibited. Cato Street conspiracy.]

1821 *Table-Talk* (I). [J. Scott killed in duel; Keats dies.]

1822 27 Jan., to Scotland. Feb.–Mar., at Renton Inn. Difficulties of Bell and Henderson. May, lectures at Glasgow; excursion to Highlands. *Table-Talk* (II). July, divorce final; Sarah refuses him, preferring his rival, Tomkins. [Shelley drowned. Suicide of Castlereagh.]

1823 Contributes to *Liberal*. The test on Sarah. *Liber Amoris*. Is pilloried in *Literary Gazette*, and in *John Bull*. *Characteristics*. At Fonthill, meets cousin, Kilner Hazlitt. Notices of Fonthill pictures in *Morning Chronicle*. Helps Bewick. [*Essays of Elia*; Lamb's 'Letter to Robert Southey'. French invade Spain to restore absolutism.]

1824 Marries Isabella Bridgwater. Meets Bewick and Knowles at Melrose, but not W. Scott. *Sketches of the Principal Picture Galleries*. Attacks Croker. To Paris; finally meets Stendhal; negotiates Paris *Table-Talk* and *Spirit of the Age* with Galignani. [Early months, or late 1823, son, Frederick, b. to Sarah Tomkins.]

1825 Italian tour. At Vevey rereads W. Scott, sends 'Bad English in the Scotch Novels' to *Examiner*; reads *Edinburgh* review of *Spirit of the Age*, sends 'The Damned Author's Address to his Reviewers' to *Morning Chronicle* and *London and Paris Observer* (printed by latter). Oct., back in England.

1826 Returns to Paris (to fashionable Chaussée d'Antin) and begins life of Napoleon. *The Plain Speaker*; *Notes of a Journey through France and Italy*. 'Boswell Redivivus' series begins. Money troubles. Frequents the French Liberals. [Bank failures at home; Constable and Scott ruined.]

1827 (A second Italian tour, for son's benefit?) Returns to England. Wife leaves him.

1828 *Life of Napoleon*, vols. i and ii. Dramatic critic of the *Examiner*. Much at Winterslow.

1829 Writes for *London Weekly Review*, *Atlas*, and again for *Edinburgh Review*. Shares lodgings with his son, Bouverie St.

1830   *Life of Napoleon*, vols. iii and iv; *Conversations of James Northcote, Esq.,*
         R. A. A last attachment. Dies at Frith St., 18 Sept. [July Revolution
         in France.]

Died thereafter: Lamb, 1834; Coleridge, 1834; Godwin, 1836; John Hazlitt,
            1837; Sarah Hazlitt, 1840; Frederick Tomkins, 1852; John
            Tomkins, 1858; Isabella Hazlitt, 1869; Sarah Tomkins,
            1878; Micaiah Walker (b. 1803), 1899.

I

Withdrawal from London

Enough my soul, turn from them
and let me try to regain the
obscurity and quiet that I
love . . .

Hazlitt

AFTER William Hazlitt and Sarah Stoddart were married at St Andrew's, Holborn on Sunday, 1 May 1808 they remained for several
months in or near London.[1] That autumn they were living at Camberwell Green, three miles south of St Paul's (and a short step beyond the
Montpelier Tea Gardens that had so vividly impressed Hazlitt's
boyhood).[2] The Green was regarded at that time as 'a pleasant retreat
for those citizens who had a taste for the country whilst their avocations
daily called them to town'.[3] His own avocations called him to publishers'
offices, to John Murray's at 32 Fleet Street and no doubt to those of
others whom he tried to interest in a translation of Bourgoing's recent
Tableau de l'Espagne moderne, which had been made topical by the French
usurpation, the insurrection, and the British expedition of that summer.
The project failed, and there was nothing further to detain the Hazlitts in
town.[4] Sarah was well advanced in pregnancy and in need of rest. They
probably also needed to economize. Soon after 16 November they
packed up and went down to live in Sarah's cottage at Winterslow.

In retreating to Wiltshire Hazlitt was once more veering away from
the flame of London life to which, moth-like, he had returned again and
again for sixteen years, but this time the renunciation seemed a final
one. It was no longer excursions into the provinces as an itinerant

[1] The accepted date for their departure from London is 1 May. For my suggestion that they did
not leave until late Nov. see 'The Hazlitts at the Mitre Court "Wednesdays" in 1808: Hidden
Implications of a Mary Lamb Letter', *Charles Lamb Bulletin*, NS, 57 (Jan.1987), 17–19. My
deductions were subsequently corroborated by C. E. Robinson's discovery of a letter of late 1808,
from Camberwell Green, to John Murray (see following note).

[2] The letter mentioned in the previous note, which proposes a translation of Bourgoing, *Tableau
de l'Espagne moderne* (Paris, 1807; London, 1808), is dated [? 3 Sept. – 26 Nov. 1808] by Professor
Robinson. It appears in his article, 'William Hazlitt to His Publishers, Friends and Creditors:
Twenty-seven New Holograph Letters', *The Keats–Shelley Review*, 2 (August 1987).

[3] P. Wakefield, *Perambulations* (1809), 433.

[4] The translation was completed and was apparently still in existence in 1867 (*Memoirs*, i, p.
xxiii).

portrait-painter, nor more or less protracted visits to his family in Shropshire: he was deliberately choosing to settle down as a married man and householder in the depths of the country. There was contradiction here. He went down to the Wiltshire village, as he said, 'a willing exile'.[5] He chose to go, certainly, but to what he still called exile. He left the theatres, Drury Lane, Covent Garden, the Haymarket, Sadler's Wells; he left the picture-galleries and the art-dealers, the sale-rooms of Pall Mall and the print-shops of St James's Street; the great buildings, the noble squares,[6] the mansion-rimmed curve of the river from Westminster to London Bridge, the water-steps and the plying boatmen, and, over them all, the serene presiding dome of St Paul's; he left the parks, the bow-windowed shops of Fleet Street, Long Acre, and the Strand; the press of stage-coaches arriving from all parts of the kingdom at the White Horse, the Bell and Crown, the Bull and Mouth, the Swan with Two Necks; the quiet backwater of his beloved Southampton Buildings, lost between the greater solitudes of the Inns of Court although but a stone's throw from bustling Holborn; he gave up the leisurely browsing over the bookstalls of Paternoster Row, the hot excitement of the fives-court in St Martin's Street, Leicester Fields, and left the swarming life around the statue of Charles I at Charing Cross, the carts and carriages squeezing through Temple Bar, the sinister vitality of St Giles's and Seven Dials, the coffee-houses, the taverns, the night-cellars, the street-walkers of a hundred lanes and alleys, the studios of artists, the conversations of writers, the political discussions of his elder brother John's radical cronies, the more youthful companionship of his own contemporaries. All this, to live in a hamlet on the edge of a plain that a foreign visitor described as drearier than the Libyan desert.[7] Why did he go? And why did he say he went willingly?

The answer lies in the duality of his own nature. Alternately indolent and violently active, at once abstracted and strongly responsive to the claims and pull of reality, formed for a life of contemplation and yet aspiring, though without any great conviction, to the fame attendant on a life of artistic creation, intermittently self-distrustful and persuaded of the vanity of all ambitions except those of the philosopher and the artist, it seemed that he was no better fitted for the struggle for existence than his father before him, and as little endowed with the will and capacity for the subtle deferences that bring worldly success and the insidious and progressive abdication of principle or stifling of passion that ensure its continuity.

---

[5] 12. 121.

[6] He was fond of quoting Cowper, *The Task*, 1. 643, 'Our palaces, our ladies, and our pomp' (see 1. 225; 8. 69; 9. 126).

[7] Anon., *Londres en mil huit-cent vingt-trois* (Paris, 1824), 155.

His father, the Revd William Hazlitt, was a Unitarian minister who had briefly known, at the promising start of his career, the most distinguished members of the intellectual aristocracy of Dissent, but who, after vainly seeking a congenial church in the four corners of the Kingdom and then in America (that far haven of civil and religious liberty), had finally returned to England at the age of fifty and buried himself, a failure, in the little Shropshire town of Wem, remoter than ever from the company and conversation he loved — and even this obscure charge he had only obtained, humiliatingly, because it had just been vacated by the father of a young man of twenty-nine recently appointed to the very charge in the thriving city of Norwich for which earlier he himself, a veteran in the cause with a strong claim to recognition, had also applied: if this young man's father had not gone to live with him the Revd William Hazlitt would not even have had Wem.[8] He served out the remaining twenty years of his ministry in the claustrophobic atmosphere of a small town, preaching to a sparse congregation in a back-street chapel no bigger than a byre, a chapel dwarfed by the massive looming tower of the disdainfully secure Anglican church the other side of the main street. The contributions of the congregation were not sufficient to keep him and his family, and most of his stipend came from a Unitarian trust together with £12 a year from the Presbyterian Fund founded by a Wem glover a century before.[9]

The reasons for his ill fortune do not clearly emerge from his son's account of him (the portrait is no doubt modified by filial piety), nor yet from his surviving letters, which were all addressed to this son as a young boy. These letters suggest a mild and conciliatory man, but may simply illustrate attitudes he hoped to implant in an indocile son whose career he feared, in moments of discouragement, to see repeating his own. Indeed, one reason may have been the passionate hit-or-miss argumentative vehemence attributed to him in a graphic eighteenth-century Wem poem giving a portrait-gallery of local worthies, where we find:

> Hazlitt the rights of man explaining
> And his ugly jaws a-straining
> As if he'd swallow up the Ring.[10]

We have no comparable picture, in the reminiscences of the essayist's friends, of Hazlitt laying down the law in political discussion, but it is not difficult to recognize here an analogue to the furious cascading prose of

[8] Moyne, 168.
[9] Shropshire County Library, The Hill Trust papers; *Christian Reformer*, Nov. 1838, 764.
[10] I. Woodward, *The Story of Wem* (Shrewsbury, 1952), 56. The Ring was the old market-place.

the essay on Scott in *The Spirit of the Age* or of the *Letter to William Gifford, Esq.*

The old minister, we can be sure, was the dominant figure in Hazlitt's early life, and although his image as hero-against-the-world was elaborated, confirmed, and also qualified as the years passed, it had been fixed in childhood and persisted to the end. The year before Hazlitt died his pride in his father shone forth as plainly as his reticence allowed in a letter on the sermons of Channing, the American Unitarian: 'I do not know whether I have much natural piety in my constitution, but Dr. Channing preaches at an Unitarian meeting in Boston the liturgy of which (formerly Trinitarian) was drawn up by my father forty years ago and upwards, who went to America to plant Unitarianism there.'[11]

In both Hazlitt and his father there was a fundamental intransigence, a stubborn idealism, a perverse zeal to relinquish the profitable delusion and embrace the dowerless truth, which produced in both the same pattern of behaviour: initially, a rejection of the career their parents had intended for them, and then a failure to attract adequate patronage or support in the one they chose; at all stages, a revolt against the established authorities in Church and State; the publication of unsuccessful didactic or polemical tracts; a tacit admission of defeat; and ultimately a withdrawal from the world (never quite final, however, with the essayist). An astute man once refused to put his son to school to the Revd William Hazlitt because (said he) the lad would be given an education that, by teaching him always to tell the truth, would unfit him for getting a living.[12] Hazlitt, to the end of his days, ruefully remembered this incident and its implications for his own life: in the year he died he said, 'Education (which is a study and discipline of abstract truth) is a diversion to the instinct of lying and a bar to fortune'.[13] This was the education he himself got.

It was fortunate for him that the early years of his second career, when he found a living in journalism, were marked by a violent intellectual and political ferment, involving the most representative figures of the age, which made the contentious and unprofitable reflections he had been elaborating in solitude without thought of gain peculiarly acceptable to the more enlightened of the two main warring camps, otherwise he might never have emerged from obscurity. His genius first declared itself in occasional criticisms, political, dramatic, and literary, provoked by the events and exigencies of the times.

And that was not all. In addition to unfitting him for getting a living,

---

[11] A PS first printed in the writer's 'Two Hazlitt Quotations', *NQ*, Oct. 1976, 456. The version subsequently published in *Letters*, 366 is defective.

[12] 8. 288–9.   [13] 20. 340–1.

his education, by confirming him in 'an obstinate constitutional preference of the true to the agreeable',[14] unfitted him for maintaining his equilibrium on the polished floors of society. A certain awkwardness in company arose from his narrow provincial origins, from the isolation of an unpopular religious sect, and the further isolation within that sect of the family of its pastor who were obliged to maintain dignity without money and prestige without benefit of protocol (they apparently saw more of their opposite numbers, the Jenkinses of Whitchurch and the Rowes of Shrewsbury, than of their own townspeople[15]). It was often aggravated by a muteness imposed by a clash between sympathetic feeling and a theory of truth-telling, or by an equally embarrassing earnestness resulting from an abrupt sacrifice of urbanity to plain-speaking. Henry Crabb Robinson, who met him at Bury St Edmunds in 1799, tells of the striking contradiction between this petrified gawkiness and the active intelligence he (and he, as he claims, alone) discerned in him:

. . . at the time I first knew him he was struggling against a great difficulty of expression, which rendered him by no means a general favourite in company. His bashfulness, want of words, slovenliness in dress, etc., made him the object of ridicule. . . . The moment I saw him I saw he was an extraordinary man. He had few friends and was flattered by my attentions.[16]

In a letter of six years later (1805) he fears that not even Hazlitt's gifts would enable him to rise above his circumstances. The association which at that moment suggested Hazlitt's name to him is itself significant:

Talking of mystics has put me in mind of William Hazlitt. Of all the young men of my acquaintance in England, I consider him as incomparably the first in point of intellect. I am inclined to think that in the whole stock of my friends, he is the only one who promises to be a distinguished and original character; tho' on the other side for various reasons I fear that he will never be able to show himself advantageously in life, but perhaps be another sad instance of Genius sinking in its struggle with fortune and the world.[17]

Hazlitt was confirmed in this uncomfortable see-saw of polite lies and impertinent truths in the 1790s by the precepts of William Godwin, who was a friend of his elder brother John, the miniature painter, and who had proposed in his *Enquiry Concerning Political Justice* a naïvely idealistic imperative of candour: 'How great would be the benefit if every man were sure of meeting in his neighbour the ingenuous censor who would tell him in person, and publish to the world, his virtues, his good deeds,

---

[14] 5. 111, where Hazlitt applies the phrase to Swift.
[16] E. J. Morley, *The Life and Times of Henry Crabb Robinson*(1935), 9.
[15] Moyne, 103.
[17] Ibid. n. 2.

his meannesses and his follies.'[18] Hazlitt was impressed, and did violence
to his tender heart in essaying to adopt this principle, which ran counter
to his whole instinct: however doggedly he proclaimed unpalatable truth
in his writings he was always the soul of watchful consideration in his
dealings with those about him, and his readiness to postpone his personal
claims in deference to the comfort of others was a constant threat to his
own well-being. 'Delicacy, moderation, complaisance, the *suaviter in
modo*,' he wryly said towards the end of his life, half in jesting
self-depreciation, half in sober earnest, when he had got a severe illness
from not protesting against an open window on a winter coach-journey,
'whisper it about, my dear Clarke, these are my faults and have been my
ruin'.[19] And again, in an endlessly significant though trifling remark,
characteristically concrete and immediate, and betraying a curiously
modern consciousness, 'If I see a person coming at the other end of the
street, I am not easy till I have taken my own side of the pavement, lest it
should be thought possible I do not mean to take it.'[20] His reputation for
misanthropy, which rests largely on certain powerful but isolated
passages in his writings, is a paradoxical proof of this: it was the
misanthropy of a Timon or an Alceste, disappointed in a world he had
approached openly, without suspicion, and which he conceived, not
without reason, to have persistently taken advantage of his initial
disinterestedness and candour, and dealt badly with him. It would be a
strange cynic indeed who could claim, as he did, 'I place the heart in the
centre of my moral system.'[21]

Godwin was probably an old family acquaintance, since his father, a
dissenting minister, had immediately preceded the Revd William
Hazlitt in his early charge at Wisbech. Hazlitt had met him at John's
studio in Long Acre in September 1794, at the impressionable age of
sixteen, when he was a pupil of Hackney New College visiting his
brother on day-leave.[22] This and later meetings with Godwin confirmed
Hazlitt in the path of stubborn opposition to authority first traced out by
his father and then dramatically signposted by a cardinal event in
English history and in the annals of his brother's circle. 1794 was the
year of the historic Treason Trials which, if the accused had not been
acquitted, might well have swept England into the vortex of a revolution,
certainly into a repressive Terror, and John Hazlitt was on terms of
friendship with at least two of the twelve prisoners, prominent members
of the foremost reformist societies of the day. These societies, the
Constitutional and the London Corresponding, were popularly supposed to be revolutionary and were in fact sustained by the flood of

---

[18] bk. iv, ch. vi (referred to by Hazlitt (1. 251 n.) as 'Mr. Godwin's plan of plain speaking').
[19] *Letters*, 349–50.   [20] 17. 337.   [21] 12. 193.   [22] Baker, 70.

enthusiasm loosed by the publication two years earlier of *The Rights of Man*, Tom Paine's reply to Burke's *Reflections on the Revolution in France*. Godwin's measured *Enquiry Concerning Political Justice*, published in 1793, was sold at three guineas, and was therefore, in Pitt's estimation, although subversive, not dangerous. Certainly it had been less widely diffused than the popular *Rights of Man*, of which some of the numerous editions were priced as low as sixpence, but its author was nevertheless indirectly threatened by the trials. He was also anxious on account of his friends among the accused, particularly the playwright Thomas Holcroft, and it was the anonymous appearance in the *Morning Chronicle* of his 'Cursory Strictures on the Charge delivered by Lord Chief Justice Eyre to the Grand Jury, 2 October 1794' that undoubtedly saved the day. This was an open letter to the Lord Chief Justice demonstrating that the evident intention of the Prosecution, with the blessing of the Court, was not to attempt to prove treason but to infer it by construction.

The trials, although they ended in acquittal, sent overt revolutionary activity underground for many years, but they did not quench revolutionary thought, nor even propaganda: one of the accused, John Thelwall, for years afterwards travelled the North and East of England dealing freely with contemporary politics under the cloak of lectures on Roman history.[23] His thunder and lightning oratory had already captivated young William Hazlitt, no doubt at meetings the boy attended with his brother.[24] His acquittal was celebrated on 5 December 1794 by a party at his house, at which the John Hazlitts as well as Godwin were present. Godwin's diary shows John Hazlitt moving in this circle at various other times in late 1794 and early 1795.

Friendship with such men, however inspiring and however creditable to one's principles, was, from a worldly point of view, more likely to compromise than to further a career so dependent upon the patronage of fashionable society as that of miniature-painter, and indeed John Hazlitt, handicapped also by a perverse obstinacy to neglect his professional commitments in order to attempt uncommissioned oil-paintings for which he had no aptitude, was, like Godwin, never free from financial worries. The unprofessional wrong-headedness that so exasperated his friend Charles Lamb[25] is in reality a striking if remote instance of the distinterestedness which Hazlitt made the cornerstone of a whole theory of 'metaphysics', ethics, and aesthetics (in opposition to eighteenth-century sensationalist systems), and which in its own admirable image as well as under the less impressive guises of obstinacy,

---

[23] Maggs cat. 583 (Spring 1933), item 374, letters from Thelwall to Thomas Hardy, 1796–1806.
[24] 12. 264–5.   [25] Marrs, ii. 208.

carelessness, and perversity, marks the whole history of the Hazlitt clan.

The pattern of the father's life, manifestly repeated in the essayist's, must also have reappeared with variations in the life of the elder son, about whom, significantly enough, we know so little, but who evidently shared the inherited temperament. It certainly reappears sixty years later in the letters sent home by *his* son, the essayist's nephew, 'cousin Will', from Mauritius to the essayist's grandson, which, in their scathing condemnation of the island legislature, suspicion of spies, and accusations of corruption in high places, momentarily recall, although in a debased (or perhaps simply in a non-literary form), the old, awkward, independent family spirit.[26] These letters were written to the grandson, W. C. Hazlitt, when he was seeking material for his first biography of his grandfather, a book which itself illustrates this obstinate attachment to the truth. Where else in Victorian biography, not to say in Victorian family biography, do we meet with such plain speaking, such reckless frankness? Its publication outraged the Mauritian cousin, who (evincing this time the Hazlitt individualism) protested: 'I am afraid the work has darkened rather than illuminated the name of Hazlitt.'[27] At about the same time Sam Hazlitt, a distant Irish relation, writing from Tipperary to the same grandson, was pronouncing the Hazlitts to be all 'oddities'.[28]

It was owing to this proneness to contumacy, more apparent however than real, but presenting a strange outcrop in the Mauritius letters, that the old preacher and his eldest son the miniaturist were withheld from prosperity all their lives long. It played its part also in denying reputation to the essayist for half a century or more after his death. Families no less than individuals can be wise or unwise in their generations, and that cumulatively down the years in one life after another: 'So if we look back to past generations (as far as eye can reach) we see the same hopes, fears, wishes, followed by the same disappointments, throbbing in the human heart; and so we may see them (if we look forward) rising up for ever, and disappearing, like vapourish bubbles, in the human breast!' 'There is', as he said elsewhere, 'an involuntary, unaccountable family-character, as well as family-face.'[29]

Hazlitt's life at Winterslow must have been narrow and skimping indeed. His brother John viewed his prospects at the time of the marriage with a misgiving which no doubt reflected his own disappointments. Speaking from years of trying to make ends meet, and perhaps also from a recognition of a hint in his younger brother's character of the flaws of improvidence and waywardness he had for so long observed in his own, he warned him that he 'must have furniture & be clear in the

---

[26] BL Add. MS 38898, fos. 144, 157.   [27] BL Add. MS 38899, fo. 462.
[28] Ibid. fo. 71.   [29] 17. 110; 12. 232.

world at first setting out', otherwise he would 'be always behindhand',[30] a modest enough ambition and eloquent of the family's slender means. It was eloquent also of the precariousness of John Hazlitt's own situation (which twenty years later, when positions were reversed, would in turn be a source of anxiety to the essayist on his brother's account[31]). The miniaturist's own fecklessness, and the desperation to which his difficulties often drove him, are briefly revealed in a letter of 1809 from Mary Lamb to Sarah Hazlitt announcing the safe arrival of a present for the John Hazlitts' infant son:

[Mr John Hazlitt] got fifty four guineas at Rochester and has now several pictures in hand. He has been very disorderly lately. I am going to tell you a secret for Mrs. H says he would be very sorry to have it talked of. One night he came home from the Alehouse, bringing with him a great rough ill-looking fellow whom he introduced to Mrs Hazlitt as Mr Brown a gentleman he had hired as a mad keeper to take care of him at forty pounds a year, being ten pounds under the market price for keepers, which sum Mr Brown had agreed to remit out of pure friendship. It was with great difficulty, and by threatening to call in the aid of watchmen & constables that she could prevail on Mr Brown to leave the house.[32]

In the remoter corners of Hazlitt's experience too, nourished by obscure inner conflicts, there was undoubtedly something of the same temptation to self-destructive abasement, the vertiginous plunge into the whirlpool of abandonment, the fatal pull of the gutter. There lurked within him what he discerned in the character of Hamlet: 'bitterness of soul makes him careless of consequences'.[33] His sympathy with outsiders, from players down to gypsies and tramps, and with the weaknesses of a Kean or a James Barry, finds its echo here, as it does also in a pathetic letter written during the worst year of his life, in 1822, when in despair he said that 'the only thing that soothes or melts me is the idea of taking my little boy whom I can no longer support and wandering through the country as beggars'.[34]

The dreaded consequences of self-abandonment may have loomed larger in 1808, and the fear of what might happen if he let go may in part explain his decision first to marry Sarah Stoddart and then to move out of London. The reality of the danger is suggested by a 'conversation' with James Northcote years later, where he puts in the old painter's mouth this judgement on his own conduct: 'You turn your back on the world, and fancy that they turn their backs on you. This is a very dangerous principle. You become reckless of consequences. It leads to

[30] Marrs, ii. 264.   [31] Le Gallienne, 331; Bonner, 249.   [32] Marrs, iii. 31.
[33] 4 .232.   [34] *Letters*, 265.

an abandonment of character. By setting the opinion of others at defiance, you lose your self-respect.'[35] There was probably not at that time, in 1808, the wild desperation threatening the borders of sanity which is implied in John Hazlitt's escapade, but we shall see it pursuing the reeling William Hazlitt through the midnight streets of London in the summer of Waterloo, or later still, haunting his skulking shadow at the corner of Southampton Buildings on windy autumn nights in 1823, the year after his divorce. In 1808 the germ of it was certainly present. The little we know of his courtship of Sarah Stoddart, who was three years his senior, hints that in marrying her he was in an important respect seeking refuge from the world.

What kind of woman had he married? One who was perhaps clumsily, even at times crudely, unconventional, but who was also cheerful, practical (except in a foreseeing sense), and possessed of a little money, with the prospect of more to come. The letters of her excellent friend Mary Lamb show her in reflection as a very pleasant companion, fun-loving and high-spirited, but in addition eccentric, erratic, and strong-willed, with little of the demure refinement expected of the well brought-up young ladies (the Miss Bennets) of those days.

She had, however, two valuable qualities in the eyes of the withdrawn and agonizingly sincere Hazlitt: sociability and forthrightness. He had had a painful misadventure a few years earlier when he was an itinerant portrait-painter in the Lake District. A combination of romantic illusions and strong passions had delivered him up — a trusting victim — to a local hoyden who had deceived his ardent simplicity and led him on to emulate the fine speeches of the hero of his admired *Nouvelle Héloïse*, and then made a fool of him and turned him over to the menfolk of the village. This had confirmed him in a morbid horror of the duplicity of women, as well as in a belief in the fatality of the ill fortune that dogged his love-affairs ever after. It always seemed to him inevitable that he should not be liked, and although he ruefully searched for clues in his mirror it was on paper that he most effectively explored and rehearsed the reasons. For the most part he there forgot his own case, and his speculations rapidly widened to embrace human nature in general, but the point of departure remained discernibly his own misadventures as the type of the studious romantic alienated from life.

But this despondency was still largely in the future: for the present he was content. Sarah's honesty was patent; she was so apt to blurt out the first thing that came into her head that Mary Lamb despaired of scolding her into a little dignity.[36] When she was a child her father, a naval officer, sipping his grog, would say to her prim older brother John, 'John, will

---

[35] 11. 318.   [36] Marrs, ii. 124, 211.

you have some?' — 'No, thank you, father' — 'Sarah, will you have some?' — 'Yes, please, father'. Years later, writing to ask a favour of an early friend of her brother, now become the great Lord Brougham, she very innocently began her letter with 'My dear H[enry]', to the unutterable scandal of those about her.[37] At her first entry upon life her unpolished manners so worried her brother that he told her, abruptly and pompously, but not unkindly (according to his lights he was not being unkind, although it was an abruptness Hazlitt would never have been guilty of), that she was 'deficient in those minutiae of taste and conduct which constitute elegance of manners and mental refinement'.[38] This is the language of a vulgarly ambitious Revd Mr Collins, bowing and scraping before the condescending Lady Catherine de Bourgh, and young Stoddart's insistence in his undergraduate letters to his sister on the need to 'make valuable friendships', his congratulations when she unintentionally did so, as well as his evident determination to worm himself into the acquaintance of dukes and countesses do indicate that he judged from an artificially rigid point of view. Nevertheless the disobliging inference is clear, and her ungracefulness is borne out by such incidents as her clumsy reception of Coleridge on his unexpected arrival at her brother's house in Malta in 1804.[39]

The little we know of their parents' circumstances suggests that the childhood they passed in Salisbury was dull and unprosperous. The father, according to his great-grandson, W. C. Hazlitt, who cites no documents and was doubtless repeating a family tradition, was a disappointed man, and in this respect, if in no other, he resembled the Revd William Hazlitt. Born in 1742, he went to sea as a midshipman in the Seven Years War and was present at the capture of Quebec in 1759 and at Belle Isle, Louisburg, and Havana, but this brisk beginning proved deceptive.[40] His career did not fulfil its early promise. When he passed lieutenant in July 1763 the Peace of Paris had already been signed many months since, and thereafter, apart from a brief spell of four years in the late 1770s in a humble bomb-ketch (ingloriously named the *Carcass*) on the New York station during the War of American Independence, the rest of his service — the major part of it, forty years — was passed in the doldrums.[41] Like most officers between the wars, when

[37] *Memoirs*, i. 166.

[38] Pinney Papers, Bristol University Library, letter of 24 Nov. 1797. Attention was first drawn to the Pinney papers by J. R. Barker, 'Sarah Stoddart and the Lambs', *Huntington Library Quarterly*, 24 (1960), 59–69. Further details were given in R. S. Woof's excellent lecture 'John and Sarah Stoddart: Friends of the Lambs', *Charles Lamb Bulletin*, NS 45 (Jan. 1984), 93–109.

[39] *The Notebooks of S. T. Coleridge*, ed. K. Coburn (1957–), ii, entry 2099.

[40] *The Commissioned Sea Officers of the Royal Navy, 1660–1815* (1954).

[41] PRO, Adm. 107/6; Adm. 6/20; Adm. 6/213–20 (half-pay registers); Adm. 36/8873 (muster book, *Carcass* bomb); Adm. 36/9478 (muster book, *Monarch*).

the bulk of the fleet was laid up under skeleton crews in the naval ports, he remained ashore long years, kicking his heels. There is however something suspicious in the consistency with which he was passed over whenever a sea-going opportunity presented itself. A few months in charge of a press-gang in Southampton in 1790 (the same year in which the twelve-year-old William Hazlitt wrote to his father from Liverpool, 'They were pressing on Saturday evening. The world is not quite perfect yet; nor will it ever be so whilst such practices are reckoned lawful'[42]), and, after the outbreak of war with France, a longer stretch on the same tack at Poole from 1795 to 1801, when he left his wife and daughter in Salisbury, complete the brief story of his active professional career.[43]

His last period of duty made no difference to his rank and little to his pay. Nor can it have improved his spirits: in that distasteful job of impressment you were not only loathed on the one hand by the merchant seamen and fishermen whose mates you abducted but on the other disliked and snubbed by local society, especially by the business folk whose commercial affairs you dislocated, and on top of that you were also isolated within the service, cut off from your fellow officers and on uneasy terms with your men. So that from the age of thirty until he died aged sixty-one at Salisbury in 1803 he was in effect continuously on half-pay, the proverbial poor lieutenant, frustrated by inactivity, immobilized by lack of means, existing on the fringes of society, or, when occasionally employed and relatively in funds, even more literally an outcast. A sad career. Hazlitt cannot have met him and he is probably echoing his wife's childhood memories when he muses in an essay on the fate of the 'many discontented half-pay lieutenants [who] are in vain seeking promotion all their lives, and obliged to put up with "the insolence of office, and the spurns which patient merit of the unworthy takes" '.[44]

It seems that it was at Salisbury that Lieutenant Stoddart met Sarah Brown, who became his wife — at least it was there, according to the *Dictionary of National Biography*, that their first child, John, was born on 6 February 1773. The date and place of birth of their second, Sarah, are not known, but the year was 1774, and on 5 June of the same year the parents were married by special licence at Chalfont St Giles, distant by the width of three counties from Salisbury.[45] This is a mystery: nothing else connects either of them with this village or any other near it.

[42] *Letters*, 49.   [43] PRO, Adm. 6/217; and Pinney Papers.   [44] 8. 93–4.

[45] PRO, Adm. 6/340 (Widows' pension papers, 1803), and Buckinghamshire Record Office, Aylesbury. The parish registers of neighbouring Chalfont St Peter, Beaconsfield, Iver, Amersham, Chesham, Fulmer, Hedgerley, Denham, and High Wycombe do not suggest any family connection. John Hunt in 1824, angered by John Stoddart's attack on the dead Byron, makes a point of his illegitimacy: 'this bastard Englishman, this illegitimate stickler for legitimacy . . .' (*Examiner*, 23 May, 322).

His wife was of weak intellect, and indeed became insane a few years after his death.[46] Her daughter's erratic behaviour may have been a worry to her. But there was more. An even greater worry to both parents was the brother's unaccountable flirtation with the political and social black sheep of the 1790s, the Jacobins, an unfilial deviation from the path of rectitude that must have affronted the Lieutenant, a pensioner and sworn liege of the King, and certainly shocked the Stoddarts' friends at Salisbury. John Stoddart's career in the years after Hazlitt's return to London in 1812 will clash with his brother-in-law's at almost every turn, so that some account of his life will be useful at this point, quite apart from the light it throws on his sister's.[47]

In 1796, after an exemplary career of conformism and social climbing, first as an undergraduate at Christ Church, and then as a tutor to the titled nephew of the patron who, from a connection we cannot now discern, had paid for his education, the Bishop of Durham, he abruptly abandoned his pupil, donned the Phrygian cap of revolution, and plunged into Grub Street in an agony of political idealism and literary ambition.[48] This short-lived error in his ways came, no doubt, of reading the *Enquiry Concerning Political Justice*, the common source of the staggering illumination that turned the heads of most of the literate young men of the day, however ephemeral and unfruitful it proved in most cases. Whereas in the spring of 1794 he had sent a copy of what he called 'Burke's famous pamphlet', the *Reflections on the Revolution in France*, to his parents, and had 'feared that Sarum was not as loyal' as it ought to have been, by 1798 he had swung round to an opposite extremity of views and blandly assured them that 'an invasion of the country was not much to be dreaded, since peace must appear, to the people on both sides, of more advantage than victory'.[49] A common enough opinion among the patriots of the day: we find it expressed by John Thelwall; but John Stoddart the younger had the fatal gift of absurdity. As extravagant in his republicanism at this time as he was later to be in his '*ultra*' loyalty after his violent reaction from it, he affected the conspicuous short-cropped haircut of the sansculottes, and became derisively known as 'Citizen' Stoddart, just as later, after his volte-face, he became derisively known as Dr Slop. He seemed to invite derision. By the beginning of

---

[46] PRO, Adm. 22/130 (Widows' pension pay-books, 1806), where a certificate of insanity is attached. Similar certificates were submitted annually until her death in 1811.

[47] My account of the Stoddarts, based on the Pinney and other papers, was completed before Dr Woof's lecture was delivered, and it did not seem advantageous to recast it, although Dr Woof's is at some points more expansive (see n. 38 above).

[48] This explains the obscure allusion in 7. 99, 'robbed "Durham's golden stalls" of their hoped-for ornaments, by sending our aspiring youth up to town to learn philosophy of the new teachers of philosophy'.

[49] Pinney Papers, 26 Feb., 14 Mar. 1794; 11 June 1798.

1797 he had met Godwin, Holcroft, Hardy, Horne Tooke, Thelwall, and probably also John Hazlitt, but he had an obstinate turn for incessant and pugnacious disputation which wearied them almost as quickly as his inability to see when he was not wanted. William Hone, a hostile witness writing twenty years later, shows that he had changed little when he says that 'he mistook passionate heat for the enthusiasm of genius, a habit of loud talking for talent, a ranting way of writing for reasoning, and pertinacity of manner for firmness of character'.[50] No doubt there is exaggeration here: Lamb, for all his indulgence, would not have stayed thirty years the friend of such a man; but even Lamb took some time to warm to him, as may be inferred from his first reaction in 1796 when he described him as 'a cold hearted well bred conceited disciple of Godwin'.[51] It is easy to see how such a man would jar on the nervous and self-effacing Hazlitt. In these republican days Stoddart translated a history of the French Revolution, and then Despaze's chronicle of the Directory, and Schiller's *Fiesco* and *Don Carlos*, and either wrote or translated a work called *The Orphan*. Through his Oxford friends Allen and Southey he met Coleridge, and through Lamb and him, Wordsworth, and he also met at this time their lawyer friend Basil Montagu, at whose house in Lincoln's Inn he soon went to live economically. Whether it was with literature or with politics that he first grew dissatisfied we do not know. In 1797–8, urged probably by Montagu, he took his LL B at Oxford and kept his terms at Gray's Inn, but he looked upon law as a stand-by. At the turn of the century he was writing his *Remarks on the Local Scenery and Manners of Scotland*, which he dedicated, no doubt with appropriate self-congratulation, to the eccentric Duchess of Gordon.

He had scraped acquaintance in Edinburgh not only with the Duke of Gordon but also with Lord Perth, Lord Balgonie, and Sir Harry Moncrieff-Wellwood, whose interest he attempted to enlist in furthering his father's promotion. During a month at Margate in 1801 he triumphantly reported to Sarah that he was living near the Duchesses of Gordon and of Manchester: he had avoided, he said, the expensive dissipations of the place, but had been to two balls where he had danced with Lady Georgiana Gordon and a niece of the Duke of Portland. The Revd Mr Collins himself could not have done better. He had in the meantime begun to dabble in investments. Both he and his sister seem to have inherited money from grandparents (probably on the distaff side) or other relations, and also in his case even from outside the family, since the sum of 500 guineas was bequeathed him by a 'benevolent benefactor', an interestingly named Mr Earl, of Salisbury, in 1796. He invested

---

[50] W. Hone, *A Slap at Slop and the Bridge Street Gang* (1821), 4.   [51] Marrs, i. 22.

his sister's money with his own and would have liked to involve his father's also in his affairs, but the Lieutenant proved deplorably cautious. In 1802 he claimed to be worth £1,500 (of which £300 was his sister's), a sum largely made up of the reversion of insurance policies on the lives of friends, but including £200 which his rich Bristol friend Azariah Pinney had promised to *give* him (his italics: Wordsworth was not the only professional beneficiary in the field in those days). By this time he had become engaged to Isabella Wellwood (although Sir Harry was not enthusiastic about the match) and shortly before his marriage was reluctantly obliged, so ingrained had the habits of opportunism become, to extricate himself from a prospective alliance with another Scottish family, the Fullartons of Ayrshire, whose daughter had a dowry of £20,000. These transactions help to explain Hone's later and otherwise obscure jibe about his being particularly interested in Godwin's chapter on Property in *Political Justice*, and also his later disagreement with and contempt for the unendowed and unacquisitive Hazlitt. On his way north to Tullibole Castle to claim his bride in July 1802 he celebrated his return to the political fold after his Jacobinical aberration by a reconciliation with the Bishop of Durham.

The papers in the Pinney family archive that enable us to trace his early career cease abruptly in 1803, the year Azariah died. A few months after his death Stoddart, writing from Salisbury, where his father was ill, violently recalled Sarah from a visit to London, but it was not because of a worsening of the sick man's condition. She was staying at the Wimpole Street house of Azariah's brother John Frederick and Stoddart seems to have discovered that he was as little inclined to continue contributing to support his late brother's friends as his own father had been some years before to indulge him, John Frederick, in his rent-free loan of Racedown Lodge to Wordsworth. He wrote in a great fright for Sarah to come home: 'You have no time to lose because I learn from Mr. Pinney that your further residence at his house will be productive of inconvenience to his family.'[52]

An embarrassing moment. 1803 was a bad year for the Stoddarts. The Lieutenant's death on 17 July seems to have thrown the family into disarray.[53] His will must have been a disappointment: within twenty-four hours John Stoddart applied for the post of King's Advocate in Malta.[54] Accounts in the Lamb letters of his departure and early days in the island refer to his uncertain finances and his sister's 'probable life of poverty.'[55]

---

[52] 5 Apr., the month in which the letters come to an end.
[53] Obituary, *Salisbury and Winchester Journal*, 25 July 1803.
[54] PRO, Adm. 1/3896, 18 July 1803.

[55] Marrs, ii. 133, 141, 143.

To turn to Sarah, it is certain that she did not get on well with her
mother, and this was one reason why she went out to Malta in the year
after her father died (no doubt using part of her small legacy) to join her
brother. Another reason was that she hankered after matrimony, an
ambition so pressing that at the age of twenty-nine she had already made
an embarrassingly large number of futile and humiliating efforts to
procure herself a husband.[56] Like Tabitha Bramble 'she had left no stone
unturned to avoid the reproachful epithet of old maid'. Her artless
gambit was not to wait on events but simply to do the proposing herself.
A letter to her from Mary Lamb, dated September 1803, dwells with
perplexed, perplexing, and perhaps significant ambiguity on her behaviour towards her current swain. In 1822 during the divorce proceedings
Hazlitt, defending his own conduct, told the wife of his confidant, J. R.
Bell, without apparent bitterness but as though believing himself thereby
exculpated, that 'she was no maid when he married her', and when she
noted this remark in her journal she showed no indignation and did not
deny the fact.[57] Whatever mystery is concealed behind the 'jewish
bargain' with her suitor of 1803 to which Mary Lamb refers (possibly
some form of dowry provided by herself), we may be sure that her
neglect of convention and her free and easy attitude struck men as
equivocal if not provocative. While in Malta she was the victim of some
kind of sexual molestation, and years later, in 1822, she was the object of
the drunken ardour of the unpleasant Bell.[58]

Not that she was scatter-brained, nor yet stupid. It is evident from her
remarkable Scottish journal of 1822, describing the books she read, the
pictures she saw, the places she enthusiastically visited and the people
she talked to in her brisk bird-like way, that she was a cultivated woman
with considerable intellectual curiosity. As much, indeed, may be
inferred from her intimacy with the quicksilver-witted Lambs. Her early
letters to her brother, written when she was reading Rousseau and such
works as Marmontel's popular *Bélisaire*, were in French (as part of an
exercise in which they were conscientiously corrected by him).[59] We are
told that she had by heart a good deal of contemporary poetry — Scott,
Byron, Wordsworth, and others, and had gone to the pains of copying
into a commonplace-book extensive extracts from Chapman, Bacon,
Jeremy Taylor, Sterne, Thomson, Dr Johnson, etc.[60] Hazlitt's complaint after their separation that she was indifferent to his writings lacks

[56] Pinney Papers, letters of 11 Sept., 12 Nov. 1794; 28 Sept., 5 Oct. 1795; 13 May 1796, etc. Dr
Woof (n. 38 above) gives a judicious account of Sarah's manoeuvres.
[57] Le Gallienne, 326; Bonner, 245.
[58] Le Gallienne, 294; Bonner, 223.
[59] Pinney Papers, letters of 3 May 1792; 30 May 1793; 16, 20 May 1795, etc.
[60] *Memoirs*, ii. 268-9.

support: it was she who checked his references and looked out the quotations he wished to insert in the blanks he left in his manuscripts, and even after his marriage to another woman she carefully collected and preserved what he published. When he was a parliamentary reporter he made notes on some of the speeches in an old commonplace-book of hers into which she had copied an early version of a poem by a friend of her brother's: the poem was *Christabel*.

All this suggests a frank, downright woman, nothing averse to coaxing a reluctant lover into wedlock, and who, however unlucky with the stolid farmers of Wiltshire or the conventional garrison-officers of Valletta, could well be successful with someone as far out of step with the world as she herself was. Through her brother she had met many of Hazlitt's circle of acquaintances, although she does not appear to have met Hazlitt himself until 1806 or 1807. It was natural that he should feel at home with so close a friend of Mary Lamb, whom he admired, and particularly with a woman who had no nonsense about her, whom we glimpse so comfortably installed at the Lambs' fireside in the Temple, with her feet on the fender and a glass of brandy and hot water in her hand. It seems from Mary's letters that she was affectionate and sensible, and we may judge of her warm heart by Hazlitt's having already at the beginning of February 1808, when he was very ill, abruptly hurried down to her at Winterslow, either because he needed nursing or was anxious at not hearing from her.

We cannot be sure that the old minister and his wife knew enough of their daughter-in-law to be pleased with the match. They could hardly have heard of her slightly tarnished reputation, but if all they were told was that she was a Church of England woman older than their son, and the daughter of a press-gang officer married to a woman of unsound mind, then their discussion of the proposed marriage must have occupied many a long evening in the winter and spring of 1808 in the little wainscoted parlour. It is possible that Hazlitt and his bride paid a brief visit to Wem between the May and the November but more probably it was not until the old couple moved to the neighbourhood of London in 1813 that they saw anything of Sarah, although their daughter Margaret was at Winterslow the year after the wedding.[61] It says much that when the divorce came in 1822 it was to her side that Hazlitt's widowed mother and his sister rallied, and not to his.

Hazlitt, then, when he took out a Special Licence on 26 April 1808[62] to be married in St Andrew's, Holborn (rather to Mary Lamb's disappointment, who was looking forward to a village wedding) was no

---

[61] Moyne, 109.
[62] The licence survives, at Buffalo; Maggs cat. 441 (1923), item 1616.

stranger to narrow circumstances, nor even to poverty. Not much money came into his father's little house, and if the old man tended the garden so assiduously it was because it provided a substantial part of the household victuals.[63] Hazlitt's early letters home mention food-hampers sent up by carrier, in the frugal tradition of the country boy come to live in the big city; and in the essay 'On the Want of Money', remembering his journeyman-portraitist years in the provinces before and after his residence in Paris at the end of 1802, he tells us (claiming that it was an experiment, which no doubt it partly was) that in Manchester he once lived for a fortnight on coffee while executing a commission, and celebrated being paid by treating himself to a humble yet ambrosial meal of sausages and mashed potatoes.[64] His temperament, like his father's, was unsubdued by material circumstances. He would have no difficulty in reconciling himself to the absence of luxuries, to a straightforward, outwardly monotonous existence, so long as he was sure of the bare necessities, and of books, paper, and artists' colours, to continue to read, write, and paint. Between 1799 and 1804 he had earned a barely adequate living with his brush, and in the end he seems to have lost confidence and given up because he could not paint like Titian. The portrait of Lamb in the absurdly improbable but significantly Titian-esque dress of a Venetian senator, painted in October 1804, was apparently his last professional work for seven years, although we do not know that he was paid for it.

Nor do we know how he lived for the next three, unless the help of his father and brother, who themselves had little to spare, kept his head above water. Between 1805 and 1808 he published five books, three of them original works in which he strove to express, as he said, 'the truth and nothing but the truth' of views based upon carefully worded premises and painfully extracted conclusions. The first, the *Essay on the Principles of Human Action*, published 19 July 1805,[65] was the most important. It had finally emerged from the 'gulf of abstraction' in which he had floundered in the years between Hackney New College and his meeting with Coleridge in 1798. Hobbes, Locke, Berkeley, and Hume had led him to the French Encyclopaedists, and it was while reading D'Holbach's *Système de la nature* that he made the discovery which he held to constitute the entire originality of his work as a 'metaphysician', and which he set out to embody in this essay.[66] It came at the start of his career as a writer, and it was fundamental to it. It essentially implied the whole paradox of his character as a realistic romantic idealist, born into

[63] 17. 110.   [64] 17. 180.

[65] *MC* advt. A list of publication dates of most of Hazlitt's works will be found in the writer's review of G. Keynes, *Bibliography of William Hazlitt*, in *Analytical and Enumerative Bibliography*, 6 (1982), 276.
                                                              [66] 9. 56.

the Christian faith and still retaining after that faith had evaporated an
indelible dye of unworldliness. It was the work of a man who, whilst
conscious of the omnipresence and permanence of self-interest and evil,
was seeking (like his ex-pastor friend Godwin) a logic that would justify,
in the teeth of all the evidences and all the pressures of reality, his own
incorrigible bias towards that ethic of altruism in which he had been
nurtured, while at the same time (unlike Godwin) recognizing the
necessary existence and inescapable importunity of those evidences and
those pressures. Shaped by the debate between self-interest and benevolence which had exercised English philosophy for the whole of the
eighteenth century, from Shaftesbury and Butler onwards, his whole life
was to be a sequence of variations on this dichotomy, an unceasing
dialogue between the ideal and the real, the one and the many, the
abstract and the concrete, the self and others, the potential and the
actual, optimism and pessimism, the scholar's study and the market-
place, a renewal of what someone he once quoted called 'the old quarrel
between speculation and practice',[67] a dialogue he was constantly trying
to resolve by an appeal from imagination to reason, whose inadequacies
he would then correct by an appeal from reason back to imagination. In
the *Essay* he elaborates a purely abstract and 'metaphysical' theory of
human action, in a general statement which deliberately neglects the
phenomenological or even the existential experience of the individual
alive in a particular time and place. This paradoxical refusal of
individuality was a limitation amply to be corrected in his later work.
The pamphlets *Free Thoughts on Public Affairs* (1806) and *A Reply to the
Essay on Population by the Rev. T. R. Malthus* (1807)[68] were in a sense
corollaries to the *Essay*, extensions or applications to politics of the
principles there laid down in 'metaphysical' terms. But its true corollary
or complement was not to begin to be developed until he turned to the
familiar form of essay in the *Round Table* in the years before Waterloo,
and though largely immanent in the inspiration behind the critical essays
and lectures that came later, on Shakespeare, the English poets and
comic writers, and Elizabethan literature, it was not to achieve its true
concrete correlative until the first volume of *Table-Talk* in 1821.

In the same year as the *Reply to Malthus* he had published two volumes
intended to make money: the abridgement of Abraham Tucker's *Light of
Nature Pursued* and an anthology of parliamentary speeches with biographical notes, the *Eloquence of the British Senate*.[69] They did not sell well.
Hazlitt seems always to have written only what he chose to write, and
this is true even of these lean years. The *Eloquence of the British Senate*, a

---

[67] 1. 182.   [68] Both advertised *MC*, 2 July 1806 and 6 Aug. 1807, respectively.
[69] The first advertised as due 'to be published in a few days' on 3 Feb., the second as 'published',
on 4 July, both in *MC*.

by-product of his passion for the prose of Edmund Burke, is a companion-piece to the *Free Thoughts on Public Affairs*, which was published — to Lamb's scandal — at author's expense. The abridgement of Tucker's book (a book he esteemed so highly that he gave it great reputation among his friends, including Leigh Hunt at a later date) may equally be regarded as a companion-work to the *Essay on the Principles of Human Action*. None of these books, however, made money. He was publishing in a depressed market.[70] He found that those in the second category, where as editor he relied for his sale on the fame of his authors, fared only a little better than those in the first, the original writings which were the fruit of what he wryly called with his customary self-depreciation, his '*logical way of writing*',[71] so that his marriage followed upon another period of discouragement in which, having failed as a painter, he found that he could not rely upon earning his living by his pen either.

He must at first have been confident of some kind of positive response to the *Essay*: the expectation seemed justified by the excitement of his discovery and the rigour of his demonstration. Instead he found incomprehension and indifference. The reviews with one exception had been bad. The most discouraging was in the formidable organ of the Church of England, the *British Critic*, where he was accused for the first time of illiteracy and impiety. But not the last: illiteracy, an indispensable corollary to the accusation of impiety, was later to be regularly invoked in the furious assaults of the Government press on his writings. This first attack cannot have surprised him, appearing as it did in a Church and State journal founded in the year the French King was guillotined, and just before the Treason Trials, expressly to combat irreligion (and by extension, since any organ of the Church was *ipso facto* an instrument of the State, political disaffection). Philosophy had long been distrusted by Church and State, especially since Hume. And it was no doubt significant also in the eyes of the *British Critic* that the *Essay* was published by Joseph Johnson, a Dissenter.

Nor can Hazlitt have failed to expect the antagonism of the *Anti-Jacobin Review*, although he may have been startled by its brutal contempt, but the almost unanimous hostility of the other journals, the *Critical*, *Eclectic*, and *Monthly* Reviews, was depressing. The exception was the *General Review* of March 1806 which declared him to be 'undoubtedly a most profound and able writer', observing that 'the apparent ease with which he penetrates into the most difficult subjects and conducts the most subtle discussions shows that his mind has been long habituated to those operations which form the talents of a great

[70] See Farington, 5 Dec. 1807.   [71] *Letters*, 159.

metaphysician'. But any pleasure this afforded must have been diminished by its hollowness: it was evident that the reviewer had not understood — had made no attempt to understand — his argument.

The 'invisible link that connected literature and the police' was entirely invisible in contemporary reviews. Godwin's part in the Treason Trials had not been forgotten, despite his continuing studious and unobtrusive mode of life; indeed when he set up his publishing business he prudently used his wife's name. London was then a small place and the friends and agents of those in power were vigilant, and had long memories. Even the retiring Charles Lamb was not exempt from suspicion; his connections with Godwin and Holcroft had not lapsed, and distant echoes of the jibes aimed at him in the *Anti-Jacobin Review* in the 1790s may be detected in the otherwise courteous tone of notices of his work at this time.[72]

The lone voice of the *General Review*, although mildly gratifying, was not convincing enough to prevail against the combined opposition of the orthodox. And if the *Essay* met with incomprehension, *Free Thoughts on Public Affairs* which followed in 1806 was greeted with silence. The notice it might have been expected to receive from those in the same political camp, but did not, ought equally to have been elicited the following year by the anthology with commentaries, *The Eloquence of the British Senate*, but not a word of effective support was uttered in its favour either. The Whig *Critical Review*, nervous of forthright opinions, considered that the compiler had been 'betrayed into great error by the love of appearing as the arbiter of fame and "disposing of honour and of scorn" among the first characters of our age and country'. Uneasily conscious of a disturbing originality in his 'florid' style, it went on to murmur that 'the endless multiplication of his words in attempting to refine his ideas to an excess of subtlety destroys the value of his best observations'.[73]

Nor were his potential allies the Dissenters any more cordial or encouraging. The editor of the *Annual Review* was Arthur Aikin who had entered Hackney New College a few years before Hazlitt. He belonged to a prominent Unitarian family who managed the *Review* amongst themselves: the most prominent contributors were Aikin himself, his father the Revd John Aikin, his sister Lucy, and his aunt Mrs Barbauld. It was as little indulgent as the *Critical Review* to *The Eloquence of the British Senate*. Deploring the 'often unintelligible and therefore unwelcome language', it singled out for special reprobation what it called 'a laboured delineation of the rhetorical taste of the Long Parliament'.[74] By

---

[72] *British Critic*, July 1809, 73 ('in one or two instances alluding unnecessarily to sacred subjects, in a way which we could wish had been avoided'), and see also B. R. Pollin, 'Charles Lamb and Charles Lloyd as Jacobins and Anti-Jacobins', *Studies in Romanticism*, 12 (Summer 1973), 633–47.
[73] 13 (Mar. 1808), 298–9.   [74] *Annual Review for 1807*, 6 (1808), 609.

this was meant Hazlitt's prefatory note to a speech by Bulstrode Whitlocke, and it is curious that the very same passage was later picked out by the *British Critic* as 'a most egregious example of bad taste in writing'.[75] The coincidence perhaps indicates an unwilling recognition by both reviewers of something bafflingly significant in Hazlitt's words at this point, misconstrue them though they might. And indeed the passage they seized on seems to me a remarkable and unique anticipation of the existential basis of the whole of his subsequent work. He always claimed that what he had once made up his mind to he abided by to the end of the chapter. The claim is proved here. The Whitlocke note might have been written on the same day as the quintessential 'On Reason and Imagination' of fifteen years later. In it he deprecates the '*vacuum* of abstract reasoning' and 'artificial logic', and insists that 'the understanding [should be] invigorated and nourished with its natural and proper food, the knowledge of things without it', and, again, that 'ideas' should not be sundered from their 'natural objects'.[76]

As to the *Reply to Malthus*, there was even more likelihood that it would prove politically combustible and there are hints that Hazlitt was particularly watchful of its reception.[77] It got only two notices before he departed a year later for Winterslow. The *Annual Review* reproduced Hazlitt's arguments at length, in ambiguous terms which suggest that the reviewer himself subscribed to them. The convention of the time required non-committal wishy-washy summaries rather than criticisms, but here the abstention (if it was such) is unexpected, since Malthus was a life-long friend of the Aikin family. Lucy Aikin remarked of him in after years that 'no one who knew him could help loving him', going on to ask, apparently oblivious of this notice of 1808 in the family journal, 'but what author of our day has been so much maligned?'[78]

On the other hand, in the *Monthly Review* (in a notice covering two books on Malthus, with Hazlitt's relegated to a perfunctory second place) the reviewer was convinced that Malthus's 'grand proposition' was impregnable and his deductions impeccable. This reviewer 'moderately censured' Jarrold, the author to whom he principally attended, but reserved stronger language for Hazlitt's 'medley of coarse abuse that can boast of neither logic nor learning' from this 'unknown and nameless writer . . . who set modesty, breeding and a sense of decency at defiance', whose 'manner is disgusting and preposterous', marked by 'rudeness and vulgarity', and whose 'style seems to have been formed on that of the most offensive of the daily prints which disgrace the times, degrade the public mind, and pervert its views and feelings'.[79]

---

[75] Aug. 1810, 131.   [76] 1. 147.

[77] See the writer's 'Hazlitt, Cobbett, and the *Edinburgh Review*', *Neophilologus*, 53 (1969), 69–76.

[78] P. H. Le Breton, *Memoirs of the Late Lucy Aikin* (1864), 152–3.   [79] 56 (1808), 53–62.

This twilight mood of doubt and lassitude to which the ill success of his books contributed was not new. It is reflected between jest and earnest (on his friends' part as on his) in the curious 'suicide' hoax of early 1808, in a correspondence in which it was pretended that he had put an end to his life, a hoax suggested, no doubt to its perpetrators, his friends Charles Lamb and Joseph Hume, by the imminence of his marriage, which to them gave it excruciatingly symbolic point.[80] Hume was eight years older than Lamb (who himself was three when Hazlitt was born) and worked like Lamb in a counting-house. Chief Clerk at that time to the Accountant of Stores at the Admiralty Victualling Office, Somerset House, where he had served for over twenty years, but certainly possessed of private means, he lived in some comfort at Montpelier House, Notting Hill, with his wife, a loud-talking, loud-laughing, bustling woman,[81] and their numerous daughters. Again like Lamb, he nursed literary ambitions (the only ambitions the two ever nursed) but, unlike him, he could not write; nor, said Crabb Robinson, would anyone suppose him to be a literary man unless it were from his physical resemblance to Gilbert Wakefield.[82] He was one of the hundred obscure competitors, aspirants to the laurel-wreath, who submitted to the managers of Drury Lane an address for its reopening after the fire in 1812.[83] This rejected address, with its magniloquent exordium, 'The world! a mimic stage!', is only slightly less bad than the execrable translation of the *Inferno* he published in the same year, which his friends swiftly interred in an embarrassed silence rarely broken since (and even then the discredit of it has been implausibly given to his namesake Joseph Hume, MP, the Benthamite radical).[84] He was one of the 'Mitre-courtiers' who attended Lamb's Wednesday nights, along with Captain Burney, brother of the author of *Evelina*, who had sailed with Cook to the South Seas, but whose horizon was now circumscribed by the whist-table; and the Captain's son Martin, Mary Lamb's pet, aged twenty. Among the others were John Rickman of the robust figure and hoarse voice, whose portrait by John Hazlitt was shown at the Royal Academy in 1807,[85] the ebullient, unpolished, well-informed, disputatious, and humorous Secretary to the Speaker of the House of Commons, the friend of Southey's republican youth, who was not yet the reactionary he increasingly became when gagged by overwork on the one hand and rendered speechless with rage on the other by the inefficiency and corruption of the House. He had recently married a wife

---

[80] It is given in *LH*, 61–96. [81] Robinson Diary, 24 Sept. 1812. [82] Ibid.

[83] BL Add. MS 27899, fo. 197.

[84] There is no copy in BL. A copy in the Mitchell Library, Glasgow, has an engraving of Joseph Hume, MP, tipped in against the title-page.

[85] Royal Academy, annotated cat., 1807, no. 830.

who was 'sensible, prudent and genteel-mannered, but without modern
accomplishments'.[86] Edward Phillips, his clerk, was an ardent card-
player, possessed of a '£100 a year and desperate debts', whose sporting
loyalties (and calculations) veered between whist and cribbage. The
learned, slovenly, absent-minded, and child-like George Dyer, historian
of the University of Cambridge, was a Unitarian, a poet, and a radical.
James White, Lamb's schoolfellow, the open-hearted joyous author of
Falstaff's Letters, was also the open-handed founder of the annual feast of
chimney-sweeps. The easy-going William Ayrton, who always lagged
behind in conversation, and was often caught with his mouth open, had
much in common with Will Honeycomb of the *Spectator*. But he had other
talents: the son of a musician, he married the daughter of another, who
was also the sister of S. J. Arnold, manager of the Lyceum Theatre, then
called the English Opera House, and was soon himself to become one of
the foremost musical impresarios of the age. Lastly, although a less
frequent visitor than the others, there was Coleridge, Lamb's other
schoolfriend, who as a talker never seemed to abandon the measured
tireless pace of the long-distance runner (with lamentable effect upon the
nerves of poor Mary Lamb).

To return to the suicide joke of 1808, the pretence that Hazlitt had
died by his own hand was a jest of an order recurrent in Lamb's writings,
where it seems to have served his fissured personality as a protective
device, an amateurish psychiatric homeopathy, following on his
mother's terrible death. Reading between the jokes in Hazlitt's contribution (his letter of 10 January 1808 to Hume) we can guess at his
dissatisfaction with the rewards of his painting and writing. The banter
does not conceal the unpromising state of his purse. His brave words to
his parents twelve months earlier, 'I have done what I wanted in writing
and I hope I may in painting', were meant to convey that he was pleased
with himself: they do not assert that he was pleased with the profits.[87] We
gather from the 'suicide' correspondence that '[borrowing] money from
his friends . . . was at all times his constant custom' and that he was then
living 'upon the well-timed contributions of a few chosen friends who
knew his worth'.[88] We find him supping with Holcroft, Godwin, the
Lambs, Hume ('all the world know that he could always smell an
Aitch-bone 4 miles off', said the last),[89] and while we know from Godwin's
diary that he entertained a number of friends at his lodgings on at least
two occasions that winter this may have been to impress his prospective
brother-in-law who was present and whose disapproval of the marriage
he apprehended. His own anxiety about his marriage was, however,

---

[86] BL Add. MS 60373, at 12 May 1836.   [87] *Letters*, 91.   [88] *LH*, 71, 80–1.
[89] Ibid. 94.

stronger than his uneasiness about the disapprobation of Stoddart (and of Stoddart's Scottish wife, as is hinted by a reference in his letters to 'the prudence of Edinburgh')[90] and when he failed to hear from Sarah at the beginning of February he hurried down to Winterslow, though he had not yet recovered from his illness, and without waiting to tell his close friends the Lambs.

All that we learn of his behaviour at this time suggests that in retiring to the Winterslow cottage he was sheltering from ill health and poverty, that he was seeking a retreat, or at least a respite, from a nagging sense of failure.

This turning away with distaste from the rough and tumble of life, this inclination to abandon the unequal struggle of existence and withdraw into the dreaming life of the mind was inevitably and at all times characteristic of Hazlitt. He inherited from his father, according to his grandson, 'a constitutional tendency to meditate and brood over questions, and to prefer to think about a subject than to write about it'.[91] Even when he was pleased with his thoughts he felt no urge to impart them to others. In 1823 at Winterslow, remembering his first arrival in the village, he said, 'Here I came fifteen years ago, a willing exile; and as I trod the lengthened greensward by the low wood-side, repeated the old line, "My mind to me a kingdom is!".' The thought, as he says, was constantly present in him: 'I found it so then, before, and since.'[92] In 1820, when he reflected on his by now well-established career as a writer, he found the inward activity of the mind richer and more satisfying than its external manifestation in writing:

I have more satisfaction in my own thoughts than in dictating them to others: words are necessary to explain the impression of certain things upon me to the reader, but they rather weaken and draw a veil over than strengthen it to myself. However I might say with the poet, 'My mind to me a kingdom is', yet I have little ambition 'to set a throne or chair of state in the understandings of other men.' The ideas we cherish most, exist best in a kind of shadowy abstraction,

'Pure in the last recesses of the mind;'

and derive neither force nor interest from being exposed to public view. They are old established acquaintance, and any change in them, arising from the adventitious ornaments of style or dress, is hardly to their advantage.[93]

This was at a time when he had already published much, but most of it was published to earn a living, and the sentiment itself was true at all times, as his grandson attests: 'It was offering a violence to his nature to

---

[90] *Letters*, 105.   [91] *Four Generations*, i. 146–7.   [92] 12. 121.
[93] 8. 6–7.

enter the arena of action and compete for the prizes of life.'[94] The theme
echoes from end to end of his writings. From one point of view it was a
Romantic withdrawal from the bustle of cities to a private and ideal
country peace, and certainly it was often purely that. From another it
was Augustan, in that what he busied himself with, what he constantly
reverted to in his solitude (whether rural or even urban), while
remaining apparently *au-dessus de la mêlée*, was the study of man, a
'looking abroad into universality' (in the words of a favourite quotation
from Bacon, whom he placed very high, as one of the four greatest
Englishmen[95]), a meditation upon the conditions of man's existence in
society and upon the realities of the *mêlée*, and behind all this the
ever-remembered principles of civil and religious liberty. This is the
essential polarity of Hazlitt's nature, a polarity not perhaps effectively
represented by the figure of Bacon (although Bacon's essays may span
themes as widely sundered as 'Of the Greatnesse of Kingdomes' and 'Of
Gardens') but rather by two other gods in Hazlitt's pantheon, Rousseau
and Burke; his life was a recurrent tension in which he was drawn first
towards one of these two opposed symbolic figures and then towards the
other.

   'For many years of my life I did nothing but think', he said in the essay
'On Living to One's Self'. Indeed, he never willingly did anything else,
and the necessity to get a livelihood was an irksome drag on the real
business of existence. And when, over and above the need to earn his
bread, the urgencies of civil and religious liberty forced him to descend
into the arena, it was only into the annexe arena of print, not of action
proper. Even then his thoughts on public affairs remained free thoughts:
as he said of Holcroft, perhaps with less reason, he was 'a purely
speculative politician'.[96] He never meddled with any kind of party
activity, nor even with any party, unless it was for a brief moment in the
Westminster election of 1819, in which, as a householder, holding for
once in his life the franchise, he supported the candidature of John Cam
Hobhouse and cast 'an ineffectual' vote for him. His antipathy to party
organization, lobbying, and horse-trading, is confirmed in his condemnation of Francis Place's part in this campaign.[97]

   His colleague of 1813 in the House of Commons gallery, Peter
Finnerty, might harangue electors from the hustings when off duty, but
not he. Instead he remained as remote from the 'vulgar democratical

---

[94] *Four Generations*, i. 147.

[95] The others, he said in his lecture on Bacon, were Newton, Shakespeare, and Milton: this is
not in the text of the *Lectures on the Dramatic Literature of the Age of Elizabeth*, but is to be found in the
Examiner report, 26 Dec. 1819, 831. The suppression has escaped Hazlitt's eds.

[96] 3. 132.

[97] Westminster Poll Books, Westminster Public Library; 12. 379; 12. 382.

mind' (the phrase he used to Coleridge in 1798) as he did from Tory prejudices. He 'lived to himself', an existence he defined as living in the world as in it, not of it. This definition is so evident a key to his character that, familiar though it is, we had better quote the passage that completes it:

. . . it is as if no one knew there was such a person, and you wished no one to know it: it is to be a silent spectator of the mighty scene of things, not an object of attention or curiosity in it; to take a thoughtful, anxious interest in what is passing in the world, but not to feel the slightest inclination to make or meddle with it. . . . He who lives wisely to himself and to his own heart, looks at the busy world through the loop-holes of retreat, and does not want to mingle in the fray. 'He hears the tumult, and is still.' He is not able to mend it, nor willing to mar it. He sees enough in the universe to interest him without putting himself forward to try what he can do to fix the eyes of the universe upon him. Vain the attempt! He reads the clouds, he looks at the stars, he watches the return of the seasons, the falling leaves of autumn, the perfumed breath of spring, starts with delight at the note of a thrush in a copse near him, sits by the fire, listens to the moaning of the wind, pores upon a book, or discourses the freezing hours away, or melts down hours to minutes in pleasing thought. All this while he is taken up with other things, forgetting himself. He relishes an author's style, without thinking of turning author. He is fond of looking at a print from an old picture in the room, without teasing himself to copy it. . . . He hardly knows what he is capable of, and is not in the least concerned whether he shall ever make a figure in the world. . . . he looks out of himself at the wide extended prospect of nature, and takes an interest beyond his narrow pretensions in general humanity.[98]

This passage, with its lingering echoes of Wordsworth's contemplation of the world at a distance, in the blessing of the gentle breeze, was written at Winterslow in 1821. As we shall see in a moment it is only a little wider and a little more explicit than what he had said in the closing paragraph of the *Reply to Malthus* just before withdrawing there in 1808. The expressions 'loop-holes of retreat' and 'hears the tumult and is still' are taken from a favourite, oft-quoted passage on the retired life in that breviary of repose, Cowper's *The Task*, while the phrase 'discourses the freezing hours away' comes unacknowledged from a passage of the same tenor in *Cymbeline* where Belarius, among the Welsh mountains (the mountains at which Hazlitt's boyish eyes had gazed from Wem, blue against the western sky), tells his young companions Guiderius and Arviragus how fortunate is their solitary country existence compared with what he has known — the city's usuries, the sly arts of the court, and the uncertain favour of princes. Of some forty quotations from *Cymbeline* in Hazlitt's works, over half are taken from this short passage

[98] 8. 91–2.

where Belarius lauds the satisfactions of the quiet life away from cities, like Amiens in the Forest of Arden singing of the winter wind and man's ingratitude: we are not surprised when Hazlitt at the end of his life tells us that he has never owned a watch ('there's no clock in the forest').[99] The dialogue of Belarius and the youths was one of the extracts included in the schoolbook, Enfield's *Speaker*, which Hazlitt studied as a boy at Wem and from which he evidently received the proverbially indelible impression common to earnest minds in their first contact with the printed word. That popular eighteenth-century anthology of prose and verse (including parliamentary oratory) also contained a passage which was an antidote to this counsel of withdrawal. It was a call to duty which, as we shall see, made an equally ineffaceable mark on Hazlitt's memory.

The theme of withdrawal is implicit in the fragment 'On the Love of Life' published in the *Morning Chronicle* upon his return to London in 1813. It later achieved the significant prominence of the first page of his first book of collected essays, where his analysis of the permanence of hope as a mainspring of human action paradoxically reposes on a constant disappointment and mistrust of life (a prelude to his work that symmetrically balances the unexpected concluding paradox of his death-bed words, 'Well, I've had a happy life'):

The love of life is, in general, the effect not of our enjoyments, but of our passions. We are not attached to it so much for its own sake, or as it is connected with happiness, as because it is necessary to action. Without life there can be no action — no objects of pursuit — no restless desires — no tormenting passions. Hence it is that we fondly cling to it — that we dread its termination as the close, not of enjoyment, but of hope. The proof that our attachment to life is not absolutely owing to the immediate satisfaction we find in it, is, that those persons are commonly found most loth to part with it who have the least enjoyment of it, and who have the greatest difficulties to struggle with, as losing gamesters are the most desperate. And farther, there are not many persons who, with all their pretended love of life, would not, if it had been in their power, have melted down the longest life to a few hours.[100]

The attractions of peace and solitude had already appeared, before this *Morning Chronicle* article, in a wry, self-mocking guise, but coloured by a deep yearning, almost at the end of the settled Winterslow years, in the peroration to the last lecture on philosophy delivered in London in April 1812, when he related (as we are told by Crabb Robinson — the lectures were not published) the Indian legend of a Brahmin so devoted to meditation that he neglected to wash and was transformed into a monkey but even then retained his original propensities and kept apart from the other monkeys: his sole delight was in studying metaphysics,

[99] 17. 245.   [100] 4. 1-2.

while subsisting on coconuts. ' "I, too," said Hazlitt, "should be very well contented to pass my life like this monkey, did I but know how to provide myself with a substitute for coconuts." ' The example of the Brahmin-monkey was remembered in 1815 in his 'Mind and Motive' in the *Examiner*, where almost the same words appear as in Crabb Robinson's version.[101]

In 1818, in *Lectures on the English Poets*, he envies the 'learned leisure and contemplative indolence' implied in Gray's *Letters*, and the way the poet 'moralises through the loopholes of retreat, on the bustle and raree-show of the world . . . His life was a luxurious, thoughtful dream. . . . What a happiness never to lose or gain any thing in the game of human life, by being never any thing more than a looker-on!'[102] And in the same lectures, although deprecating the indolence of Thomson, no doubt because he feared it in himself, he yet envied his courage (or nonchalance) in 'dreaming out his existence'. It is the burden of the introductory lecture that we all live in a world of our own making. 'If poetry is a dream, the business of life is much the same. If it is a fiction, made up of what we wish things to be, and fancy that they are, because we wish them so, there is no other nor better reality.'[103] Thus in 1818. At other times the fear of being mistaken comes uneasily to the surface, as in 1822, when the contemplation of his own past life gave him a troubled suspicion of failure:

In looking back, it sometimes appears to me as if I had in a manner slept out my life in a dream or trance on the side of the hill of knowledge, where I have fed on books, on thoughts, on pictures, and only heard in half-murmurs the trampling of busy feet, or the noises of the throng below [an image possibly inspired by his stay at Winterslow, on the slope overlooking the high road to London]. Waked out of this dim, twilight existence, and startled with the passing scene, I have felt a wish to descend to the world of realities, and join in the chase. But I fear too late, and that I had better return to my bookish chimeras and indolence once more![104]

But even when the urge to live seems about to reassert its sway, as at this time in 1822, where he is describing his pleasure in going on a journey, no aim or destination is proposed, it is really the note of withdrawal that is being sounded: 'Oh! it is great to shake off the trammels of the world and of public opinion — to lose our importunate, tormenting, everlasting personal identity in the elements of nature, and become the creature of the moment, clear of all ties.'[105] The theme reappears in January 1823 in 'My First Acquaintance with Poets', where he speaks of the obscure

---

[101] Morley, 70; 20. 51–2. The essay may have been based on the notes of the lectures.
[102] 5. 118.   [103] 5. 3.   [104] 8. 326.   [105] 8. 185.

existence his father spun out, poring endlessly over the Bible and the Commentaries:

. . . though the soul might slumber with an hieroglyphic veil of inscrutable mysteries drawn over it, yet it was in a slumber ill-exchanged for all the sharpened realities of sense, wit, fancy, or reason. My father's life was comparatively a dream; but it was a dream of infinity and eternity, of death, the resurrection, and a judgment to come.[106]

It appears unchanged in 1827, on the last page of 'On a Sun-Dial', where he likens his life to his father's, and his thoughts revert to Winterslow, and to the same image of the crucible of happiness:

'What I like best is to lie whole mornings on a sunny bank on Salisbury Plain, without any object before me, neither knowing nor caring how time passes, and thus "with light-winged toys of feathered Idleness" to melt down hours to moments.'[107]

And again in 1828, during a further stay in Winterslow, the epigraph to 'A Farewell to Essay-writing', 'This life is best, if quiet life is best' from the old familiar passage in *Cymbeline* introduces recollections of earlier days at Winterslow, when he had first come to live there, and had no suspicion, in that dreaming world, far from the tumult of the streets and the febrile hurry of newspaper-offices, that he would ever go back to London to be pitchforked into journalism, still less that he would ever become a successful author. It makes a further appearance in 1830, in 'The Free Admission', when it is the theatre that provides the 'loop-holes of retreat'. And then a final appearance, in which he no longer strives to keep nostalgia in check. To yield to it was always a temptation, but he always resisted it as a betrayal of life, and we should add to the number of the polarities of his nature this alternation between a lapsing into the past and a welcoming of the future. Hence his describing Crockery, a character in the play *Exit by Mistake* who, returning from abroad, regrets all the vanished inconveniences and groans over all the improvements, as one of the drollest inventions in the contemporary theatre: it was one he was peculiarly fitted to recognize and relish.[108]

After the public and private disappointments of 1814–15 he had turned more and more to the past: all that is worth remembering in life, he said, is the poetry of it, and this feeling underlies his last essays, which are the true prose-poems of English Romanticism. Later in the same year, then, in 'The Letter-Bell', the pastoral theme finally merges with the themes of time and memory, and spanning a period of thirty years and more, in a meditation on hope and regret, on the love of life, on the

[106] 17. 110.   [107] 17. 245.   [108] 7. 280.

allegorical progress of the weary traveller, it becomes etherialized in a Romantic Proustian reverie, into the pure music of the spirit:

As I write this, the *Letter-Bell* passes: it has a lively, pleasant sound with it, and not only fills the street with its importunate clamour, but rings clear through the length of many half-forgotten years. It strikes upon the ear, it vibrates to the brain, it wakes me from the dream of time, it flings me back upon my first entrance into life, the period of my first coming up to town, when all around was strange, uncertain, adverse — a hubbub of confused noises, a chaos of shifting objects — and when this sound alone, startling me with the recollection of a letter I had to send to the friends I had lately left, brought me as it were to myself, made me feel that I had links still connecting me with the universe, and gave me hope and patience to persevere. At that loud-tinkling, interrupted sound (now and then), the long line of blue hills near the place where I was brought up waves in the horizon, a golden sunset hovers over them, the dwarf-oaks rustle their red leaves in the evening-breeze, and the road from [Wem] to [Shrewsbury], by which I first set out on my journey through life, stares me in the face as plain, but from time and change not less visionary and mysterious, than the pictures in the *Pilgrim's Progress*.[109]

And so in 1830, in the closing stages of the dialogue between hope and disappointment, we find the final recognition of what had for so long been instinctively assented to, by the son as by the father before him, of the superiority of the life of contemplation over the life of action. He reverts to the position briefly stated in 1824: 'It is better to desire than to enjoy — to love than to be loved', a position unconsciously echoed at the end of the century by Proust, the late flower of Romanticism, who might well, if he had ever heard of Hazlitt, have been paraphrasing these very words when he said, 'L'ambition enivre plus que la gloire; le désir fleurit, la possession flétrit toutes choses; il vaut mieux rêver sa vie que la vivre, encore que la vivre ce soit encore la rêver, mais moins mystérieusement et moins clairement à la fois, d'un rêve obscur et lourd . . . .'[110]

Let us go back twenty-three years to the first false start of his career as a writer, where the permanent state of conflict between the two opposing forces within him is unconsciously but vividly illustrated by his ironical assertion that the only advantage of Malthus's argument is that 'it would be the most effectual recipe for indifference that has yet been found out. No one need give himself any farther trouble about the progress of vice, or the extension of misery.' He reflects bitterly that he of all people ought to be the first to welcome it as an unhoped-for liberation from the eternal corrosive brooding that filled his days: '. . . no one would have more cause than I to rejoice . . . for I have plagued myself a good deal about

---

[109] 17. 377.
[110] 'Characteristics', No. ccv (9. 198); *Les Plaisirs et les jours* (definitive edn., Paris, 1924), 185.

the distinctions of right and wrong'.[111] The passionate unremitting pursuit of these distinctions must often have worn him out. He said the same thing years later: 'There is nothing more likely to drive a man mad, than the being unable to get rid of the idea of the distinction between right and wrong.'[112] During one of the bouts of depression frequent in the bad year of 1822 he became convinced that thinking beyond his strength, and feeling more than he needed to, about so many things, had withered him up and made him old before his time.[113] Now in 1807 at the end of the *Reply* he appears to sink back in lassitude, to abandon both the struggle and the irony. In a peroration of extraordinary and significant ambiguity, in a tangle of accusation, self-criticism, envy, and desire, he commends (condemns?) the tranquillity of 'living to one's self':

To persons of an irritable and nervous disposition, who are fond of kicking against the pricks, who have tasted of the bitterness of the knowledge of good and evil, and to whom whatever is amiss in others sticks not merely like a burr, but like a pitch-plaister, the advantage of such a system is incalculable.—
    Happy are they, who live in the dream of their own existence, and see all things in the light of their own minds; who walk by faith and hope, not by knowledge; to whom the guiding-star of their youth still shines from afar, and into whom the spirit of the world has not entered! They have not been 'hurt by the archers,' nor has the iron entered their souls. They live in the midst of arrows, and of death, unconscious of harm. The evil thing comes not nigh them. The shafts of ridicule pass unheeded by, and malice loses its sting. Their keen perceptions do not catch at hidden mischiefs, nor cling to every folly. The example of vice does not rankle in their breasts, like the poisoned shirt of Nessus. Evil impressions fall off from them, like drops of water. The yoke of life is to them light and supportable. The world has no hold on them. They are in it, not of it; and a dream and a glory is ever about them.[114]

This passage so exactly expressed a complex attitude constant to his mind that he used it again in an *Examiner* essay of 1815 and more significantly as conclusion to an essay of 1818 in which he was thinking of his father.[115] The image of the fatal archers, borrowed like so many others from Cowper's *Task*, reappears in a variant in 1823 when, passing by Dulwich College, he thinks of his own days at Hackney and of the happy ignorance of the schoolboy: 'He sees not the archers taking aim at his peace; he knows not the hands that are to mangle his bosom.'[116] Its recurrence may well hint at a profound and bitter and specific reason for

---

[111] 1. 283.   [112] 5. 111.
[113] *Letters* 267 (Hazlitt to P. G. Patmore (9 June 1822)).   [114] 1. 284.
[115] 'Mind and Motive', 9 Apr. 1815 (20. 53), and 'On Court Influence', *Yellow Dwarf*, 10 Jan. 1818. The par. was omitted (it had already served three times) when the piece was reprinted in *Political Essays* (1819).   [116] 10. 18.

his retirement from the world, a mysterious secret hurt, a wound inflicted by the deadly marksmen, such as was intimated in Joseph Hunter's notes on the Hazlitt family, and spoken of as if from knowledge after Hazlitt's death by Mrs Montagu, and a hundred years later guessed at by the discerning Middleton Murry, a secret wound somehow related to the violence of his passions, to an inner warfare hauntingly shadowed forth in an obscure passage of *The Plain Speaker* where the hunters again emerge:

The 'winged wounds' that rankle in our breasts to our latest day, were planted there long since, ticketed and labelled on the outside in small but indelible characters, written in our blood, 'like that ensanguined flower inscribed with woe:' we are in the toils from the very first, hemmed in by the hunters; and these are our own passions, bred of our brain and humours, and that never leave us, but consume and gnaw the heart in our short life-time, as worms wait for us in the grave![117]

Or perhaps his retirement to Winterslow may have been no more than a lull in the struggle constantly renewed by the contradiction between the two eternally opposed factors: on the one hand his unshakeable conviction (at once temperamental and reasoned) of the natural disinterestedness of human behaviour, and on the other the unmistakable evidence that in the world the principle of obstinate selfishness was the stronger. He had not yet acquired patience to recognize that 'evil is inseparable from the nature of things',[118] nor arrived at his conception of the paradoxical but manifest 'theoretical benevolence, and practical malignity of man'.[119]

At any rate, by November 1808 when he and Sarah left London for Winterslow he had made a choice. He was abandoning the literary and artistic market-place, but he did not intend to wallow in bucolic sloth like his brother's friend, the gifted, inglorious Dan Stringer, who threw up his career after a brilliant start in the Royal Academy School in the 1770s to retire to his native Knutsford, and whom he was shocked to find there in the early 1800s, a Tony Lumpkin, heedlessly drowning his talent in Cheshire ale and rejoicing in the company of the lubberly, patronizing, local squires. Twenty years later he was still saddened by Stringer's wilful, wasteful renunciation: '[He] seemed to have given up all thoughts of his art. Whether he is since dead, I cannot say: the world do not so much as know that he ever lived!'[120]

[117] Howe, 11–12; NLS MS 665, fos. 61–2 (Mrs Montagu to Jane Carlyle, 13 Oct. 1830, '. . . a bitter sense of injury when a boy . . . curdled all that should have been milky in his nature'); *Athenaeum*, 30 Sept. 1922, 855; 12. 234.

[118] 11. 79, where he says this of Southey.   [119] 20. 343.

[120] 8. 21; see also 11. 199–200. Stringer was dead. Having opted for the humbler destiny of a provincial Professor of Painting (as he is described on his tombstone at Knutsford), he had died at

Hazlitt had not relinquished the aspiration to fame which was still to be the long-term objective of his simple if not narrow country existence. He still hoped to achieve a more comprehensive, a more persuasive formulation of his 'metaphysical system'. It had taken up so many years of his life and brought so many anxious thoughts, that he would grieve to see it entirely lost. He dared also hope that success would eventually crown his efforts with the brush and his search for 'the true secret of Titian's golden hue and the oleaginous touches of Claude Lorraine'.[121] Claude was the presiding genius of this period of his life. In his room at Winterslow there hung a print of the *Landscape with the Arch of Constantine*, which he used to contemplate by the hour together day after day, 'sighing out his soul' into it. Eighteen years later he chose it to epitomize his journey to Rome as frontispiece and sole illustration of his *Notes of a Journey through France and Italy*. When he walked out of a summer evening with his Claude glass in his pocket it was to seek under the calm of the slow-fading sky the pure remote distances of those idealized landscapes. He was content. 'A life', he said later, 'spent among pictures, in the study and the love of art, is a happy noiseless dream: or rather, it is to dream and to be awake at the same time; for it has all "the sober certainty of waking bliss," with the romantic voluptuousness of a visionary and abstracted being. They are the bright consummate essences of things.'[122] And he goes on to quote Milton's sonnet 'To Mr. Lawrence', the apologia of pastoral and domestic tranquillity, 'and we may say "that he who of these delights can judge and knows to interpose them oft, is not unwise." ' However, in painting he remained to the end a sentimentalist; his feeling was all derivative; he never threw off the paralysis of imitation, nor moved forward to the frontiers of originality that he later successfully crossed in his writing.

His ultimately untroubled confidence in his mode of life at this time, his secure if muted faith, and his modest ambition, were already anticipated in a paragraph of the *Essay on the Principles of Human Action* where he celebrates the obscure intellectual victories of the *promeneur solitaire*:

There are moments in the life of a solitary thinker which are to him what the evening of some great victory is to the conqueror and hero — milder triumphs long remembered with truer and deeper delight. And though the shouts of multitudes do not hail his success, though gay trophies, though the sounds of music, the glittering of armour, and the neighing of steeds do not mingle with his joy, yet shall he not want monuments and witnesses of his glory, the deep

---

the age of 54 on 2 Aug. 1808 (not some years before 1796, as Farington believed (Farington, 27 Oct. 1796)). There is an impressive self-portrait in E. Waterhouse, *Dictionary of British Eighteenth Century Painters* (1981), 361 (which gives the date of Stringer's birth, 14 June 1754, but not of his death). [121] *Letters*, 120. [122] 8. 173.

forest, the willowy brook, the gathering clouds of winter, or the silent gloom of his own chamber, 'faithful remembrancers of his high endeavour, and his glad success,' that, as time passes by him with unreturning wing, still awaken the consciousness of a spirit patient, indefatigable in the search of truth, and a hope of surviving in the thoughts and minds of other men.[123]

[123] 1. 46.

2

The Loopholes of Retreat

My mind to me a kingdom is.

Edward Dyer

In the three years following his marriage Hazlitt published only one book for money, the *New and Improved Grammar of the English Tongue* (1809). Not that it was mere lowly hackwork. Like everything he wrote it bore the stamp of originality, and thirteen years later he still thought well enough of it to give it as a Christmas present to his son. Disappointed in the autumn of 1808 over the Bourgoing translation, he accepted Godwin's invitation to contribute an English grammar to the Juvenile Library. But in the event, if he had been vexed by the ill success of his earlier writings, he was shocked by the fate of this new venture, which was scuttled even before it was properly launched, and, worse still, scuttled by the very man who had commissioned it.

The story is curious. It seems to have happened somewhat as follows. Between 25 May and 18 June 1809, before the illness of Hazlitt's child recalled him to Winterslow, he and Godwin discussed his draft, from which he then made out a final version. This he delivered about mid-September, and on 5 November told his father that it was printed and would soon appear. In the mean time Godwin, in a flash of inspiration, had 'made an entirely new discovery as to the way of teaching the English language', which produced his 'New Guide to the English Tongue', included in a new edition of Mylius's *School Dictionary of the English Language* on 22 September. This guide he now republished within the same covers as Hazlitt's *Grammar* on 11 November. A favourable notice of the *Grammar* in the *Critical Review* of December did not satisfy him. He itched to improve the sales of the volume by improving the *Grammar* itself, and in March 1810 wrote to Hazlitt asking if he would mind his abridging it so as to make it more acceptable to 'Heads of Seminaries of Education'. On the 30th of that month Hazlitt replied.

Dear Sir,

I was not at all offended, but a good deal vexed at the contents of your former letter, having had three books which I have written suppressed, and as I had taken some trouble with the grammar, I thought it might answer the purpose, and as you seemed to approve of what I had done to it, I was sorry to be dashed in pieces against the dulness of schoolmasters.[1]

However, he accepted Godwin's decision to abridge, and Godwin set to work. On 29 June 1810 the *Morning Chronicle* advertised the abridgement with the following note:

Published This Day: Many of the most respectable Heads of our Seminaries of Education having expressed a desire that the improvements contained in Mr. Hazlitt's *Grammar* should be brought out in a form better adapted for the use of the Junior Classes of their Pupils, the following is now given to the public in compliance with this suggestion.

Hazlitt's forbearance may perhaps be explained by his having probably sold the copyright to Godwin outright, but his disappointment is conveyed by the curious word 'suppressed', which can hardly mean 'rejected'. Only one of his books was turned down, that we know of, and that was the Bourgoing, which he cannot be said to have *written*. The 'suppression' must refer to the hostility or neglect of reviewers which dashed the sale of the *Essay, Free Thoughts on Public Affairs*, and the *Reply to Malthus*.

He had already by this March of 1810 engaged to write the life of Holcroft who had died twelve months earlier, but for the rest of his stay at Winterslow he undertook no more commissioned books. He and his wife were content thereafter to live upon her small annual income of £80. Its bare sufficiency will be apparent if we remember that although Lamb had a salary of £170 a year, exclusive of bonuses and overtime, and also a small private income, he and his sister were unable to save a penny: Mary was not even able to give Sarah a wedding present.[2] We can be pretty sure that Hazlitt spent some of his time working in the garden and that if he cultivated anew the flowers he had once tended so carefully at Wem he conformed also to the husbandry of country-dwellers in all times and planted at least the humble undemanding potato and cabbage, and his favourite lettuce, even though he may not have risen to the broccoli and kidney-beans that had been his father's pride.

Sarah's finances are obscure despite the survival of her father's will, dated 2 January 1795, in which she was left two houses in Salisbury and an unspecified amount in 5 per cent annuities, to come to her when she was twenty-five. She reached that age in 1799, before her father's death, and either these legacies were thereupon inexplicably assigned to her, or she inherited money from another source. At all events, by 1802 she had

---

[1] Dated '[Latter part of June 1809]' in *Letters*, 108. I give my reasons for proposing 30 Mar. 1810 as the date in 'The "Supression" of Hazlitt's *New and Improved Grammar of the English Tongue*: A Reconstruction of Events', *The Library*, 6/9 (1987), 32–43.   [2] Marrs, ii. 282.

£300 in five per cents available for investing in her brother's schemes.[3]
After her father's death she must have spent a good deal of her
inheritance during her stay in Malta, in an effort to dazzle the garrison
officers. No Winterslow property is mentioned in the will; the two
cottages she owned there may have been bequeathed to her by her
grandparents on one or other side. In 1806 Sarah was back in England
and by October had settled in Salisbury. It was probably there, in the
town, perhaps in some neighbour's house on a market-day, that she met
young William Dowling,[4] the farmer to whom she was engaged before
marrying Hazlitt; or it may have been during a visit to her property in
Winterslow, where he lived on his uncle's farm, and where, at some
point after the engagement, she contemplated setting him up on a farm
of his own.[5] The match fell through in 1807, but she seems to have 'laid
out so much money on her cottage',[6] apparently in repairs, in anticipation of setting up house there, that she had none left to buy furniture. In
January 1808 Hazlitt refers with mock pathos to the 'ruinous state of a
cottage or tenement which he hope[d] one day to call his own'.[7] On the
other hand, Peggy Hazlitt tells a different story: according to her the
cottage her brother and his wife occupied at Winterslow was 'offered' to
the happy couple by John Stoddart.[8] Whether this means a gift, a
rent-free tenancy, or a nominal rent, is again not clear, but although
Stoddart had a hard head for money he was a good brother, as the
Lambs recognized.[9] Yet another, and equally unclear version comes
from William Hazlitt the younger who years later told his son William
Carew Hazlitt that the cottage 'was bought in after years by her brother,
who invested the money with what money was left her by her parents and
paid her the interest up to her death'.[10] Whatever the truth of the matter
John Stoddart comes out of it well. We find him in 1806 sending his
sister £50 'to make her comfortable', and there were probably other such
gifts. And years later we find him making provision for his nephew,
Hazlitt's orphaned son.[11] But it seems that at this moment, in 1808, he
was not eager to contribute to Hazlitt's support, and the kind-hearted
Mary Lamb passed some anxious moments over his discouraging
attitude to the match.[12] Hazlitt was aware of his reluctance but remained

---

[3] Pinney Papers, letters of 21, 29 Aug. 1802. The preceding facts I take from Howe.

[4] A good Wiltshire name: it is the Salisbury attorney's in *Tom Jones*.

[5] No William Dowling appears in Winterslow Parish Register. A William Dowling was married
in neighbouring Broughton, 23 July 1798, and is shown also in the West Winterslow Land Tax
Assessment Register (County Record Office, Trowbridge) as a tenant-farmer. It is probable that
Sarah's William Dowling was his nephew (see Marrs, ii. 242).

[6] Marrs, ii. 264.   [7] *Letters*, 97.   [8] Moyne, 109.   [9] Marrs, ii. 263.

[10] BL Add. MS 38913, fo. 72.   [11] BL Add. MS 38898, fo. 17.

[12] Marrs, ii. 263–4, 268–9.

unperturbed. He knew that his future brother-in-law disliked him (this was almost certainly from political feeling), and suspected that his brother-in-law's Edinburgh wife disapproved of him too, but 'was not terrified'.[13] The plans he proposes at the end of a pre-marriage letter to Sarah are full of optimism but at the same time are ominous of shifts and devices impending: 'I think you had better sell the small house, I mean that at 4.10, and I will borrow a £100. So that we shall set off merrily in spite of all the prudence of Edinburgh.'[14]

Almost as little is known of their life at Winterslow as of their finances. The ten letters that survive from this period are mostly about business. We must depend largely upon what we can piece together from his essays and upon reflected glimpses in the letters or journals of others. All we have in that way for Winterslow is a few Lamb letters and a page of Peggy Hazlitt's journal.

It is a pity the wedding did not take place as planned in the village church: the Rector, Matthew Marsh, himself in those months about to embark on a late marriage, might then have left some mention of the Hazlitts.[15] We cannot hope to discover how they were viewed by the various tribes of the village, the Reeves, the Yeatses, the Sheers, the Annetses, or even by those raised a notch above total anonymity by their occupation, like Bell the baker, Fiander the exciseman, Judd the butcher, or Rogers the blacksmith, but it is odd that such unusual cottagers as the Hazlitts should never figure in the numerous surviving letters of the Rector, and odder still that Hazlitt himself never refers either to the Rector, or to the Rector's patron, the lord of the manor, who was none other than Lord Holland.[16] Henry Vassall Fox, 3rd Baron Holland, favourite nephew of the great Charles James Fox who was, during Hazlitt's boyhood, at once the glory of the Whig party and the idol of the Dissenters, had been born in the family mansion, Winterslow House, just outside the village, in 1773. When the local living that was in his gift fell vacant in 1804 he presented it to his former tutor at Christ Church, Oxford, the Revd Matthew Marsh, with whom he maintained a close tie.[17] He frequently went down to Wiltshire, and we find him in October 1808, when he was passing through in his carriage on his way to Spain, briefly pausing at Winterslow Hut (the inn a mile below the village on the main road to Salisbury) to pen a brief word of apology to Marsh for not having time to call on him.[18]

---

[13] See *Letters*, 102, 105.   [14] Ibid. 105.

[15] Marrs, ii. 281; BL Add. MS 51711, letter of 14 Sept. 1808.

[16] Diocesan Record Office, Salisbury: Bishop's Transcripts; there may be an oblique reference to the Rector in 4. 122.

[17] BL Add. MS 51711 and 51714, *passim*.

[18] BL Add. MS 51711, letter of 31 Oct. 1808.

Marsh was the type of old-fashioned don, fond of his wine (much of it supplied by Lord Holland) and of his classics, swapping Greek epigrams with his patron, quoting Horace at him in praise of the country life, tenacious of his ease, but in the interest he took in the life of the village not unlike Gilbert White (he advises Lord Holland on the choice of woodmen, describes George Judd as a drunkard, and Dowling — whether uncle or nephew does not appear — as a very good-natured man). He is anxious for his parishioners, especially in the hard times of 1817–18; and when there was a question of cutting down the trees in 1824 he protested: 'I do . . . not know what the poor will do. As it is, they are wretchedly off, having nothing whatever but the refuse and dead wood of these coppices . . . The Squire, the Parson and the Blacksmith are the only persons in this part of the country who ever use a coal fire.'[19]

Marsh never mentions current publications in his letters, and he does not sound likely to have been interested in Hazlitt's books.[20] But his amiability makes it surprising that he took no interest in Hazlitt the man. He can hardly not have known he was there, especially after the spectacular visit of the chartered eccentric Lamb in 1810, in high holiday humour and breeches of 'lively Lincoln-green'.[21] He cannot certainly in such a small place have failed to notice Hazlitt's absence from church. The Hazlitts' neighbours John and Maria Armstead who came to Winterslow in April 1810, were church-goers, and Sarah may also have attended from time to time. She and Mrs Armstead became very intimate, but although her neighbour's children were christened by Matthew Marsh young William Hazlitt was not: he had to wait until his parents moved to London.[22] As for Hazlitt himself, we can be pretty sure that he never heard a single sermon from the Winterslow pulpit.

This was long before *Characters of Shakespear's Plays* brought him a certain celebrity, and in the villagers' eyes he was simply an oddity who went mooning about the woods with paint-box and canvas under his arm: Marsh does not mention him in his letters to the Hollands, gossipy though these were. More surprisingly, later on, after Hazlitt's return to Winterslow to spend the summer of 1818, when he had long been a contributor to the *Edinburgh Review* (the organ, almost, of the Holland House circle through a whole network of contacts including Sydney Smith, John Allen, the Hollands' librarian, Brougham, and of course the editor himself, Francis Jeffrey), when *The Round Table*, the *Characters*

---

[19] Ibid, letter of *c.* Sept. 1824.

[20] Ibid., letter of 25 Aug. [1807]: '. . . I did not understand the metaphysical part of it, and indeed I seldom derive much knowledge or satisfaction from either reading or hearing what are called metaphysics in these times.'   [21] 17. 66.

[22] The Armstead children, John and Cecilia Maria, were baptized on 25 Sept. 1810 (Parish Register).

of Shakespear's Plays, A View of the English Stage (Lord Holland was a committee-member of Drury Lane), and *Lectures on the English Poets* had been published as well as a long series of resounding anti-Tory articles in the *Examiner*, and when the lively young Henry Edward Fox, the Holland heir, who had recently stayed at the rectory to be coached for Oxford (he went up in the autumn of 1819), must have brought his host and crammer a good deal of news about what was being published in London, Marsh still did not break his silence.[23] It probably did not even occur to him to do so. He asks no questions of Lord Holland, and indeed seems unaccountably narrow in his views. Sydney Smith wondered at this blinkered outlook in a letter to Lady Holland: 'How can [Marsh] have lived with such men as he has lived with without catching one atom of their liberality and Wisdom?'[24] Like Hazlitt the Hollands were unshakeable admirers of Napoleon; and in this regard Marsh seems to have taken a stand of his own: in his letters to them he declared the great man to be 'the greatest scourge that ever afflicted the civilised world'.[25] From all this it is plain that sympathy between Hazlitt and the Rector was hardly to be looked for.

Not that the names of the significant writers of that age often figure in the Holland House papers. The style of the coterie, a Whig aristocracy of blood and intellect, lay embalmed in the measured and polished urbanity of the past age, the conventional late eighteenth-century style of calculated impersonality and sounding generalities, and to them the familiarity, the directness, and above all the vehemence of Hazlitt's manner would appear inexpressibly vulgar. When Lord Holland turned from the debates at Westminster to seek relaxation he turned to the higher and more gentlemanly forms of journalism (and a journalism itself stiff with the dignified abstractions and elegant clichés of the eighteenth century). He turned to the lawyer Francis Jeffrey, the parson Sydney Smith, and the lawyer-politicians James Mackintosh and Henry Brougham. In a few years' time when Hazlitt became an *Edinburgh* reviewer, his editor Jeffrey, much as he valued the shrewdness and energy of Hazlitt's contributions, deprecated what he saw as his rashness, carelessness, extravagance, and bias to paradox, and was frequently at the labour of ironing out the asperities, or indeed excising whole passages, of his barbarous prose.

It is true that the Hollands would invite to their table anyone of note. Kean dined with them as soon as they judged his reputation secure, some three weeks before they actually went to see him themselves, and so

---

[23] BL Add. MS 51711, letter of 3 Feb. 1818.

[24] *The Letters of Sydney Smith*, ed. N. S. Smith (1953), i. 416.

[25] BL Add. MS 51711, fos. 101, 105; 51714, letter *c*. Oct 1814.

did the gawky Scotsman, Wilkie, once he had been safely elected to the
Royal Academy. Certainly Moore (translator of Anacreon and Whig
court-jester to boot) was a frequent, and even Wordsworth an occasional, guest at their board, and there was always a place for exotic
notabilities like Schlegel, Foscolo, and Mme de Staël (opponent of
Napoleon though she might be), but they can hardly have given a second
thought to inconsiderable vulgarians like Hunt, Keats, or Hazlitt, and
although there were several Lambs among their regular visitors they
were not of Charles's branch of the family.[26] Howe's enthusiasm and
admiration for Hazlitt misled him to believe that the author of the
Political Essays (which was totally ignored by the *Edinburgh Review*) 'could
easily, had he so wished, have found acceptance in the Holland House
circle'. He is never mentioned in their voluminous surviving papers. In
any case, after the plainness of his country upbringing, the negligence of
his Bohemian youth, and his solitary years at Winterslow, he was not
always comfortable even in the relatively modest salon of Mrs Basil
Montagu in Bedford Square: by 1817, when he had been visiting the
Montagus for several years, he had, according to Crabb Robinson, only
just arrived at a tolerable degree of polish and social grace ('He talked
much at his ease to Mrs. Montagu and is greatly improved in his
manners. He is almost gentlemanly compared with what he was.').[27] But
he would as little have relished Lady Holland's upstart-aristocratic
insolence as she his half-indifferent want of gentility. He was no Tom
Moore; he had nothing of the dancing-master, nor of the amiable rattle.
And even the amusing Moore's drawing-room skills and charm could
not apparently in that age of supercilious arrogance redeem his plebeian
origins as a Dublin grocer's son nor camouflage his fatal parasitism in
the houses of the rich and great.[28] So far from being a Moore, Hazlitt
was not even a Crabb Robinson, and it is astonishing with what
adroitness the normally sociable and ubiquitous Robinson tried to avoid
being presented to Lady Holland when he was the *Times* war correspondent in the peninsula in 1808 and she arrived at Corunna apparently
intent upon founding a salon.[29]

   The tranquillity of Winterslow must have seemed like a return to the
ideal existence of Wem, to which he looked back with so much regret,
when he was in his father's house and his 'path ran down with butter and
honey', and when for many years of his life he was blessed in having
nothing to do but think.[30] Just as he seems to have been the only member
of the family with any feeling for Wem, so also he loved Winterslow and
Salisbury Plain. Twenty years later an acquaintance in Paris heard from

---

[26] Although Lord Melbourne, years after his death, questioned Haydon about him: Pope, iii.
647.   [27] Morley, 211.
[28] See Moore, iv. 60.   [29] Sadler, i. 276–7.   [30] 12. 222; 8. 92.

his lips 'a long, eloquent, and enthusiastic dissertation' on the Plain so radiant that he never forgot it.[31]

It is easy to picture their existence in those Winterslow years: on cold winter mornings, the tinder-box and the brushwood and sticks in the ashy grate, water from the village pump for the tea-kettle, or, for washing, from the butt under the drain-pipe; the low light to the east above Tytherley Woods as he went down the muddy garden path alongside the flattened potato-patch to collect the eggs from the uneasily-stirring hen roost; and at the end of that path, the jakes screened by a privet-hedge or an elder-tree; for dinner, a couple of rashers of bacon from the flitch hanging in the kitchen, or a rabbit, and beer from the Lion's Head in the village (more convenient than the Hut, a mile away), and bread from Bell, or perhaps, though less probably, from Sarah's own oven; and of a Sunday, a chicken, or some roast beef from Judd's, and sherry; in the evenings, with the curtains drawn against the rainy dark, the table pulled closer to the fire, and the bundle of mended pens, with the books and the quire of long paper spread white under the well-trimmed lamp, and a silence broken only by the scratching of the crow-quill or the settling of a log, or out on the road the creaking axle of a late cart. In the spring, the first lark-song on the downs to the north, then daffodils, and bluebells in the wood. In the summer the parlour would shimmer with the yellow-green light ceiling-reflected from the garden, where the pink and white or brown and white blossoms of bean-row or potato-plant trembled in the breeze, and above them billowed the clothes-line full of the week's washing, and all the doors and windows of the cottage would be open the livelong day.

The Hazlitts' main occupations were no doubt those familiar to the countryside: wrestling with the vagaries of the weather, and watching the slow return of the seasons. The summers of 1811 and 1812 were exceptionally fine but they had been dearly paid for in advance: the great floods of January 1809 were a staggering event, when the heaviest rains for forty years put the cathedral-stalls at Salisbury under water, and submerged acre after acre of field and meadow as far as Wilton, the next village to Winterslow. There were other gloomy events, of human agency, when local or neighbouring or even remote militia, the Wiltshire Yeoman Cavalry, the Dorset Volunteers, the Radnorshire Militia, or the Salisbury Volunteers, manœuvred on the Plain; or when the regular troops went marching by, to embark at Plymouth for Portugal, or straggled back exhausted after service in the Peninsula — the faint sound of the cavalry's hoof-beats or the tramping feet of the infantry borne on the wind from the high road in the hollow below the village — the King's

[31] Howe, 354.

German Legion (hateful name!), the 66th Regiment, the Light Dragoons; or sometimes again it might be several hundreds of French prisoners toiling beside their captors through the pouring rain to gaol (a sight to arouse pity, as Hazlitt was probably gratified to find humanely recognized in a report in the local news-sheet).

All his life he was an enthusiastic playgoer. The delight that in the year of his death inspired the essay 'The Free Admission' had been apparent forty years earlier in a letter describing his first visit to the theatre, in Liverpool, when he was twelve, and saw Kemble, Suett, and Dignum on a tour after the end of the London season. It appears in 1820 in a reminiscence of the modest-flamboyant manager of the strolling players (twin brother to Vincent Crummles) who, in the early 1790s, put on *The West Indian* and *No Song No Supper* in the market-halls of Wem and Whitchurch,[32] and later in that decade he must have tramped in to Shrewsbury to the little theatre on St John's Hill, as well of course as attending, on visits to his brother in London, the more impressive, the dazzling, productions of Drury Lane and Covent Garden. So we should not suppose that he neglected such minor pleasures of that order as Wiltshire afforded. He and Sarah no doubt paid an occasional visit to the dingy little Salisbury theatre, unless they were put off by the chill of its small open grates, inadequate to dry out the damp walls, or by the garrison officers who were its desperately uncouth principal patrons. If they did, they saw Mrs Shatford's company, fresh from Poole or Blandford, in *The Honeymoon*, or perhaps in their friend James Kenney's *Raising the Wind*. The references in *Table-Talk* to 'the theatre at ——, a little country town in the West of England', to 'the officers of the regiment quartered there', and to 'the comic strength of the company at ——, drawn up in battle-array in *The Clandestine Marriage*' are inapplicable to Bath and probably refer to the theatre at Salisbury (where in fact *The Clandestine Marriage* was performed by Mrs Shatford's company on 1 April 1811).

Like Liverpool and most other provincial towns Salisbury was occasionally visited by the celebrities of the London stage. In 1809 Jack Bannister came down, and in 1811 Mathews, and also Incledon, who played Young Meadows in the ever delightful and ever popular *Love in a Village*. The playgoing season in Salisbury was late winter and early spring, and for the Hazlitts it meant a walk home through the darkness up the hill from Winterslow Hut after being dropped by the coach or the carrier. Other seasons too had their appropriate attractions: in summer there were 'pink-feasts' (flower-shows), sheep fairs, wrestling matches, and single-stick contests, some held no further away than the forecourt of

---

[32] 18. 294.

the Hut; and once during those years, in July 1810, the Triennial Music Festival took place in Salisbury Cathedral, a festival at which Mme Catalani sang (that 'ocean of sound', as Hazlitt called it, moved by her 'divine power and impassioned tones').[33] They could hardly have been unaware of this event after the 'powerful impression made on the public mind' the previous autumn by the singer's engagement at Covent Garden at the staggering salary of £4,000.[34] Sarah, an ardent and determined sightseer and gadabout, may well have carried him, perhaps even despite himself, to many of these entertainments, but all that we have documentary evidence of is that he wrote, and painted, and visited the private collections of pictures in county seats.

The dense woodlands south-east of Winterslow even now offer a rich contrast to the bald north-western slopes beyond the Salisbury road, and in those days when almost the whole of the Plain was a barren desert (not quite the 'huge waste' traversed by Wordsworth in 1793 on his return from revolutionary France, but almost as desolate)[35] the difference was much sharper, so that when Hazlitt sallied forth with paint-box and canvas it was seldom to the downland beyond the Hut and nearly always to the woods above Tytherley. Landscapes, however, were a luxurious summer relaxation from the real business of his brush. No longer needing to paint portraits to live, he had turned again to the grand dramatic subjects in history painting that had long haunted his imagination and were always to remain a passion — the story of Ugolino, and the legend of Jacob's Ladder. He had sketched out by January 1808 his version of the Ugolino, a subject which he calls, in speaking of Sir Joshua Reynolds's interpretation of it, 'one of the most grand, terrific and appalling . . . in modern fiction'. As to Jacob's dream, this was a theme which occupied him in March 1810, at a time when he had recently been writing the life of Holcroft, where an image in one of the chapters shows that for him it was an obscure symbol of the unspoken yearnings of a lonely wanderer on the face of the earth, athirst for a spiritual lodging-place, for the companionship of allied souls, for posthumous fame.[36] There can, I think, be no doubt that the legend stood for the aspirations of his earnest youth, that it consoled his fading hopes after the false dawn of the French Revolution, as well as acting as a surrogate for those childhood religious beliefs that had early kept him warm on the benches of his father's chilly little chapel at Wem, but which were later

[33] 8. 95; 5. 193 (cf. 12 161). The details in this and the two preceding paragraphs are taken from the contemporary *Salisbury and Winchester Journal* and esp. from the issues of 13, 20, 27 Feb. 1809; 16 July, 3 Dec. 1810; 4, 25 Mar., 1, 29 Apr. 1811.

[34] *Literary Panorama*, Sept. 1809, col. 1116.

[35] Some improvements had been made in the previous decade; see *GM* 75 (June 1805), 574.

[36] *LH*, 73; 18. 58; *LH*, 101; 3. 155.

put to dogmatic rout at Hackney. That ladder stretching up from earth to heaven became the hidden link between religion and politics. In 1817 he spoke with regret to young William Bewick, the painter of the abandoned picture, and soon after, in his last lecture on the English poets, he used the legend to symbolize a dream of the progress of humanity and of mankind's eternal aspiration to happiness.[37] Elsewhere in describing a version of this theme in the Dulwich Gallery (believed to be by Rembrandt), and another by the American artist Washington Allston, he dwells upon 'the solemn awe', the 'steps of golden light', the 'emanations of flame or spirit hovering between earth and sky', and 'the glory, the intuition, the amenity' of this 'dream of love, of hope, and gratitude'.[38] For him it always remained a type of epiphany. But at this time in Winterslow, despairing of the skill to finish these pictures, he turned them to the wall, and a similar want of confidence halted also the completion of his landscapes. His sister in her account of his life at Winterslow does not propose this explanation but there can surely be no other:

Yet painting was his great delight, and while at Winterslow it was his custom to go forth into the woods to sketch views from nature, taking with him his canvas, paints etc., an hammer and a nail, which he used to drive into a tree which served him as an easel to hang his picture on. A couple of eggs, boiled hard, and some bread and cheese for his dinner, but never any liquor of any kind. Usually coming home at four (except when we went to meet him in that beautiful wood), bringing with him some promising beginning of the beautiful views around, but fated, alas, never to be finished. Several of these I remember, one in particular, the view of Norman Court, the seat of —— Wall Esqr. I wish I had these rejected landscapes.[39]

Years later the Bois de Boulogne carries him nostalgically back through two decades to Tytherley: 'Some of the woods on the borders of Wiltshire and Hampshire present exactly the same appearance, with the same delightful sylvan paths through them, and are covered in summer with hyacinths and primroses, sweetening the air, enamelling the ground, and with nightingales loading every bough with rich music.'[40]

  Rejected the landscapes may have been, but there is no doubting the intensity of feeling that went into them, the passion and excitement of his efforts during those Winterslow years, and which were later so vividly expressed in his own description: 'How often have I looked at them [the landscapes of Rubens and Rembrandt] and nature, and tried to do the same, till the very "light thickened," *and there was an earthiness in the feeling of the air*!'[41]

[37] Bewick, i. 135; 5. 167.
[38] 17. 62. See also 10. 21 (this painting is now ascribed to Aert de Gelder); 12. 120; 18. 123.
[39] Moyne, 109–10.   [40] 10. 158.   [41] 8. 7–8 (the italics are mine).

The isolation of village life was beguiled not only by his own painting but also by his portfolios of prints; and within a half-day's journey of Winterslow were great country houses with rich collections to which he doggedly found his way — Stourhead, Longford Castle, Wilton House, Blenheim. We can have but a faint notion of the urge that sent him tramping the roads from one mansion to another and gave him — a timid enough man in the commerce of daily life (although not in his opinions on paper) and made yet more timid, nervous even, by country solitude — the boldness to demand, and the tenacity to insist upon, admittance, and to argue with supercilious or refractory major-domos. Except for inadequate prints or copies he had reached the age of twenty almost entirely ignorant of the Italian masters: his dazzled initiation into the mysteries of art did not come about until the first months of 1799, when at the sale of the Duc d'Orléans's collection he was staggered by the Titians, the Raphaels, the Domenichinos. Like everyone else he was familiar enough with their names, but to see them face to face was like the breaking of some mighty spell: 'A mist passed away from my sight; the scales fell off. A new sense came upon me . . .'[42]

And three years after this revelation, when he crossed the Channel at the Peace of Amiens, it was not as a tourist like the thousand others at that time but as an apprentice painter, eager to study what was non-existent at home in England — a public collection of these masterpieces. It is difficult to realize that he could not, as things then were, have copied any of the old masters without going abroad. The Royal Academy held no classes in painting, only in drawing, and no help came for the English artist in his own country until 1806, when the wealthy amateurs and connoisseurs who had recently founded the British Institution in Pall Mall lent old masters from their own collections for the apprentice artists to study. (Among those who applied for permission to copy was John Hazlitt, by now over forty and who had never, so far as we know, been abroad.) After that first step forward it was nearly ten years before the Academy roused itself to provide comparable facilities, and by then communication with the Continent had been re-established. In the meantime the Institution's successful innovation had inspired one of its founders, the Marquis of Stafford, purchaser of many of the Orléans pictures, to throw open his collection to the public; his ticket-days at Cleveland House became famous and he soon found many imitators, some among the owners of country houses, but for the most part these still remained inaccessible to outsiders. But these innovations were insufficient to break down the English artists' parochialism. Wilkie's experience was typical: after he had visited the

Louvre in 1814 he was comically dismayed to find that the annual
exbibition of the Royal Academy 'looked' (as he said with devastating
but characteristically cautious understatement) 'very odd after what we
had seen in Paris'.[43] Not that England's isolation during the Napoleonic
wars was greatly to blame: Wilkie's remark echoed Romney's artist
friend's grotesque exclamation forty years earlier when they both
entered the Sistine Chapel: 'Egad! George, we're bit!'[44]

Hazlitt also kept in touch through Lamb with the print-shops and
sale-rooms. In the early months of 1810 when Sarah went up to town to
spend a few weeks with Mary he commissioned her to buy a number of
prints for him. Some items were meant to complete his set of the Raphael
Cartoons (from the originals at Hampton Court): he blames his failure
to join her and to attend the widely advertised Walsh Porter sale at
Christie's on 14 April on the vile weather, the discomfort and expense of
the journey, and his snug situation: 'here [he says with glee, quoting *The
Beggar's Opera*, but referring with gay blasphemy to the Cartoons] I sit
with my doxies surrounded'. On receiving the catalogue from Sarah he
had been discouraged by the flood of Dutch genre-paintings that
swamped the Old Masters he had hoped to see. His scorn of the 'parcel
of Metsus and Terburghs and boors smoking and ladies at harpsichords'
and his appealing to the *St Cecilia* of Domenichino, the *Pan and Syrinx* and
the *St. George and Holy Family* of Rubens, to the *Danae* of Correggio, and
the *Ariadne in Naxos* of Titian anticipates his violent reaction thirteen
years later to the cabinet pictures in the Fonthill Abbey sale.[45]

In Claude it was not technical perfection alone that appealed to him.
The delightful pastoral heaven that breathed in those enchanted
landscapes, the legends they enshrined, like the marriage of Isaac and
Rebecca, Hagar and the angel, the meeting of Jacob and Laban, all
plucked at a threefold strand in his memory: his boyhood among the
fields and flocks of Shropshire with the distant view of the blue,
air-drawn hills of Wales; his father's ponderings over Genesis and the
'glimpses, glimmering notions of the patriarchal wanderings, with
palm-trees hovering in the horizon, and processions of camels at a
distance of three thousand years'; and finally his own youthful journeyings through the English countryside around the turn of the century,
when he was twenty, through the Vale of Llangollen, on the Severn at
Llanymynech, in Somerset at Nether Stowey and Alfoxden, in the Fen
Country, and in the Lake District.

This enthusiasm was no doubt linked to the Romantic dream of
simple idyllic happiness on the slopes above Chambéry or in the Pays de

---

[43] Cunningham, i. 432.   [44] 11. 200; see also 12. 89.
[45] See the writer's 'Hazlitt and the Walsh Porter Sale', *EA* 26 (1973), 452–4.

Vaud that had beckoned to him in his early reading of Rousseau's *Confessions* and *La Nouvelle Héloïse*, memories perhaps also of the humbler but not less enchanting *Love in a Village*, as well as to the picaresque novels of the eighteenth century and *Don Quixote*, a pastoral dream of love and country peace. His advances to the village girl in Keswick in 1803 conformed to the quixotic pattern in their tangle of realism and idealism, of sacred and profane love, in which he was at once the Don and Sancho Panza, but he was no match for her wiles and mockery. The minx made a fool of him and the episode ended in ludicrous catastrophe and ignominious flight. It might have ended in worse if he had fallen into the hands of the pursuing villagers, but the lesson for him remained unlearned, and was bound so to remain, as far as it regarded women. There his illusions were perpetually reborn. But to his view of country folk in the mass the episode had simply brought confirmation.[46] He loved the country but was repelled by the 'hard coarse edge of rusticity'[47] in the people, with a repulsion that probably dated back beyond 1803 to the petty jealousies and cabals of the inhabitants of Wem.[48] He was too quick, subtle, and timid ever to feel at home with ignorant, stolid villagers, and what we are able to infer from his writings suggests that at Winterslow his mistrust and their equally natural dislike reciprocally intensified each other. At best there was uneasiness: he knew why it was impossible to find common ground with the villagers and why he could do little to improve matters:

They see nature through their wants, while you look at it for your pleasure. Ask a country lad if he does not like to hear the birds sing in the spring? And he will laugh in your face. 'What is it, then he does like?' — 'Good victuals and drink!' As if you had not these too; but because he has them not, he thinks of nothing else, and laughs at you and your refinements, supposing you to live upon air.[49]

In Winterslow he was more than a stranger, he was worse: a gentleman without a gentleman's means, and, unacknowledged either by squire or parson, half-writer half-painter, utterly unbucolic, he was an oddity. When he sketched a beautiful yew-tree and a local wag mock-innocently asked if he knew how many foot of timber it held, he had neither the knack to parry nor the composure to absorb the thrust; he could only

---

[46] The facts in this case are discussed in Maclean, 196–201; *TLS*, 25 July 1968, 789; 15, 29 Aug. 1968, 873, 928; 12, 19 Sept. 1968, 1032, 1062; in Ch. 12 below; and in J. Beer, 'Coleridge, Hazlitt, and "Christabel" ', *RES* ns 37 (1986), 40–54. The Keswick episode was a particular instance of 'the *ferae naturae* principle', which, he came to believe, 'is within us and always craving its prey to hunt down, to worry and make sport of at discretion, and without mercy' (20. 323).

[47] 5. 290.

[48] The expressions are his sister's (Moyne, 106 and 102–7 *passim*). She does say that unlike the rest of the family he was happy at Wem, but he can hardly have been ignorant of their feeling towards the townsfolk. Her words probably echo her mother's, who contributed her own recollections to Margaret's journal.   [49] 17. 272.

resent it.[50] This antipathy is not apparent in the few surviving letters of the period, where he seems on excellent terms at least with his immediate neighbours (who were, however, strangers, newcomers to the village, like himself), but it is plain from his essays that it existed. It re-emerged years later when he came down to Winterslow Hut to get through some solid reading for his lectures on Elizabethan literature, and overheard a conversation in the kitchen between his landlord James Carter (who had himself just asked him if he had any object in reading all the books he had brought) and a yokel who aggressively questioned whether it was right that people should be able to make a living by the pen.[51] He loved the country around Winterslow, but not the country people who, there as in other places, were always itching to hoot at or pelt any harmless stranger, and drive him off like a stray cur, with the same delight in inflicting pain as they got from bull-hanking or cock-throwing.

His happiest times there, or at least almost the only times that have left a trace, appear to have been the summer visits of the Lambs in 1809 and 1810. Lamb was instantly aware of the gulf between him and the villagers: 'Even his person', he says, 'in their rustic eyes, [had] somewhat strange in it.'[52] And Lamb's reaction was characteristic, 'mad wag' as he was, he immediately as his friend's champion exaggerated the disparity between *himself* and the villagers, by grotesqueries both of dress and conduct, no doubt to the agonized embarrassment of his host.[53] The Lambs were to have arrived at Winterslow early in July 1809 with young Martin Burney and Ned Phillips the good-natured card-playing sportsman and 'Oxonian of idle renown', whose gun was expected to furnish forth part of the table. The arrangements proposed by Mary Lamb in a letter to Sarah in May (Martin to sleep on the floor; the Lambs to contribute £5 to expenses) make it clear that none of them had much money to spare.

However, these plans came to nothing. On 29 May Mary Lamb's insanity recurred; she was confined for a month; and at the end of that time the Hazlitt's six-month-old baby died; so that it was Peggy (Hazlitt's sister Margaret) who came to Winterslow in July, instead of the Lambs. Their visit was postponed to the end of September. This meant that young Martin could not make one of the party, because he had to prepare for a law-examination on 6 November, but Phillips with his pack of cards joined the Hazlitts and the Lambs on 17 October.

It was an exceptionally fine autumn and every day the men walked ten or twenty miles across country, to Wilton House and Stourhead to see the pictures, to Stonehenge, to Salisbury. The boisterous Lamb,

---

[50] 17. 68.   [51] Ibid.

[52] *LH*, 103–4. These lines have not been claimed for Lamb, but seem likely to be his.

[53] 12. 42; 17. 66–7.

determined to make the most of his annual holiday, may 'for the joke's
sake' have dragged the unwilling Hazlitt (who, however, was an
indulgent host before he was a frowning republican) to the celebration of
King George's golden jubilee on 23 October in Salisbury, with the
bands, the parades, the bonfires, the fireworks, and the bottled stout;
but it is more likely that they stayed to cast a mild eye on the lesser
festivities at Winterslow.[54] However, we hear nothing of this in
subsequent letters, only the quieter echoes of the private pleasures of
those 'four weeks of uniform fine weather, the only fine days which had
been all the Summer'.[55] For the two women it was a delicious lazy month
that did Mary Lamb a power of good; on their walks she was too wise to
attempt to keep up with the nimble Sarah's rapid ascent of the hills. She
was content to enjoy the fresh country breezes after her confinement.
Evidently it also did Sarah good, helping to distract her after the shock of
the death of her child in July. Hazlitt no doubt also found relief from a
wretchedness that touched his sister Peggy when she arrived at Winterslow shortly after Hazlitt had followed the little coffin into Salisbury to its
resting-place in St Martin's churchyard in the same grave as the
grandfather, old Lieutenant John Stoddart, at the end of the child's few
brief months of existence; a cruel blow; nearly thirty years later she had
still not forgotten his anguished face.[56]

After they had got back to Middleton Cottage from their day's
excursion they would all, men and women together, take another brief
stroll outside the village to pick mushrooms for supper, and the unique
splendour of the evening skies in that month was something that Hazlitt
remembered for years:

I used to walk out at this time with Mr. and Miss L—— of an evening, to look
at the Claude Lorraine skies over our heads, melting from azure into purple
and gold, and to gather mushrooms, that sprung up at our feet, to throw into
our hashed mutton at supper. I was at that time an enthusiastic admirer of
Claude, and could dwell for ever on one or two of the finest prints from him
hung round my little room: the fleecy flocks, the bending trees, the winding
streams, the groves, the nodding temples, the air-wove hills, and distant sunny
vales; and tried to translate them into their lovely living hues.[57]

With Lamb too the memory still remained five years later of a sunset so
gorgeous that it drew even the fanatical Ned Phillips from the card-
table.[58]

The virulence of his subsequent picture of the blinkered life of country

---

[54] See *Salisbury and Winchester Journal*, 30 Oct. 1809.
[55] Marrs, iii. 33.   [56] BL Add. MS 38898, fo. 3; Moyne, 109.
[57] 17. 319–20. The insufficiency of the prints of the day is apparent from his last words here.
[58] Marrs, iii. 95.

folk, in his review of Wordsworth's *Excursion*, is owing either to the distortion of a writer economically intent on the unity of his subject, or perhaps to the memory of an unfortunate love-affair, a variation of the Keswick rebuff. In the spring of 1810, after another miscarriage, Sarah went alone to London to spend a month with the Lambs, and it may have been then that Hazlitt, loitering in the woods above Tytherley, fell in with the Sally Shepherd whose identity has baffled his biographers.

This is the woman mentioned in Sarah Hazlitt's journal of the divorce in 1822, when Hazlitt was in the middle of the most intense love-affair of his life and Sarah recalled his earlier infatuation: 'I told him it was like his frenzy about Sally Shepherd; he said *that* was but a fleabite, nothing at all to this, for she had never pretended to love him; but all along declared she did not.'[59] This was clearly an infatuation which Sarah had seen with her own eyes and had an evident right to resent, which suggests that it must have happened during their married life.

It was the frequency of such infatuations, as well as his recourse to women of the town, that led Sarah ten years later to accept the Sarah Walker affair and resign herself to a divorce. P. G. Patmore, who made Hazlitt's acquaintance in 1817 after his return to London, describes the kindliness of his demeanour, evidently born of a long familiarity, towards the street-women whose beat was in Whitehall on his homeward route from Fleet Street. He also says that he never knew Hazlitt out of love.[60] Hazlitt himself confessed that he preferred 'humble beauties, servant-maids and shepherd-girls, with their red elbows, hard hands, black stockings and mob-caps' to ladies, women of quality, or others in the public eye, opera-singers, actresses.[61] Not that, as between husband and wife, the faults were all on Hazlitt's side. We have seen how free Sarah's behaviour was, and in what a vital sense his marriage had from the very beginning been a disappointment to him. She too, apparently, since their marriage, had given him cause for jealousy — the occasion was probably provided by the last of the long string of prospective husbands who got away, the William Dowling to whom she had been engaged in 1806–7. His uncle, at one time landlord of the Winterslow Hut, was now prosperously installed in what had been Lord Holland's home farm.[62] There was plenty of room for a nephew, and there is no reason why young William Dowling, who was nearly ten years younger than Sarah, should not still have been at Winterslow when the Hazlitts arrived newly married, from London. At the time of her engagement to

[59] Le Gallienne, 329; Bonner, 247–8.
[60] Patmore, ii. 275–7, 301–2, 335, 343; iii. 88–9.   [61] 8. 236.
[62] He appears in the Wiltshire Land Tax Assessments as a very substantial farmer, and in the correspondence of Matthew Marsh as 'a very good-natured man'. See also J. Rogers Rees, 'Hazlittiana', *NQ*, Apr. 1908, 292.

young Dowling Sarah had arranged to bring him up to London to spend the Christmas of 1806 with the Lambs; the collapse of this plan, their subsequent parting, and her marriage would be no bar to Hazlitt's jealousy.[63]

The Hazlitts' married life, after the summer and autumn of 1808, was spent in only two places: first Winterslow, and then York Street, Westminster. No Shepherds appear in the rate-books of York Street or its vicinity between 1812 and 1819.[64] On the other hand Hazlitt alludes to an unexplained love-affair in Winterslow,[65] and there *was* a Sally (Sarah) in the village when the Hazlitts lived there, who became Sally Shepherd by marriage in 1811. She was twenty when the Hazlitts arrived and her name then was Sarah Baugh; in June 1811 she married William Shepherd, labourer (a man nine years older than herself and only two younger than Hazlitt) so that this was the surname by which Sarah Hazlitt would naturally refer to her in 1822. She was widowed two years after the Hazlitts left Winterslow. She must have been extremely well-favoured: in 1825, three years after the conversation of 1822 recorded in Sarah Hazlitt's journal, she, a widow of thirty-seven and mother of an eleven-year-old boy, became the wife of a Winterslow man nine years her junior, one Richard Judd, bachelor. She died in her native village in 1864.[66]

While we may be right in concluding that this was the Sally Shepherd that Sarah meant, we need not also suppose that she encouraged Hazlitt or in any way responded to him. She was probably as little moved and not much less sympathetic than that other country girl in Keswick.

The 'frenzy' would no doubt have come to a head in 1811 or 1812 and contributed to the Hazlitts' departure from Winterslow, through irritation or disgust, in Sarah's case, and, in her husband's, pique at Sally Baugh's marriage. But his involvement could well have begun in the spring of 1810, in the empty period of Sarah's absence. There is small suggestion of romantic love between Hazlitt and his wife; Lamb's ungallant reason for wishing to see them married was 'for the joke's sake'; Hazlitt's only extant courting-letter to her is ponderous with un-loverlike badinage. The tone of their exchanges seems to have inclined to the facetious or matter-of-fact rather than the romantic; it had a very modern freedom; and at all times they seem to have been entirely frank with each other. There was never a more sensible, clear-sighted, and realistic journal than Sarah's during the divorce and it is unlikely that she had not already acquired the same level-headedness by the time of her marriage. In the comradely letter written to Sarah in

---

[63] Marrs, ii. 241.   [64] Westminster Public Library.   [65] 10. 27.
[66] Winterslow Parish Register: b. 3 Feb. 1788, m. Shepherd 2 June 1811, widowed 17 Mar. 1814, m. Judd 28 Aug. 1825, d. 17 Feb. 1864.

London on the morrow of the Walsh Porter sale, prompted perhaps by a
wry twinge of conscience, Hazlitt says, somewhat oddly after the three
days' unremitting brush-and-palette toil he had just reported, 'I must be
a good boy. I have not been very good lately.'[67] However, these are frail
clues, and until further evidence comes to light the suggestion remains
no more than an interesting possibility.

If this was indeed the Sally who declared all along that she did not love
him, her marriage may well have had an influence on the Hazlitts'
return to London, but the major cause was no doubt the death of old Mrs
Stoddart in February 1811 and the financial position it revealed,
together, perhaps, with some kind of officiousness on the part of Sarah's
ambitious brother.[68] According to Lamb, Sarah's £80 a year was to
become £120 on her mother's death.[69] The Lieutenant's will did indeed
say that when his widow died Sarah was to get the family home and half
the residue of the estate, but what the residue was is not known. Mrs
Stoddart's annual pension of £40 was dependent on her formally
declaring her total income to be less than £30. If her death would
increase Sarah's income by £40 then the residue of the estate must have
been large enough to provide not £30 but at least £80 a year. Of course
there may have been concealment. There was in fact some difficulty over
the pension in 1805 when Mrs Stoddart became permanently insane just
before her daughter's return from Malta.[70] On the other hand, when her
right to the pension was established in 1806 payment was regularly made
thereafter to an attorney named Robert Wray, and it seems likely that
the pension was just as regularly absorbed by asylum, or nurses',
charges.[71] Mrs Stoddart's absence from home in confinement would
account for Sarah's living alone in the cottage at Winterslow before her
marriage, unless their earlier failure to get on together may have been
sufficient reason. In Hazlitt's few surviving letters of this time there is no
mention of his mother-in-law, and it is curious to reflect that although
she lived for nearly three years after her daughter's marriage he may
never have laid eyes on her.

Of course, much may have happened to the Stoddart estate between
1795 and 1803, and in any case Lamb in 1808 may have got his figures
wrong. But the sudden disruption of the subdued but even tenor of the
Hazlitts' existence following upon Mrs Stoddart's death does hint
strongly that it brought about a dismaying alteration in their prospects.
The balance of her pension up to February 1811 (£30, representing the
last nine months of her life) was paid to John Stoddart as administrator

[67] *Letters*, 120.

[68] Her recent illness and death are mentioned in an unpub. letter of her son's: BL Add. MS
41962, fo. 249.

[69] Marrs, ii. 272.   [70] PRO, Adm. 6/340.   [71] PRO, Adm. 22/130.

of the estate on 7 November 1811.[72] When the estate was finally settled he must have had disappointing news for his sister, and his irritation at his brother-in-law's inactivity would not thereby be lessened. Sarah's grandson seems however to imply that he remained the solicitous brother we have observed him to be: it was probably now, when Sarah and her husband were in difficulties, that he arranged for the sale of her 'small property' and thereafter paid her an annuity 'rather exceeding [what] would have been realised by the ordinary rate of interest' so as to ensure some kind of income for herself and the child.[73]

At all events, either a dimunition of income, or increased expense, or both, drove Hazlitt back to London in the month of his mother-in-law's death to attempt portrait-painting once more. Perhaps the shadowy remembered form of Titian's *Man in Black* beckoned also; and, having partly recovered from his discouragement of 1808, he may have hoped that something more than a mere livelihood in art was possible. The saying of his brother's, 'No young man thinks he shall ever die', impressed him no less than the dictum of Young which he had read as a boy in Enfield's *Speaker*, 'All men are about to live, forever on the brink of being born.'[74] The temptation of the quiet life was a danger of which he had always been aware, even at the moment of retiring to live to himself behind the loopholes of retreat. The *Speaker* also provided a judicious antidote to the beguiling passage from *Cymbeline* and the seductions of ease, in an eloquent appeal by Sneyd Davies against indolence and 'infamous delay' of which he remembered and later quoted in his essays whole lines, notably a minatory image of 'the cloyster'd heart [sitting] squat at home, like pagod in a niche obscure'.[75] He too, no less than Young, must often have felt that life, active life as well as the contemplative, is for the most part waiting to live, but felt also at the same time that it ought not to be so. Now the obscurity had become wearisome; he had been out of the squirrel's cage of the busy world long enough to surrender once more to the illusion of action, and the siren echo of the post-horn winding faint from the mail-coach on the London road below Winterslow village may at last have cried to him that life was passing him by.

Another factor in these obscure decisions must have been his wife's new pregnancy and the birth of their first surviving child on 26 September 1811.

Hazlitt, returning to London, put up at his old lodgings, 34 Southampton Buildings, where he found himself next-door neighbour to

[72] PRO, Adm. 22/140–2.   [73] *Memoirs*, i. 259.   [74] *The Speaker*, bk. 3.
[75] *The Speaker*, Bks. 6 and 3 respectively; for the Davies quotation see 5. 10; 11. 127; 18. 78; 20. 52, 95. Two other lines he remembered and quoted even in the last months of his life are, 'What avails from iron chains / Exempt, if rosy fetters bind as fast?' (17. 370).

Coleridge. Henry Crabb Robinson, who had recently left *The Times* to
read for the bar, met him at Lamb's on 10 February, where also was
Serjeant Rough, a bibulous improvident barrister who moves in and out
of the annals of the Robinson circle in the next five years,[76] and whom we
find in Hazlitt's company on at least one other occasion, in 1815, just
before he left England for the Colonies. At Lamb's also that night was
Thomas Clarkson, the steadfast labourer in the cause of the abolition of
slavery to whom Hazlitt paid tribute in *The Spirit of the Age*. He was a
native of Wisbech, the home town of Hazlitt's mother's family the
Loftuses which Hazlitt had visited, perhaps often, as a young man.
Either now or very soon after, Clarkson sat to Hazlitt for a portrait.

The amiable, generous, and energetic Robinson seems to have
procured him at least two commissions for portraits. He was still as
impressed as he had been at Bury St Edmunds in 1798 by Hazlitt's gifts
and as anxious as in London in 1805 that he should not sink into
obscurity and want. If friendship be reckoned by intellectual sympathy
and identity of interests, Lamb was Hazlitt's greatest friend at this
period; if by practical assistance, then it was Robinson, and he remained
so for many years.

Hazlitt knew him to be a good friend, and in the previous year when
he had been shocked to find that his Life of Holcroft was unlikely to be
published for a long time, if at all, he turned to Robinson who not long
since had invited him to contribute to a newly founded journal in which
he had an interest, Richard Cumberland's short-lived quarterly, the
London Review.[77] The purpose for which the *Memoirs of the late Thomas
Holcroft* were projected makes it plain that no payment would have been
made to the author until well after publication, if even then: the widow
had no money, Godwin (a prime mover in the scheme) certainly had
none, and Robinson wryly remarked that the committee of the dead
man's friends, so far from contributing to the relief of the widow, looked
more in need of relief themselves.[78] Nevertheless Hazlitt's disappointment must have been great when the objections raised by Godwin to the
inclusion of Holcroft's diary held the book back, and when the widow's
subsequent marriage to James Kenney the dramatist improved her
situation, and it was shelved entirely: with the emergence of the first
difficulties he approached Robinson with a suggestion that Tipper, the
publisher of the *London Review*, might be interested in a translation of the

[76] Robinson Diary, 10 Feb. 1811. This is a week earlier than the first date hitherto given for his
return to town. See also Morley, 161–2, 5 Feb. 1815.

[77] The title hints a challenge to the success of *ER*, but the *London Review* had the ill luck to be
launched in the same month as the Government-sponsored answer to *ER*, the *Quarterly Review*. See
Smiles, i. 98.

[78] Morley, 11. See D. McCracken, 'Hazlitt: a Case of Charitable Journalism', *KSJ* 28 (1979),
26–7.

recently appeared *Les Martyrs* of Chateaubriand (whose *ultra* politics, more royalist than the King, were not yet notorious), or, failing that, in a rearranged version of his projected work on philosophy.[79] It was not Robinson's fault that the suggestions came to nothing. Hazlitt was always grateful for Robinson's early regard and help. Years later when his attacks on Wordsworth's politics, and Robinson's loyal and resentful defence of the poet, had driven a wedge between them, he said to Mary Lamb, 'Robinson cuts me, but in spite of that I shall always have a kind feeling towards him for he was the first person that ever found out there was anything in me'.[80]

Helping Hazlitt practically, finding him patrons, was perhaps rather more difficult than acknowledging his brilliance, as an anecdote he told Robinson at this time implies. Recent events — the journey to town to seek commissions, Robinson's helping hand, Coleridge as next-door neighbour — all these things must have brought to mind his hopes and disappointments years before when tramping the North of England:

Hazlitt spoke of Coleridge with the feelings of an injured man. Hazlitt had once hopes of being patronized by Sir George Beaumont. Coleridge and he were dining with him when Coleridge began a furious attack on Junius. Hazlitt grew impatient at Coleridge's cant and could not refrain from contradicting him. A warm and angry dispute arose. The next day Coleridge called on Hazlitt and said: 'I am come to show you how foolish it is for persons who respect each other to dispute warmly, for after all they will probably think the same.' Coleridge produced an interlined copy of *Junius* full of expressions of admiration, from which it appeared that Coleridge himself really agreed with Hazlitt. 'But,' added Hazlitt to me, 'Sir George Beaumont is a High Tory and was so offended with me, both for presuming to contradict and interrupt Coleridge and for being so great an admirer of Junius, that in disgust he never saw me afterwards. And I lost the expectation of gaining a patron.'[81]

Naturally it did not enter Hazlitt's head that he might with advantage have held his tongue or wormed his way round to Coleridge's side.

This was not the only time that his unworldliness hampered his artistic career. When he was in Liverpool in the early 1800s he had painted a portrait of the renowned Whig banker William Roscoe, historian and patron of the arts, author of lives of Lorenzo the Magnificent and Leo X, supporter of liberal causes, of Catholic emancipation, and opponent of the slave-trade. It had seemed likely that this would lead to useful contacts for Hazlitt but he had forfeited his sitter's wider recommendation by a piece of youthfully ingenuous swank he later wryly regretted:

---

[79] *Letters*, 117. Pub. in London by Dulau, Soho Sq., 1809; reviewed *Critical Review*, 17 (Aug. 1809), 489–98.

[80] Howe, 196.

[81] Morley, 24.

I remember well being introduced to a distinguished patron of art and rising merit at a little distance from Liverpool, and was received with every mark of attention and politeness, till the conversation turning on Italian literature, our host remarked that there was nothing in the English language corresponding to the severity of the Italian ode — except perhaps Dryden's Alexander's Feast, and Pope's St. Cecilia! I could no longer contain my desire to display my smattering in criticism, and began to maintain that Pope's ode was, as it appeared to me, far from an example of severity in writing. I soon perceived what I had done, but here am I writing *Table-talks* in consequence. Alas! I knew as little of the world then as I do now. I never could understand anything beyond an abstract definition.[82]

Despite Hazlitt's annoyance he and Coleridge must have seen a good deal of each other at this time. Three weeks after the conversation with Robinson we find Coleridge calling on him at his 'particular request'(the reason for the request, alas, is not known), and it is impossible that on resuming acquaintance after such a long interval he should have found irksome the company of someone, now a near neighbour, who had more ideas than any person he had ever known, and who in 'conjectural subtleties . . . excel[led] all the persons [he] ever knew', however much thereafter he came to consider his behaviour spineless and his politics abject.[83]

The commissions for portraits were a mediocre artistic success and an outright financial failure. Hazlitt's hand had lost much of its knack. In his early twenties he had been hindered not so much by want of skill as by self-distrust and impossibly high standards, so that according to a friend he satisfied his sitters better than he satisfied himself.[84] By now his execution had become erratic: one of the portraits, of a Mr Howel (probably the same whom we find six years later listening with interest to Robinson reading *The Round Table* aloud[85]), was 'a good caricature likeness but a coarse painting'; a second, of Thomas Clarkson, was 'thought like', but as usual he himself had no confidence in it, possibly because he had noticed strange resemblances between his sitter and more than one of the figures in the 'Cartoons' and was overcome by the implied possibility of a disastrous comparison of his own work with Raphael's;[86] a third, of Robinson's brother Thomas, was a failure, and Hazlitt was so mortified he would not hand it over. He tinkered with it for more than eighteen months, before reluctantly consigning it to the sitter, who was so disgusted that he destroyed it. Robinson, who had every reason to be irritated, or at least embarrassed by these disasters

[82] 8. 204 n.   [83] Griggs, iii. 317; 12. 24.

[84] Talfourd, *Letters*, i. 255. Talfourd did not meet Hazlitt until 1815, and we do not know his authority for this statement.

[85] Robinson Diary, 6 June 1817.   [86] 11. 149–50.

was on the contrary full of concern: 'I fear poor Hazlitt will never succeed. With very great talents and with uncommon powers of mind I fear he is doomed to pass a life of poverty and unavailing repinings against society and his evil destiny.'[87]

He retreated to Winterslow with the troublesome portrait at the end of April or early May 1811. He had not the heart to set about 'improving' it right away, even though it was paid for and should already have been delivered, but turned instead to another plan. He spent the rest of the summer of 1811 revising for a new and special purpose the twice-projected History of English Philosophy. He had had this 'metaphysical' work in mind long before his first retreat to Winterslow, and had first outlined its scope in a letter to the *Monthly Magazine* in February 1809, and in a prospectus, of which he sent a copy to William Windham, MP, in the hope of enlisting his support to ensure that 'what [had] employed many years of his life and many anxious thoughts [might] not be entirely lost'.[88] The work was characteristically intended less as an impartial history of English philosophy than as a further attempt to establish the principle of the natural disinterestedness and freedom of the human mind in the teeth of 'The modern philosophy', the sensationalist doctrine of selfishness and necessity (soon to make still greater strides under the guise of Utilitarianism):

. . . it will be the writer's object, besides reporting the opinions of others, to . . . try if he cannot lay the foundation of a system more comfortable to reason and experience . . . [than] the material, or *modern*, philosophy, as it has been called. . . . According to this philosophy, as I understand it, the mind itself is nothing, and external impressions everything. All thought is to be resolved into *sensation*, all morality into the *love of pleasure*, and all action into *mechanical* impluse.[89]

A year later, 26 February 1810, came the proposal to Robinson to recommend it to Tipper, the publisher, but in a different form, turned now into 'a volume of Essays on the subjects mentioned in the prospectus, making the history subservient to the philosophy', which, he said, was what he thought he could do best.[90] It was in fact what he had always intended, but here he formally recognizes the cast of mind he confessed to years later in declining to write a history of drama for the *Supplement to the Encyclopaedia Britannica*. As he then said in all honesty:

The object of an Encyclopedia is, I take it, to condense and combine all the facts relating to a subject, and all the theories of any consequence already known or advanced. Now where the business of such a work ends, is just where I begin, that is, I might perhaps throw in an idle speculation or two of my own,

[87] Morley, 30; Howe, 122; E. J. Morley (ed.), *H. C. Robinson, Blake, Coleridge, Wordsworth, Lamb, etc.*(1922), 40; see also Baker, 183, and Wardle, 133.

[88] *Letters*, 106.   [89] 2. 113–14.   [90] *Letters*, 117.

not contained in former accounts of the subject, and which would have very little pretensions to rank as scientific. I know something about Congreve, but nothing at all of Aristophanes, and yet I conceive the writer of an article on the *Drama* ought to be as well acquainted with the one as the other.[91]

Since these earlier plans had not answered he now intended to take advantage of the rage for lectures then sweeping the town.[92] At the suggestion of Stoddart, who was mollified by this show of gainfully directed energy, he submitted proposals to the managers of the Russell Institution in Great Coram Street, which in the cultural life of the capital stood midway in wealth and repute between two other institutions, the inferior Surrey and the superior London.

When it came to the push he was uneasy about the size of the hall and the nature of the audience, but he was probably heartened by the kindly Basil Montagu, who was one of the 'proprietors' of the institution.[93] We do not know when he met Montagu, nor through whom. It could have been through Godwin, Lamb, Coleridge, Stoddart, Robinson, or George Dyer, but it must have been after 1807, when Montagu sent a copy of Hazlitt's abridgement of Tucker's *The Light of Nature Pursued* to Mackintosh and said, 'Search has been abridged and I am told very well, by some person unknown to me'.[94] Perhaps even after 1810: if they had met between 1807 and that year it is at least odd that Lamb does not mention his host in his letter from Winterslow to Montagu on 12 July 1810. Can it be that Hazlitt's friends kept him from the elegant drawing-room of 25 Bedford Square because his awkward manners might embarrass them? Montagu (a philanthropized courtier, Wordsworth called him), eight years older than Hazlitt, was the natural son of the dissolute Earl of Sandwich and the actress Martha Ray, who was shot dead in 1779 by a suitor crazed with jealousy, the unfortunate Hackman, as she was leaving Covent Garden stage door after a performance of *Love in a Village*.

A lawyer of literary tastes and liberal ideas, Montagu had embraced the principles of 1789, but was later (probably about the time when Stoddart was staying with him in Lincoln's Inn) 'persuaded to a more realistic view' of the future of the human race by Mackintosh. Mackintosh delivered a marathon series of thirty-nine lectures on 'The Law of Nature and Nations' at Lincoln's Inn in 1799 abjuring the French Revolution 'with its sanguinary history, its abominable principles, and for ever execrable leaders', a revolution which he had championed eight years before in his answer to Burke, the *Vindiciae*

---

[91] Ibid. 185–6.   [92] *MC*, 5 Apr. 1813.

[93] BL 8306.a.12. The only other known name among the proprietors is Francis Horner, the Whig barrister and politician.

[94] BL Add. MS 52452, fo. 20, 26 Aug. 1807. Search was Tucker's pen-name.

Gallicae. These lectures, which Hazlitt attended and a quarter of a century later still remembered (although chiefly for their crushing impact on Godwin) in *The Spirit of the Age*, had made him one of the most prominent apostates in an age of dramatic apostasies. He was as notorious for the alleged self-seeking that prompted his change of front as he was for the staunchless flow of measured discourse that made him a lesser but more intelligible Coleridge. Dr Parr's well-known jibe about the three attributes that constituted the nadir of disreputability stuck like a burr; it was repeated everywhere, and did him harm, but it was never repeated by Hazlitt.[95] Nor did Hazlitt ever allege or imply self-interest (although Lamb did, with some violence: 'Though thou'rt like Judas, an apostate black'[96]). He seems to have met Mackintosh, either at Godwin's in the nineties, or later at Richard Sharp's, or perhaps at the Montagus' later still, and seen enough of him to relish his conversation (however unoriginal he deemed it) but never to have got to know him well. Yet they respected each other. Mackintosh for his part was early impressed by Hazlitt's writings: the *Essay on the Principles of Human Action* was sent out to him in India (possibly by Sharp who was, as Howe demonstrates, Hazlitt's patron about 1806–7, and whom Mackintosh had commissioned to keep him supplied with books,[97] or else by Mackintosh's friend James Scarlett, who admired the work). There is no reference to the *Essay* in Mackintosh's papers (his pleased reception of it[98] may have been reported to Hazlitt by Sharp or Scarlett) but there is a good deal about the *Reply to Malthus* and the *Eloquence of the British Senate*, both of which he apparently came across independently in Bombay.[99]

In October 1811 Hazlitt went up to town to canvass subscribers and reconnoitre the territory, and, in November, after returning briefly to Winterslow (his first surviving child was but a month old), went back again to attend Coleridge's series of fifteen lectures on Shakespeare and Milton. They were delivered in Crane Court, Fleet Street, from Monday, 18 November. In paying his two guineas ('tickets from Godwin's Juvenile Library, Skinner St.') he was no doubt impelled not only by his respect for the lecturer's abilities but also by a curiosity to observe his mode of delivery and the number and receptivity of his

---

[95] The jibe is a brief abstract of some of the prejudices of the age; Mackintosh having said at a dinner that Quigley, an associate of the Irish rebels, seemed to be a character of the worst kind, Parr replied, 'No, Jemmy, not of the worst kind: he was an Irishman, and might have been a Scotsman; he was a priest, and might have been a lawyer; he was true to his cause, and might have been an apostate, Jemmy.'

[96] *The Works of Charles and Mary Lamb*, ed. T. Hutchinson (Oxford, 1908), ii. 655.

[97] BL Add. MS 32568, fo. 17.

[98] 11. 102.

[99] BL Add. MS 52451, fo. 239v, Mackintosh to Sharp, 8 Oct. 1808; BL Add. MS 52436, at 16 June 1808.

audience.[100] Coleridge had no such curiosity. He seems never to have contemplated attending Hazlitt's lectures either then or later.

In London in October and November he saw a good deal of the Lambs in the Temple, of Coleridge (again his neighbour in Southampton Buildings), and no doubt of others, including the Colliers in Hatton Garden. John Dyer Collier, a member of the Lamb circle of whom we hear little, was a former editor of the *Monthly Register*, now a reporter on the *Morning Chronicle* after having been on *The Times*. He had been succeeded on the last, which was then less important than the *Chronicle*, by his twenty-two-year-old son, John Payne Collier, who had inherited his father's literary ambitions; young Collier had just at this time embarked on a short-lived diary which gives us several glimpses of Hazlitt; it is well known for its reports of Coleridge's lectures, but the references to Hazlitt have been neglected.

It provides the earliest date for his arrival in town after the birth of his son, a date ten days before the oft-quoted letter of 26 October, in which Coleridge complains to Rickman of the injury his visits did to Lamb's health. Howe points out that if it was true that Hazlitt's calls were damaging, Coleridge's could hardly have been less so, and yet no one complained of him. We find in Collier's diary a Lamb and a Hazlitt who were the soul of mildness in making way for Coleridge's brilliant but interminable monologues. In any case, there was no need to apprise Rickman, who could judge for himself: he was on the spot: he was at the Lambs' on Wednesday, 16 October 1811, with Coleridge and Hazlitt. Also present were Dyer, Martin Burney, and Charles Lloyd. Collier describes the scene at Inner Temple Lane: 'Few others talked, although Hazlitt, Lloyd, Rickman, Dyer and Burney, with Lamb and his sister, now and then interposed a remark, and gave Coleridge, as it were, a bottom to spin upon: they all seemed disposed to allow him sea-room enough, and he availed himself of it, and, spreading canvas, sailed away majestically.'[101] The 'bare skeleton of what fell from him', as retailed by Collier, naturally enough bore on the theme of his forthcoming lectures, and Shakespeare led to Beaumont and Fletcher, and thence to Dryden, and to Southey's recent *Curse of Kehama*.

Hazlitt never doubted Coleridge's genius. Whenever his name was mentioned he made the same two observations: first, that he was the most brilliant man he had ever known; and then, that he was the most indolent (Robinson quotes him as damning him briefly in 1808 as 'a man without a will').[102] Hazlitt, although apparently self-absorbed, was

---

[100] We know from J. P. Collier's diary (see note following) that he attended on 21 Nov.; and from 5. 48 n., on 16 Dec.

[101] J. P. Collier, *Seven Lectures on Shakespeare and Milton by S. T. Coleridge* (1856), p. xxi.

[102] Morley, 11.

vividly and painfully alive to the world; he observed everything and took everything to heart. And it exasperated him, who, in trying to convey the exact truth of his impressions, toiled so hard over a paragraph, who delivered himself of a reflection as if with a forceps, that in his friend's mind such acute thought, such a plethora of subtle ideas, should be doomed to sterility. Remembering his own halting inarticulate admiration in 1798 of the talker on the road from Wem to Shrewsbury, 'whose conversation was the only one [he] had ever heard in which the ideas seemed set to music', he was humiliated that an intellect so much finer than his own should have produced nothing, nay, worse, that its owner seemed bent on destroying it. Believing, as he did, posthumous fame to be reserved for only a small number from among those who have achieved reputation in their lifetime, how should he have foreseen Coleridge's resurgence? Even Wordsworth had given Coleridge up. Hazlitt had doubtless been shocked by the deterioration apparent in Coleridge on his return from Malta, by the physical weakness and blackened lips (opium, said De Quincey and others) with which he had confronted his audience in the 1808 lectures, although he had been impressed by his readings from Milton in that series. (Robinson, who then saw Coleridge for the first time, was no less 'aware of the infirmities of his character' than he was 'deeply impressed by admiration of his genius'.)[103] Hazlitt's deprecation of Coleridge's loquacity was not the jealousy of egotism. He had listened eagerly enough in 1798. No, moving as he then was towards a pluralistic conception of truth as emergent from the interplay of minds, from a dialogue between individuals, deeply convinced as he was, both by temperament and on principle, of the need to 'look abroad into universality', how could he fail to be dismayed by Coleridge's inability to suspend his endless monologue, by his ostrich-like ignoring of interruption, the deaf ear turned to objection, qualification, or disproof, or by the changes of direction which were nevertheless perceptible whenever the spate of words encountered resistance or opposition? We must remember too that they belonged to a circle accustomed to harsh home truths: this was the time of the famous quarrel when Wordsworth at Montagu's called Coleridge something like 'a rotten drunkard', when even the indulgent Lamb 'expressed himself morally' (in Robinson's opaquely discreet phrase) about Coleridge and his habits (and about Hazlitt also, whatever obscure derelictions on Hazlitt's part he may have had in mind).[104]

If the following account of an evening at Lamb's, Wednesday, 27 November 1811, is read in the light of all this, Hazlitt's criticism of Coleridge will be seen to be relatively mild. From another point of view

---

[103] Ibid.   [104] Ibid, 50.

the account is unique: in all other descriptions of Lamb's Wednesdays, and especially in Hazlitt's own, Hazlitt plays a minor part, but here the neutral observer J. P. Collier ascribes to him a prominent, even a dominant, role. There is a suggestion of an aggressiveness that may simply be a shadow cast by the strong impression he made on Collier, or the backwash of an energy generated by the excitement of his recent escape from country solitude.

In the Evening went to C. Lamb's —— present —— G. Dyer —— W. Hazlitt —— C. Lamb —— Phillips —— Rickman —— M. Burney —— Capt. Burney —— Miss Lamb.

G. Dyer. The ancients called Wisdom ςοφια. Pythagoras was the first who called it φιλοσοφια —— He did not know what was meant when persons used the Phraze 'Philosophy of the mind'.

W. Hazlitt. Censured Coleridge much for not at least making his definition of Poetry distinct and clear.

Rickman did not wonder at it because he (Coleridge) did not understand it himself. He had better never have attempted to define it at all. Definitions were always difficult to make, difficult to be understood, and frequently not necessary or worth understanding.

W. Hazlitt thought that Coleridge had failed in his attempt —— To his mind Poetry appeared something between words and music, and the question was, what justified a man in combining the two. Coleridge said a state of excitement. He (Hazlitt) did not know, but it seemed to him that a Poet was like a man on horseback on a rough road, who, instead of travelling on among the ruts and ruggedness, preferred the greensward at the side.

C. Lamb remarked that with regard to one point, that respected the Unities, Coleridge was original, viz. in what he said regarding the Chorus being always on the stage, which prevented the change of the Scene. He praised the manner in which Coleridge had shown that the Unities were owing to accidents.

W. Hazlitt wished to remind Coleridge of Dr. Sam Johnson of Litchfield who, if he recollected rightly, had said something on the same subject.

It was remarked that Dr. Johnson founded his overthrow of the ancient unities of Time and Place on the point that the audience did not imagine that the proceedings on the stage were reality. In one place he had said that if the audience fancied any persons in a tragedy unhappy, it was themselves and not the actors.

W. Hazlitt did not think Coleridge at all competent to the task he had undertaken of lecturing on Shakespeare, as he was not well read in him. He knew little more than was in the Elegant Extracts, and Hazlitt himself had told him of many beautiful passages that Coleridge had never before heard of. It was owing to this ignorance that Coleridge had not exemplified any of his positions by quoting passages, and he doubted if he ever would —— Milton he was well acquainted with, and some years ago his readings of him were very fine; his natural whine gave them effect: this whine had since grown upon him and was now disagreeable. Coleridge was a man who had more ideas than any person Hazlitt had ever known, but had no capability of attending to one object; he was constantly endeavouring to push matters to the furthest, till he became obscure to everybody but himself. He was like a man who, instead of cultivating and bringing to perfection a small plot of ground, was attempting to cultivate a whole tract, but instead of accomplishing his object dug up the ground only for the encouragement of weeds.

G. Dyer thought that Coleridge was the fittest man for a Lecturer he had ever known: he was constantly lecturing when in company, only he did it better.

C. Lamb objected to what Coleridge had said on Shakespeare, when he remarked that it was the Poet we saw in every character, that though we beheld the Nurse or the blundering Constable, it was not those characters only, but the Poet transferring himself to those characters. If he had said that it was the Poet who spoke in the characters of Shallow or Slender or Sir Hugh Evans it would have been much better, as there was nothing very original in the characters Coleridge alluded to: there was in those he omitted.

W. Hazlitt put the characters of Coleridge and Roscoe in contrast. The one was a man full of ideas but of no industry, the other a man of great industry and no ideas. This was the merit of Roscoe's life of Lorenzo de' Medici, that he had collected with great labour all the materials, and had employed them with much exertion to advantage. Coleridge's mind was full of materials for building, but he had not perseverance to employ those he had, but, always fancying himself deficient, was in the constant search for more.

During the Evening the conversation turned upon Johnson.

Mr. Burney praised his *Vanity of Human Wishes* above his *London*.

W. Hazlitt thought Wordsworth's Criticism upon the two first lines,

> Let observation with extensive view
> Survey mankind from China to Peru,

extremely just, viz. that the first was wholly unnecessary, as the complete idea to be expressed was contained in the second line.

Phillips was a great admirer of Johnson. Boswell's life of Johnson would have been nearly as great a loss as Johnson's works, because all the Doctor's good sayings would have been lost. Boswell said nothing good himself.

W. Hazlitt said that Boswell was a Scotsman, and a Blackguard of course.

Capt. Burney asked if it necessarily followed that because a man was a Scotsman that he was a blackguard.

W. Hazlitt as naturally as that an African should be a black.

M. Burney praised Juvenal and quoted some lines.

C. Lamb had compared Halliday's translation of Persius and found it better than the original.[105]

The unexpected comparison between Coleridge and Roscoe is a reminder of Coleridge's moderate reputation at that time (as well as of the over-estimation in which the worthy banker was held), but it is more

---

[105] *Coleridge on Shakespeare*, ed. R. A. Foakes (1971), 60–3. Hazlitt's prejudice against the Scots, significant in the light of his later Scottish adventure, but a prejudice almost universal at the time, may be compared with Coleridge's (Morley, 28–9), Cobbett's, Lamb's, etc. It was certainly confirmed, among Radicals, by the infamous Lord Braxfield's conduct of the Edinburgh Treason Trials of 1793.

impressive as an early instance of Hazlitt's pluralism, and of his constant
bias towards dialectic: just as he prized the dramatic above the lyrical, so
also he found the comparative, in analysis and in description, to be more
fruitful, more effective, and a closer approximation to the truth, than the
positive.

For his own course of lectures to show a profit Hazlitt needed at least
forty subscribers, which would bring in eighty guineas, but even if he did
get such a sum into his hands its inadequacy to his needs becomes
apparent when we find him asking Robinson early in December, six
weeks before the series was even due to begin, for an advance on the few
tickets his obliging friend had placed for him.

At last the commencing date, Tuesday, 14 January 1812, was upon
him. Fresh from country solitude, he was so conscious of the blur of faces
before him that, devoured by nervousness, he delivered his first lecture
(on Hobbes, 'the father of the modern philosophy', and on Bacon)
calamitously, despite the loyal encouragement of his friends. He had
never before stood at a lecturer's rostrum. He delivered himself in a low,
monotonous, half-audible voice, kept his eyes glued to the manuscript,
not once daring to look at his hearers, and read so rapidly that no one
could follow. His fatal inexperience was compounded by a ruinous
miscalculation: he had evidently not timed his material, nor had he
enquired of the organizers how long he was allowed. He tried to cram
into sixty minutes matter which normally would have taken three hours.

His mortification was not lessened by a patronizing letter of advice he
received next day from his brother-in-law, nor even by a well-meant
word of condolence at Lamb's that evening from Robinson, who unlike
Stoddart immediately sensed his own tactlessness. A further daunting
moment came the following Monday, 20 January, when, asked to repeat
the first lecture in Mrs Montagu's drawing-room, he stopped abruptly
half-way through, whether from diffidence or discouragement Robinson's terse note does not say, but he could not be persuaded to
continue.[106] On the following evening, however, he got safely through
his second lecture — the material had been abridged at Stoddart's
suggestion by his friend (and Robinson's) the unsuccessful barrister
Burrell — and thereafter he turned his initial mishap to such good
account that his manner ultimately became 'very respectable' and even
'interesting and animated'. Robinson may have been an even more
diligent friend than we know. He had reported Coleridge's lectures in
The Times the previous November, contriving to get his paragraphs into
its columns despite the chilly discouragement of the owner, his friend
John Walter II,[107] and he seems likely to have been the author of two

---

[106] Morley, 60.   [107] Ibid, 51.

favourable accounts of Hazlitt's lectures that appeared on 23 January and 13 February.[108]

Most of his friends seem to have rallied to him, and it is strange to find an exception in Lamb, who had little taste for public lectures but who need not have kept so ungenerously silent when Robinson appealed for his help in interesting an important editor and prospective subscriber, Leigh Hunt, in their friend's venture and so procuring him some useful publicity. There was a certain rigidity in Gentle Charles. He stuck, unmoved, at Hazlitt's bad delivery and would not give him credit for the substance; whereas Robinson on the other hand had the patience to disentangle the matter from the manner, and recognized it as good; and now he had sufficient feeling to try to get help for him from Hunt.[109]

John Thelwall, Hazlitt's brother's old friend, and Lamb's of more recent years, was certainly in the audience, perhaps not out of friendship alone but partly from just such a professional interest as Hazlitt took in Coleridge's lectures: he had ambitions as a playwright, but in the mean time earned his living as a teacher of elocution, which meant in theory preparing young men for the Church and the Bar, but in lowlier practice running courses for pupils with stammers and other speech defects.[110] He had become as unshakeable an admirer of Napoleon as Hazlitt, whom we find dining, together with Robinson, George Dyer, and Northcote (the old Academician who had commisioned him in 1802 to copy a Titian at the Louvre) at Thelwall's house on 27 February, when his course was fairly launched.[111]

In the following month some further difficulty arose and the series was somehow interrupted by the pressure of his debts; he did in the end complete it, but at fortnightly instead of weekly intervals. The last was delivered on 28 April. The fable of the Brahmin and the coconuts with which he concluded on so personal a note and at such an incongruous

[108] In the absence of evidence Hazlitt's biographers have assumed that the papers on English philosophy published in the posthumous *Literary Remains of the Late William Hazlitt* (1836) are identical with these lectures (see Baker, 185). The required evidence is found in the coincidence of expression between the first of those papers and the first of these notices in *The Times*.

[109] Howe says that Lamb 'heard no lectures, neither Coleridge's nor Hazlitt's' and was not present on the 14th; he did, and he was. Howe's prejudice against Robinson also prevented him from understanding Robinson's diary entry of 17 Jan. recording a conversation with Lamb and Leigh Hunt: 'I spoke about Hazlitt's lecture in terms of great praise, but C.L. would not join me and I fear I did not succeed in my object.' His object was not, as Howe suggests, to get Lamb to attend the lectures, but to join him in praising the one they had already attended, so that Hunt might be persuaded, to mention them in the *Examiner*, if not as fulsomely as he did Campbell's in a few months (3 May) then at least less contemptuously than he did Coleridge's (9 Feb),

[110] Robinson Diary, 14 Jan., 25 Feb. 1812; BL Add. MS 27925 fo. 56; see also Robinson Diary, 3 July 1811, 27 Dec. 1815.

[111] Robinson Diary, 3 July 1811, 'Thelwall declaimed very furiously and very sillily on the *detestable* idea of assassinating Buonaparte, & said he would defend Buonaparte's life against his own father, etc., etc.'; Morley. 65.

remove from the impersonality of philosophy reiterates his rooted hunger for the quiet life of meditation.[112]

His lectures over, he went back to Winterslow, but not for long. He was probably enabled to clear his debts, but otherwise he was not much better off. He soon returned to town, in the first place to complete Thomas Robinson's portrait, but more urgently to find some permanent work, however inconsistent with the pursuit of philosophy. We may assume some pressure on Sarah's part here. As if her disappointment after her mother's death and her brother's interference (and perhaps added to these the exasperating business of Sally Shepherd) were not enough, times had suddenly become very bad. This had been dramatically brought home to their part of the world. They may not have had any money deposited in the Salisbury Bank but they were bound to have been dismayed by its failure in the summer of 1810, one of the most shattering in that period of dizzying commercial and economic disasters. Since almost the only paper money circulating in the city and its environs was the issue of that bank, now no longer negotiable, distress was widespread, provisions difficult to procure.[113] The year 1811 had seen the worst slump of the war and it was followed by the depression of 1812, more severe in the provinces than in London, giving rise not only to further bank-failures but also to an outbreak of Luddite frame-breaking.[114] In August the *Examiner* complained of 'the enormous price of the necessaries of life'.[115]

In the early years of her marriage Sarah, having finally secured a husband, may have been pleased enough to live quietly on her modest income, supplemented by her husband's very occasional earnings from books and portraits, but the worsening of the times coupled with the birth of her child required that the father, now thirty-three and with little to show for it, should cast about for a more regular source of income. Their life at Winterslow had been cramped, but not too narrow for them to help their neighbours, to entertain, to make occasional trips to London, to drink sherry, and to buy prints. Now all that was changed, and there was little else but debts; unless something was done quickly they were in danger of going under and would sink to the bottom of the stagnant pool of genteel provincial poverty. She was sorry to leave the Armsteads, with whom she corresponded for years after (young William seems to have stayed with them at Winterslow in 1822, and the Armstead children were in Exeter with the Hazlitt family in 1820).[116]

---

112 Morley, 70.
113 *Examiner*, 22 July 1810; Marrs, iii. 52–3.
114 J. S. Watson, *The Reign of George III* (Oxford, 1960), 468–9.
115 16 Aug. 1812, 520.
116 Le Gallienne, [302], 315–16; Bonner, [227], 237; *Memoirs*, i. 261.

But evidently there was nothing for it but to leave Wiltshire as soon as Hazlitt could gain a firm footing in London. This did not finally come about until the late autumn, but he was certainly there in May and in July; and so was Sarah — whether in order to add her efforts to his, as her brother's role in his finally landing a job suggests, or for some other reason, we do not know.

We catch a glimpse of him at Lamb's in May, on what was probably the last occasion when he and Wordsworth were in a room together. We owe this again to J. P. Collier's zeal to preserve Coleridge's *obiter dicta*. (He assigns no date to the meeting but it is bound to have been in this month.)[117] The talk was of Tasso's influence on Spenser, and soon involved Fairfax's translation of the *Gerusalemme Liberata*, which was praised on the one hand for its air of originality and ease, and blamed on the other for its infidelity. Said Lamb, 'Nothing could be more wanton than Fairfax's deviations.'[118]

Aye, (interposed Hazlitt), that is an evil arising out of original genius undertaking to do unoriginal work; and yet a mere versifier, a man who can string easy rhymes, and employ smooth epithets, is sure to sacrifice the spirit and power of the poet: it is then a transfusion of wine into water, and not of one wine into another, or of water into wine. It is like setting even a tolerable artist to copy after Raphael or Titian: every light and shade, every tone and tint, every form and turn may be closely followed, but still the result is only an unsatisfactory imitation. No painter's own repetitions are equal to his original pictures.[119]

Here speaks the writer who later demurred at supplying dry compilations and impersonal digests to the *Encyclopaedia Britannica*. What he says (and what Collier elsewhere reports of him) bespeaks a writer who expresses himself naturally in imagery. His later explanation of his style, that, having found that trying to express the truth and nothing but the truth did not answer, he mustered all the tropes and figures he could lay hands on, is misleading. It was, on the contrary, his early, austerely abstract style that was not natural to him.

In this passage too we discern the painter who was still trying to explain to himself why he was dissatisfied with his Louvre copies of 1802, and this brings us to observe that although the Thomas Robinson commission was his last, and his career as portraitist was closed, his passion for painting remained. It is somewhere about this time, and in the setting of Montagu's house, with Montagu's step-daughter, little Anne Skepper, as the model, that we should place a story of Procter's

[117] J. P. Collier had not met Wordsworth when Coleridge gave his 1811–12 lectures (n. 101 above, p. liv), and Wordsworth was in London once only between 1808 and 1815, i.e. in May 1812.

[118] See Lucas, iii. 384.   [119] Collier (n. 101 above), p. xxxiii.

which plainly says that the yearning to feel again the brush between his fingers was never long absent in the years that followed.

Although he had abandoned painting as a regular profession, his love for it — perhaps also some dormant ambition — induced him occasionally to return to his exercises in the art. He bought colours and brushes, and a canvas or two every year, set to work with ardour, sketched, rubbed out, grew dissatisfied, gave up his labours as hopeless, — and resumed them again in the succeeding year! Upon one of these occasions (in the year 1812) he said to Mrs. ——, at whose house he used to drop in occasionally, 'I have always wished to paint Cupid and Psyche, but I have never commenced, because I have never been able to see a Psyche — until today!' As he said this he pointed to Mrs. ——'s daughter, who was playing about the room, and asked if she might sit to him. 'I don't know', added he, 'that I shall ever succeed in painting a picture; but if not, I shall at least learn that I am unable to do so — and that is something.' Consent was given, and the little girl (who had a face full of expression) sate as a model for his Psyche. He did not succeed in the picture. He was not quick and dextrous enough to catch ere they vanished all the transient and playful graces of childhood, nor could he revive them by that species of amusing dialogue, adapted to all ages, which is part of the accomplishment of a portrait-painter. The consequence was that the little Psyche speedily got weary . . . , and became a model for a sleeping nymph, rather than for the young and girlish bride of Cupid.[120]

In Collier's note of the conversation at Lamb's in May there is no sign of any awkwardness between Hazlitt and Wordsworth such as might foreshadow their differences of a few years' time, but this may be because young Collier was not close enough to these older men to be in the secret of their relationships. He records another conversation which also seems to refer to this period, so we will give it here. It evidently took place in his father's house,[121] and it turned on Samuel Rogers the banker-poet; John Dyer Collier produced a manuscript copy of *The Pleasures of Hope* given him by the author:

Hazlitt contended that there was a 'finical finish' (his own words) about the lines, which made them read like the composition of a mature period; and he added his conviction that they were produced with much labour and toil, and afterwards polished with painful industry. Such was indisputably the fact; and it was generally declared that no free and flowing poet could write so neat and formal a hand: it was fit for a banker's clerk, who was afterwards to become a banker. Coleridge dwelt upon the harmony and sweetness of many of the

---

[120] [B. W. Procter], 'My Recollections of the Late William Hazlitt', *NMM* 29 (1830), 472. The anecdote, evidently deriving from the daughter herself, Procter's wife, is confirmed by two echoes: the recurrence of the same consoling reflection to Hazlitt at a later unsuccessful attempt at painting, in 1826–7 (17. 219); Patmore's independent allusion to Hazlitt's admiration of 'Mrs. M. and the Psyche-like form and features of her daughter' (ii. 333).

[121] Necessarily, unless his father carried the manuscript *Pleasures of Hope* about with him.

couplets, and was willing to put the versification about on a par with Goldsmith's *Traveller*. Hazlitt, on the other hand, protested against Rogers being reckoned a poet at all: he was a banker; he had been born a banker, bred a banker, and a banker he must remain; if he were a poet, he was certainly a poet *sui generis*. 'Aye, *sui generous* (stuttered Lamb, in his cheerful jocular way, looking at everthing on the sunny and most agreeable side), Rogers is not like Catiline, *sui profusus*, any more than he is *alieni appetens*, but he is *sui generous*, and I believe that few deserving people make appeals to him in vain.' This characteristic joke put everybody into good humour, and it was voted, almost *nem. con.*, that Rogers was a poet in spite of his purse; — 'by virtue of it', added Hazlitt, and so the matter ended.[122]

Godwin encountered Sarah in London twice that summer, on 14 May and 12 July; and Crabb Robinson, on 28 May, speaks of 'walking with Miss Lamb to Hazlitt's', and again on 30 June of being 'at Chambers in the forenoon till half past two, when [he] called at William Hazlitt's', which suggests that if the Hazlitts were so close to Robinson in Gray's Inn and to the Lambs in the Temple then they had again returned to Southampton Buildings.[123] Montagu tried to put money in Hazlitt's pocket at this time by commissioning a pamphlet for the Society for the Diffusion of Knowledge upon the Punishment of Death which he had founded in 1809, but the pamphlet did not appear; nor is it clear that Hazlitt was paid.[124] A further miscarriage that Sarah suffered on 6 September cannot have helped matters. It is possible that at this time Hazlitt was in another difficulty: he had still not repaid £30 he had borrowed 'for a fortnight', in anticipation of payment for his pamphlet, from a friend of Robinson's some six weeks before.[125] At the end of September he tried to borrow another £20 from Robinson. 'Such are the difficulties from which great talents alone, without discretion, will never relieve a man,' Robinson had noted solemnly. Everything seemed to be going wrong.

At his wits' end to find a steady livelihood, he concluded that the only hope, a fairly lowly one, was parliamentary reporting; or perhaps it was the suggestion of his wife, with her brother again in the background. Stoddart was now on the staff of *The Times*, having been appointed leader-writer, apparently on the strength of a series of political articles he

---

[122] Collier (n. 101 above), p. xlix.   [123] Godwin Diary.

[124] Vol. 2 of Montagu's *Opinions of Different Authors upon the Punishment of Death* had been published that March (*The Times*, 2 Mar. 1812). Hazlitt's pamphlet was not published until 1831, and then perhaps only in part (19. 368).

[125] Robinson Diary, 8 Sept. 1812. The friend was Anthony Robinson, one of the 'proprietors' (numbering a thousand), along with Alsager and James Perry, of the wealthy London Institution, of which Richard Sharp was one of the 'managers'. I say it is *possible* because there is some doubt about the name of the borrower. It is given in shorthand in H. C. Robinson's diary and the typed transcript in Dr Williams's Library does not give it as 'William Hazlitt' but as 'William? Holt?', which corresponds to the indications of the key.

contributed over the signature of J.S. in 1810,[126] but Hazlitt's relations
with his brother-in-law were even then sufficiently strained to make a
direct approach irksome; he asked Robinson to sound him out, and
Stoddart agreed to speak to John Walter II. This was towards the end of
September 1812. In the interval Lamb had also taken steps on Hazlitt's
behalf; he wrote to John Dyer Collier, Foreign Editor of the *Morning
Chronicle*, adducing as special qualifications the familiarity with the
history of parliamentary oratory Hazlitt had shown in *Eloquence of the
British Senate* and particularly his vice-like memory (which enabled him to
reproduce whole conversations he had heard). Collier promised to speak
to James Perry, the owner and editor. Hazlitt was interviewed, and
appointed at four guineas a week.

Whether or not the engagement took immediate effect and the salary
became immediately payable is not clear.[127] If Hazlitt received no pay
until there were debates to report, it was two months before he felt the
benefit of his good fortune: Parliament was dissolved on 29 September;
there was a general election; and the new Parliament did not meet until
30 November.

[126] A. Andrews, *The History of British Journalism* (1859), ii. 75.

[127] W. H. Watts, parliamentary reporter on *MC* in 1808, received 5 guineas a week all year
round, but John Payne Collier on *The Times* was paid only when Parliament was sitting; salaries
varied — he received 7 guineas a week, but his brother William, also a parliamentary reporter,
only got 3 (Anon., *The History of The Times* (1935–52), i. 135–6).

3
Return to Town

Le vent se lève, it faut tenter de vivre

Valéry

HE probably took lodgings near the Old Palace of Westminster from the beginning so as to be within easy walking distance of his work. But he had two places of work, and though near the one he was at some distance from the other. After the House rose he still had to walk to the *Morning Chronicle* office opposite Somerset House in the Strand, transcribe his notes, and hand over the copy to the printer. The hours were arduous and when he was on late shift (the relay system had come into operation) he did not leave St Stephen's Chapel (where the Commons met; the Lords were in the adjoining Court of Requests) until long after midnight; the House sometimes sat until five in the morning, and when he finally retraced his steps down Whitehall he must often have seen the dawn streaking the sky in the direction of St Paul's.

This was the unreformed Parliament and the experience of the newcomer was certain to be one of shocked incredulity. Looking back, one such newcomer remembered 'his first feeling of disappointment as he gazed with a sense of wounded pride around the darkened narrow room and looked in astonishment at the honourable members grouped in various attitudes of carelessness and indifference'. But this no doubt had already been Hazlitt's case many years since, if as is likely he had attended debates while compiling the *Eloquence of the British Senate* in 1806. He had been awakened to the art of oratory by the writings of Burke after the revelation of the 'Letter to a Noble Lord' at Shrewsbury in 1796; he was also the son of a preacher. It is uncertain how many orators he had heard before 1812; probably Sheridan, Canning, Horne Tooke, and William Windham, but not Curran or Erskine.

Although he was no longer astonished by carelessness and indifference, we may imagine moments when the debates he was now forced to attend for long hours made him angry as well as ashamed. A man of his political views must have found it difficult to record impartially claims and assertions that filled him with irritation or disgust. 'Public affairs vex no man', said Dr Johnson speaking, it is true, six years before the fall of the Bastille inaugurated in England an epoch of reaction that vexed a good many, and none more than Hazlitt, but speaking also a year after

Hazlitt's father was so vexed by the ill treatment of American prisoners-of-war by British officers that he risked ostracism and even violence to protest. The son must indeed have found it particularly hard to have to listen to and record the suave 'claims of barefaced power' at a time when he was in no position to 'expose the little arts of sophistry by which they were defended', as he was able to do in a few years' time. Parliament was a model academy of those arts.

There was no Press Gallery, since reporting was a breach of privilege. It was, however, unofficially tolerated (a toleration which the Tory *Courier*, in a staggering paradox, held to be the most remarkable example of 'the progress of tyranny and the encroachments upon freedom during the present King's reign'[1]), but the reporters had to crowd into the Strangers' Gallery, itself notoriously inadequate. Perhaps even more than by the lounging members with their feet on the benches, dozing with their hats over their eyes or engrossed in their wide-spread newspapers, Hazlitt was struck by the dissipated insouciance of his hard-bitten colleagues. One of the two most celebrated had died four years earlier, the legendary Mark Supple, a big-boned loud-voiced Irishman who 'reported like a gentleman and a man of genius'. The members, it was said, hardly knew their speeches when they had undergone his free and bold treatment, but none ever came to the office to complain. He was a favourite, and presumed upon it.[2] His cynicism and recklessness goaded him one night in a moment of tipsy, exasperated boredom to cut into a pause in the languid proceedings of the House by suddenly bawling out, 'Mr. Speaker! Give us a song!', an anecdote that Hazlitt, who put it in an essay many years later, probably heard from Supple's boon companion, also an Irishman, the wild Peter Finnerty, one of Hazlitt's fellow-reporters on the *Chronicle*.

Finnerty at this time was forty-six and three times already a martyr in the cause of the freedom of the press. In 1797 he had been sentenced to the pillory, two years' jail, and a huge fine for exposing the scandalously rigged trial of the United Irishman William Orr; in 1809 he had been expelled from Walcheren by the Secretary of War (Castlereagh, Chief Secretary in Ireland during the Orr affair) for reporting the appalling condition of the troops on that expedition; and when Hazlitt met him he had just served eighteen months in Lincoln gaol for a libel charging Castlereagh with cruelty in Ireland. Hazlitt probably remembered the much-publicized meeting held by the Friends of the Liberty of the Press at the Crown and Anchor the previous year, Sir Francis Burdett in the chair, to raise money for Finnerty's legal fees.[3] Nor did Finnerty fight

[1] 17 Mar. 1817.   [2] I quote and adapt *MC*, 11 Mar. 1829.
[3] *Examiner*, 17 Feb. 1811, 104. The same number has a leader, 'Prefatory Observations on the Case of Mr Finnerty', 97–8.

only with his pen: he spoke from the radical platform in the Westminster elections, and in 1810 did Cobbett a service at a critical moment in his career. This was when Cobbett thought of giving up his *Political Register*, in order to get out of the two-year gaol sentence passed on him for an article on the flogging of militiamen by a Hanoverian guard.[4] It was Finnerty who persuaded him to stand firm against the Government. Among those who spoke in favour of Finnerty's petition to Parliament in 1811 were Whitbread, Brougham, Burdett, Romilly, and William Smith. It was then that Shelley, an undergraduate at Oxford and a stranger to Finnerty, helped him with money from admiration for his stand against Castlereagh.[5] We can be pretty sure that Hazlitt drank with him and the other reporters at Bellamy's (the chop-house described by Dickens in *Sketches by Boz* that was so near the House that the division-bell could be heard and dining members would rush headlong downstairs and into the Chamber) or at Finnerty's other haunt, the Cider Cellars in Maiden Lane. He got to know him well enough to be able to call him in after years 'my old friend'. He liked him; but did not consider him in his harangues at the hustings to be anything more than a voluble, plausible, animated, and prejudiced arranger of common-places. Years later he recalled with amusement showing him a para-graph he had just written poking fun at the Scots, at which Finnerty laughed heartily, only to look grave when his eye strayed down the page to a sly dig at the Irish.[6]

James Perry, their editor, bore patiently with Finnerty's drunkenness and unreliability; he too had known what it was to be imprisoned for his convictions.[7] Perry *was* a Scot, and so far as we know, the first with whom Hazlitt had much traffic; it was he, and another of Hazlitt's colleagues, John Black, who gave Hazlitt the chance of putting to the test the anti-Scottish prejudice to which we have already seen him give utterance, a prejudice almost universal in the eighteenth century, with an edge sharpened by the two Jacobite risings, and which was most notorious in Dr Johnson. Perry and Black represented a nation which, although Hazlitt did not then know it, would throw up in a few years' time the two most malignant enemies he would ever encounter, and who almost ruined his writing career. The inveteracy of this prejudice is apparent in Black's tormentor, the intensely English William Cobbett, and in such of Hazlitt's friends as Coleridge and Lamb.[8] It was exacerbated by the invasion of the London press by Scottish journalists.

[4] *Political Register*, 4 Jan. 1817, cols. 13, 14.
[5] *Examiner*, 23 June 1811, 392–3; N. I. White, *Shelley* (New York, 1940), i. 108, 219, 597.
[6] 17. 300.
[7] P. L. Gordon, *Personal Memoirs* (1830), i. 295–9.
[8] See the writer's 'Hazlitt, Cobbett, and the *Edinburgh Review*', *Neophilologus*, 53 (1969), 69–76.

One reason for the Scots' flight south was that their poverty, incessantly chafed by their analytical ponderings, had aroused them to a raw awareness of social inequality, and they were ardent to let fresh air into the closed society of their native land which the propertied, or the feudal, or the merely pious Establishment was determined should not be disturbed, a determination personified in the sinister crude figure of Braxfield, the judge who sentenced Palmer, Muir, and Gerrald in the Edinburgh Treason Trials of 1793.[9] Nor was it likely to be disturbed, for in Scotland, ostensibly more democratic than England but in fact devoid of representative institutions and consisting of a multiplicity of co-optive oligarchies at once more timid and more oppressive than those in the South, the parliamentary reformers had no voice in the newspaper press (nor would have, until the *Scotsman* was established in 1817) and the 'lad o' pairts' found himself obliged to go four hundred miles out of earshot to make himself heard. Once in the South, however, these men spoke out loud and firm, and earned an honourable place in the history of free speech in Britain.[10] One such was the editor who guided the *Morning Chronicle* consistently, if at times confusedly, along the path of justice and humanity during thirty of the most troubled and crucial years in the annals of the country.

James Perry was born in Aberdeen in 1756, the son of a builder. He began to study law at the University of Aberdeen but his father became bankrupt and he had to earn his living, becoming first a draper's assistant, and then joining Booth's company of travelling players (among whom was the equally obscure Thomas Holcroft). However, his Scottish brogue proved ineradicable, so he gave up the stage and went to work as a clerk in a Manchester factory, read widely, and at twenty-one, in the year before Hazlitt's birth, moved on to London. Many anecdotes were told of his early struggles in journalism; one of them, an amusing story of the illegality of parliamentary reporting, was included by Hazlitt before he met Perry, in his as yet unpublished *Memoirs of the Late Thomas Holcroft*, an anecdote no doubt of intriguing point to him at the present time.[11] Perry became an industrious and skilful gallery reporter, showing his ingenuity by the invention of the relay system. He also published political pamphlets and poems, and frequented the 'spouting clubs' that flourished from the early 1780s onwards. He first became editor of the *European Magazine*, and then took over the *Gazetteer*, making it a vehicle for the views of Charles James Fox, which, as he often

[9] J. S. Watson, *The Reign of George III, 1760–1815* (Oxford, 1960), 359.

[10] *Blackwood's* complained furiously that 'the Whig news-writers in London are, for the most part, Scotchmen — more's the shame and pity', 13 (1823), 93. The *Edinburgh Literary Journal*, 19 Sept. 1829, 220, gives a long list of Scottish editors in London.

[11] 3. 92–4.

remarked later, 'from their liberality in the cause of freedom, justice and humanity, had made, on his first entering the House of Commons, an indelible impression on his mind'.[12]

When the *Gazetteer* was bought by the Tories in 1789 Perry resigned and acquired the *Morning Chronicle*. His business acumen, his cordial enjoyment of social life, which spread a wide net for news, and finally his jealousy of his own independence and the good name of his profession, which made him vigilant on the one hand to escape the snares of venality and on the other to outwit the manœuvres of the powerful and great, all contributed to a steady and ultimately massive growth of the paper's influence and wealth (when he died he left more than £100,000).[13] In particular, he set out to make his paper supreme in parliamentary reporting and political coverage, and until his death in 1821 it was by far the most influential journal the country had ever seen. He was never corruptible, although not of course impartial. During the 1807 election the *Morning Chronicle* was made a daily vehicle for the political paragraphs of Fox's favourite nephew and his circle, but not so pliant a vehicle that Brougham did not complain to Lord Holland that Perry 'had too much an opinion of his own'.[14] His leaning was constantly towards the humane and liberal view.

But the paper had a hard struggle. The Government eyed it suspiciously from the start, and in December 1793, when it reported a meeting of the Constitutional Society, Perry was charged with having 'printed and published a seditious libel'. He was defended by Thomas Erskine, who had been Tom Paine's counsel the year before, and who was to be counsel for Thelwall, Holcroft, Hardy, and Horne Tooke the year after. After a fifteen-hour all-night sitting the jury returned a verdict of 'Not guilty', said to be due to one obstinate juryman. Perry's course was now set and there was no driving him out of it. It was he who printed Godwin's anonymous 'Cursory Strictures', which effectively altered the initial direction given the 1794 Treason Trials by Lord Chief Justice Eyre, and which may fairly be called a landmark in the history of British freedom. In a few years, in 1798, he was hauled before the House of Lords for breach of privilege and committed to Newgate, where he held levees of sympathizers, just as, fifteen years later, that other jailed editor Leigh Hunt received his friends, one of whom was Hazlitt, in Surrey Gaol. It was Hunt, as it happened, who unwittingly caused

[12] *Annual Biography and Obituary*, vii. 380.

[13] *Caledonian Mercury*, 6 Apr. 1822.

[14] BL Add. MS 52178, Brougham to John Allen, 10, 16 June 1807; and BL Add. MS 52180, 9 May 1813, Francis Horner to Allen: 'I am glad Lord Holland did not omit the opportunity of keeping Perry right who is so liable to err on such occasions'. By 1811 Perry counted as 'a sort of *sous-ministre* of the Fox party' (*The Diaries of Sylvester Douglas*, ed. F. Bickley (1928), ii. 118).

Perry's third trial, before Lord Ellenborough in February 1810, for quoting from the *Examiner* an assertion that the successor of George III would have 'the finest opportunity of becoming nobly popular'. Perry conducted his own case with such vigour and eloquence that the judge charged the jury in his favour and he was forthwith pronounced not guilty (whereupon a projected case against the *Examiner*, held in reserve, also fell to the ground).

Perry's interests extended beyond politics. His keen eye for literary talent recognized and encouraged as early as 1793 the promise of Coleridge.[15] He was also a connoisseur whose name appears in the catalogue of the Lucien Buonaparte sale as purchaser of canvases by Titian and Rubens. He had a very engaging personality: Hazlitt continued to like him even after his unceremonious and ambiguous dismissal. Leigh Hunt describes him as 'a lively, good-natured man, with a shrewd expression of countenance, and twinkling eyes, which he not unwillingly turned upon the ladies'.[16] He was of a warmly hospitable turn, and most of the celebrities of the period dined at his table. He could not, of course, have been received in the houses of the nobility (except Holland House whose divorced mistress was herself ostracized) but he achieved a remarkable eminence in an age when newspaper editors were socially unacceptable.[17] Mary Russell Mitford who often stayed with the Perrys at their grand twenty-three-roomed house in Tavistock Square on her annual visits to town from 1813 until Perry's death in 1821, calls him a 'delightful person at whose house men of all parties met, forgetting their political differences in social pleasure'. She mentions Erskine, Romilly, Tierney, Southey and Moore.[18] Others name Byron, Hobhouse, Brougham, Dr Parr, Benjamin Constant, and Sir Thomas Lawrence; and there was also Professor Porson, who was Perry's brother-in-law.

Finnerty was a reporter of long standing. There was another colleague, a newcomer, who must have been closer to Hazlitt from a shared inexperience; he became a good friend, and remained so, long after Hazlitt left the *Chronicle*. This was John Black. Five years younger than Hazlitt, the son of a Scottish pedlar, he had supported himself at Edinburgh University by a clerkship and by hackwork (including translation from the German). One of his fellow-students was William Mudford, son of a London shopkeeper, with whom he collaborated in an

[15] Griggs, iv. 897; i. 138, 226–7.

[16] Hunt, 208.

[17] When Lockhart seemed likely in 1825 to be offered a newspaper-editorship, Scott warned him (with unconscious irony) against 'sacrificing . . . a considerable portion of his respectability in society . . . ' (Scott, ix. 250).

[18] M. R. Mitford, *Recollections of a Literary Life* (1852), iii. 14.

exchange of letters in which he defended classical education against
Mudford, who contrived to get the debate into the *Political Register* of
William Cobbett. Cobbett later, however, took a dislike to his earnestness and persistently ridiculed him as 'Dr. Black, the Scotch feelo-
sopher'. After Mudford had returned to London and become editor of
the *Universal Magazine*, in which he published articles of Black's on
German and Italian literature, he persuaded Black to come south. Black
arrived in 1810 and Perry appointed him to report parliamentary
debates and deal with foreign correspondence in the *Morning Chronicle*.
He was studious and well informed, solemn, rather dull, but good-
hearted and notably unworldly.

On 30 November the new Parliament opened with a debate on the
battle of Salamanca and the conduct of the war, in which both Canning
and Castlereagh spoke; in the Lords, Wellesley attacked the Government for its inadequate support of the Peninsular campaign. The
Morning Chronicle of 2 December approved this speech by a man who was,
before long, to furnish Hazlitt with the occasion for a short but
unforgettably ironical paragraph that proved the harbinger of his
political journalism. The succeeding weeks of that wet muddy month of
December 1812 saw further attacks by Wellesley and his faction on the
Ministry for its neglect of his brother in Spain. At the same time the first
news arrived of the reverses suffered by Napoleon in Russia after the
winter set in on 6 November. On 17 December the Prince Regent sent a
message to the Commons proposing financial aid to the gallant Russian
people, which Sir Francis Burdett described as not only extraordinary
but insulting to the country, whilst Whitbread said that charity should
begin at home, with their own starving factory-workers — sentiments
soon echoed in the petitions of towns and counties to Parliament for
peace with France, on account of 'the evils brought upon them by this
dismally long-protracted war'.[19] The Regent's name was embarrassingly prominent elsewhere. On 9 December after a spectacular trial John
and Leigh Hunt were found guilty of libelling that royal personage.
They had resumed the attacks that led to Perry's trial: the offending
article had appeared in the *Examiner* in March 1812, a week after Lamb's
punning verses against the Regent, 'The Triumph of the Whale', had
appeared in its columns. As we have seen, Leigh Hunt occasionally
attended Lamb's Thursdays, and it seems likely that it was in autumn
1812, almost as soon as Hazlitt returned to town, that he met Hunt
there. The trial took place before Lord Ellenborough (who had tried
Perry) and a special jury. 'The Court and Westminster Hall were
crowded to excess during the trial, and it was with the utmost difficulty

---

[19] *Parl. Deb.* 25 (1813), cols. 532, 598, 1157 (2, 6 Apr., 10 May).

that a great number of special constables and Bow Street officers could maintain order and tranquillity.'[20] The report filled eight columns of next day's *Morning Chronicle*. Hazlitt must have witnessed a great stir in the office in the Strand. The Hunts' 'hard truths' are well known, but we cannot omit to quote them briefly as typical of the turbulent polemical arena Hazlitt was now about to enter :

What person, unacquainted with the true state of the case, would imagine that . . . this *delightful, blissful, wise, pleasurable, honourable, virtuous, true*, and *immortal* prince, was a violator of his word, a libertine over head and ears in disgrace, a despiser of domestic ties, the companion of gamblers and demireps, a man who has just closed half a century without one single claim on the gratitude of his country, or the respect of posterity![21]

We have no evidence of Hazlitt's immediate reaction to these events, but they frame with sufficient drama his first entrance upon the scene of his future political journalism.

What we do hear of him at this time is peaceful enough. His relief after months of uncertainty is reflected in his high spirits at Lamb's Christmas Eve party, two days after the House adjourned, when he told Robinson that he found the work easy. Even more than of satisfaction we detect a sense of hope heralding the beginning of an entirely new life. At thirty-four he was not yet too old to start afresh, and for a while we shall find a note of youthfulness and optimism in his work and in what others say of him. On the second day of the new year Robinson was at Hazlitt's lodgings, where he found the Lambs and Martin Burney and probably also Martin's father: 'I found H. in a handsome room and his supper was comfortably set out. Enjoyments which have sprung out of an unmeaning chat with Mrs C[ollier] at Lamb's.[22] On what frivolous accidents do the important events of our lives depend!' Wherever this room may have been, it was certainly not at 19 York Street. The setting was evidently very different from what Robinson had earlier seen at Southampton Buildings and also quite unlike the accounts we later get of the house in York Street. There is no more mention of Sarah than there was on 24 December, and it is likely that she and the baby made the final move to town just after this. Robinson says that both Hazlitt and Sarah were at the Lambs' on 14 January. But they did not move into York Street for several months.

Hazlitt may have found the reporting easy and the late hours tolerable (to such a night-bird as himself), but there were disadvantages. The

---

[20] *MC*, 10 Dec. 1812.   [21] Hunt, 231.

[22] It appears from this that although Lamb wrote formally to recommend Hazlitt to J. D. Collier it was actually Robinson who first raised the matter (however casually) with Mrs Collier. Lamb was probably better acquainted with the husband.

gallery itself was awkwardly placed, on one side of the hall, so that only half the house was visible. And it was unhealthy. St Stephen's was ill-ventilated and the air in the gallery high up under the roof rapidly became foul.[23] The reporters were crammed into its back bench, and it was such a struggle out (when a suitable break in the speeches offered) that one unfortunate reporter some years before this — Thomas Campbell the poet — contracted a disease of the bladder that racked him for the rest of his existence.[24] And in addition to these physical vexations, which may have contributed to the ill health Hazlitt suffered in a few years' time, there were the speeches themselves: when not exasperating to Hazlitt from the unpalatable views advanced they were dull, and he found it hard to bear patiently with 'the well-known, voluminous, calculable periods roll[ing] over the drowsy ears of the auditors, almost before they [were] delivered from the vapid tongue that utter[ed] them', and was depressed by the 'want of originality', the universal 'suspension of thought and feeling'.[25] He was temperamentally no more docile than his brother, and as prone to reckless bouts of dissipation when in funds or out of spirits. We hear of at least one occasion when 'to the great annoyance of some of his colleagues he preferred his wine with a few friends to taking his share in reporting an important discussion in the House of Commons'.[26] The euphemism is transparent. We hear little of his drinking before 1812 but in the circle of newspapermen he now entered there was ample opportunity and constant encouragement. He had probably been initiated by his reckless brother in the late 1790s; it was difficult to remain sober in Lamb's company; both Finnerty and Black were fond of the bottle; and finally it was a hard-drinking age. It was seldom enough that he heard a speech that commanded his personal as well as his professional attention (an exception was Plunket's speech on the Catholic question on 25 February 1813, which he remembered in after years as the most brilliant he ever listened to[27]), but his lecturing had taught him that successful speaking is quite a different thing from effective writing: an audience of flesh and blood naturally tends to blunt originality of thought and inhibit discrimination. Not that he failed to recognize appropriate degrees of excellence in public speaking. His *Eloquence of the British Senate* demonstrates the contrary. But his familiarity with the best speeches of two centuries may have done him the disservice of arousing the forlorn hope of some inspired departure from the

---

[23] S. C. Hall, *Retrospect of a Long Life* (1883), i. 112.

[24] NLS MS 9818, fo. 137 and MS 3363, fo. 16.

[25] 12. 270–1.

[26] *NMM* 29 (1830), 438.

[27] 17. 13 (according to Castlereagh, 'a speech never to be forgotten', *MC*, 3 Mar.).

hackneyed arguments, the threadbare metaphors, the appeals to established maxims he was now obliged to endure night after night.[28]

In the mean time the new parliamentary session was not to begin until 2 February, and he had a whole month on his hands, time to find a place to live, to look up old friends, and to write. His former visits to Horne Tooke's fine house overlooking Rushmere Pond on Wimbledon Common, where he used to meet Godwin, Curran, Holcroft, and Fuseli, and perhaps Dr Parr, and which he describes in *The Spirit of the Age* almost as nostalgically as the evenings at the Lambs',[29] were a thing of the past. His host had died some months earlier. Holcroft too was gone. Godwin he had always called on when he came up to town. His relations with Curran were more formal, and with Fuseli's violent humours ('always swearing and straining for something that was out of his reach . . . all sound and fury — a mere explosion of words'[30]) he did not greatly sympathize. We must assume, with Howe, that the hospitality of Richard 'Conversation' Sharp's board at Fredley Farm under Boxhill was from some cause no longer open to him.[31] His greatest pleasure, as we may judge from 'On the Conversation of Authors' came from the Lambs' regular parties — now on Thursdays — where he saw once more across the hearth or the card-table the familiar faces of Hume, Dyer, White, Captain Burney and Martin, Phillips, Rickman, Robinson, and Ayrton, as well as a new face, that of Barron Field, eight years younger than Hazlitt, student at the Inner Temple but already a contributor of long standing to the *Examiner*, *The Times*, and even the *Quarterly*.[32] When the weather was fine the long rambles with Lamb he had known at Winterslow were repeated, on the outskirts of London. A notebook of Ayrton's gives a glimpse of Lamb dining at the inn at Dulwich in company with Martin Burney and Hazlitt, after walking round by 'Norwood's ridgy heights', whence they might 'survey the snake-like Thames or its smoke-crowned capital'.[33] The holiday mood of these pleasures is embalmed in a Lamb-inspired pun upon which Ayrton here congratulates himself. When the bill came, Burney, hard up as usual, put his arm around Ayrton's shoulder, 'Pray, pay for me, Ayrton!' And Ayrton made excruciating reply, 'I will, Martin, but don't fatten and lean on me too.'[34] A mile from Norwood rose the 'bloomingly sylvan' Sydenham Common with, on its far side, extensive views, as in a Claude landscape, of the fields and orchards of Kent. Here lived their friend Thomas Hill, dry-salter by trade, universal quidnunc and gossip by nature, man of letters by inclination (he was a book-collector and

---

[28] 12. 265–6. [29] 11. 47–50.

[30] His description, quoted by Procter, 481. [31] Howe, App. III.

[32] Field's bookplate is in a copy of Hazlitt's *Political Essays* in the National Library of Australia.

[33] 12. 114. [34] BL Add. MS 60358, fo. 111.

part-owner of the *Monthly Mirror*). His faithful attendance at Hazlitt's lectures in 1818–19 suggests a long-standing acquaintance.[35] But apart from such excursions, we may suppose, his steps were most often turned towards town-visits, or took him at the furthest out through Hyde Park and Kensington Gardens to Notting Hill to see the Humes at Montpelier House.

The absence of information about such friendships is a familiar and sad lack in the scanty chronicle of Hazlitt's days, and makes him appear less sociable than he was. The little we know of his daily movements we do not learn from him, but indirectly from the diaries of such as Robinson and Haydon and the correspondence of Lamb, none of whom appear to have known either Tooke or Sharp (although Robinson did meet Sharp in 1829). This is strikingly attested by what we know of another friend, James Northcote, a painter who was the antithesis of Fuseli, and who filled a big place in his existence. To the reminiscence-laden conversation of this last survivor of Dr Johnson's circle Hazlitt had been attending with relish ever since the beginning of the century. Year after year he returned to his studio to spend hour after hour in his company. But if we set aside the volume of their *Conversations*, which Hazlitt published in the last year of his life, the only other evidence that they ever met, let alone exchanged more than a few words, amounts to half a dozen brief uninformative allusions.[36] If Hazlitt had died abroad in 1825 the *Conversations* would never have been written and we would never have suspected the extent of their intercourse. It was to the old painter's studio in Argyle Street that Hazlitt repaired when he got back to town. He found him adding the last details to his rambling *Life of Sir Joshua Reynolds* (his former teacher, and, we are told, John Hazlitt's at a later date), published the following June, from which Hazlitt liked to quote anecdotes of Goldsmith and a eulogy of Titian and Claude entitled 'The Dream of a Painter'.[37]

Around the corner in Great Marlborough Street, but otherwise poles apart from the matter-of-fact and reasonable Northcote, lived another painter, the irascibly ambitious Benjamin Robert Haydon, who nearly ten years since had come storming up from the West Country, seeing

[35] BL Add. MS 20081, fo. 144; and Godwin Diary, Mar.–Apr. 1818; 5, 12 Nov. 1819. Three letters from Hazlitt to Hill, now lost, were sold in 1841 (see the writer's 'Some New Hazlitt Letters', *NQ*, July–Aug. 1977, 342). Hill employed Lamb's friend J. Fenwick in 1805 (BL Add. MS 2081, fo. 95), subscribed to Coleridge's *The Friend* in 1809 (Forster MS (V. & A.) 112, fo. 15), attended Coleridge's lectures in 1811 (Godwin Diary, 25 Nov.), and visited Hunt in gaol in 1813 (Hunt, *Corr.* i. 83).

[36] *Letters*, 85; Godwin Diary, 18 Mar. 1808, 26 Feb., 23 Mar. 1811; Haydon, 211; Morley, 65; Pope, ii. 417.

[37] Pub. 29 June according to BL M/612 (Colburn List); the Goldsmith anecdotes are in 8. 93, 12. 100, 168.

himself in his colossal vanity as the Saviour of British Art, the long-awaited Genius of Historical Painting. A paragraph in his autobiography indicates that Hazlitt and Leigh Hunt, as well as his former fellow-student the Scotsman David Wilkie, were frequent callers at Great Marlborough Street during this winter. Hazlitt had met Haydon at Northcote's in 1812 just after he had seen his Romantic canvas *Macbeth the instant before he murdered Duncan*[38] at the British Institution exhibition which ran for some months from 3 February, when he was in town delivering his lectures. Haydon rather despised Hazlitt as a failed artist who had long since abandoned all hope of ascending to the high plane on which, he, Haydon, had his being, or of breathing the rarefied atmosphere of the life of genius. He saw him as a lame dog lacking in will, application, and moral courage. Haydon was in no doubt of his own energy, resourcefulness, and tenacity, although he had already at this early date, by misapplying these virtues, committed the gigantic blunder that was to poison and clog his whole future career. He had embroiled himself in a lone vendetta with the Royal Academicians in 1809, when they deliberately, as he alleged, tried to keep him down by hanging his *Death of Dentatus* in a bad place, and, isolated as he was (his only artist friend Wilkie was a shrewd realist and must often have shaken his head over Haydon's absurdities and obstinacies, loyal to him though he remained), would have been glad to attract an ally, or more acceptably a lieutenant, in the fight. In a few years from now, when Hazlitt began to cover art exhibitions in the papers, Haydon would attempt to stifle his impatience at what he claimed to be Hazlitt's pessimism about English art, because he recognized the force and subtlety of his judgements. It would be excellent policy, he told himself, to cure Hazlitt's wrong-headedness about the prospects of English historical painting and bring him to a just perception of the genius of Benjamin Robert Haydon. He must 'make it his object to manage such an intellect for the great purposes of art'.[39] But he immediately became irritated again on realizing that any such docility in the critic would be incompatible with the rooted originality of outlook he had come to admire; that such an intellect was not to be managed; it had its own purposes.

As to Hazlitt's writing during this period, none of it immediately emerged into print. A paragraph on Pope in the *Morning Chronicle* of Boxing Day 1812, quoting not only the admired 'Epistle to Mr. Jervas' and 'Windsor Forest' but also the lines in Thomson's *Seasons* on the redbreast's winter visits to the breakfast-table later used in the *Lectures on*

---

[38] 'A truly energetic, original and grand performance' (*Examiner*, 3 Feb. 1812).
[39] Pope, ii. 65.

the English Poets, may well have come from his pen, but there is nothing else characteristic of him for a good many months. The other departments of the paper were adequately staffed, including the theatrical. We cannot doubt but that his love of the stage, dating from his first visit, his first sight of Dignum and Suett in *No Song No Supper* (that 'bright vision of his childhood' that had never ceased to 'play around his fancy with unabated vivid delight'[40]) immediately reasserted itself on his return to town. Playgoing had a unique sense for him: the privilege of inhabiting a special region of time arrested between past and future, the paradox of a real suspension of events bodied forth in an ideal succession of incidents; like novel-reading, but less solitary, more palpably related to concrete existence; when he was in the world but not of it, at once absent and present, gazing at the raree-show from the loopholes of retreat.

It was a fascination that never slackened, to the end of his life. If ever the day should arrive (he said) when he ceased to cast a sidelong glance at those pregnant abridgements, the playbills that unfolded the map of his life, anyone who pleased might write his epitaph.[41] He can have lost no time in returning to Covent Garden and Drury Lane. They were not indeed the theatres he had known; both buildings, as though to mark his departure, had been destroyed by fire in 1808. But in the corps of players, with one great exception, he found little change. His great idol, Mrs Siddons, had retired in 1812. But the majesty of Kemble, whom he hardly admired less, remained, as yet unchallenged on his throne; and in addition all his old favourites, Emery, Liston, Mathews, Elliston, Incledon, Munden, were still appearing in his favourite characters. He had missed the long siege of the O.P. riots of 1809, when the public forced the opportunist shareholders of the new Covent Garden to restore the old prices, after nearly half a season of nightly renewed para-political disturbances when the actors strove to make themselves heard while the audience whistled, mewed, crowed, grunted, barked, brayed, and bellowed, to the accompaniment of sticks, rattles, bugles, and hunting-horns.[42] He was no doubt glad to have been out of town. Even in normal times the unruliness at the theatres was notorious. One night a sweep in the gallery gave vent to his disgust at the play by emptying his bag into the pit: he was fined — this was thought to be carrying criticism a little too far. Talking during the performance was normal. Hazlitt's zest for enjoyment and his concentration must have been great for such irritations to have left so little trace in his writings. Years later he pointed to the attentiveness and decorum of Parisian audiences, whose good

[40] 12. 193.
[41] 18. 392.
[42] Naturally, fights were common, and one Peter George Patmore, 23 years old, who reappears in our story, was fined for assault (*Examiner*, 8 Oct. 1809, 671).

manners and artistic feeling raised them far above their English counterparts.

However, dramatic criticism in the *Morning Chronicle* was the domain of Black's friend Mudford, who had previously been parliamentary reporter.[43] The grey flaccid prose of the review of Coleridge's tragedy *Remorse* (produced at Covent Garden, 23 January 1813), which has been ascribed to Hazlitt,[44] almost certainly dribbled from Mudford's pen. It is not likely that Perry would either commission or accept such a notice (two and a half precious columns of his paper) from a middle-aged fledgling parliamentary reporter just up from the country, who was floundering between failure as a painter and failure as an author, when he had an accredited dramatic critic who already had a dozen books to his credit and was in any case paid for the job. Perry's new parliamentary reporter would have to surprise his editor with some exceptional evidence of ability before he would send him to the theatre. His place was in the reporters' gallery.

Two days after Parliament reassembled on 2 February the Hunts were sentenced to two years' imprisonment and a fine of £1,000, Leigh Hunt having 'romantically' turned down what might have been an effective, if blackmailing, offer of help from Perry and the Whigs to bring sinister pressure to bear on the authorities, just as he and his brother now turned down a separate offer to pay the fine.[45] We have quoted the words of the libel as illustrating the violence of contemporary journalism. The libel is significant in two other senses. First, it undoubtedly inspired the savage attacks on Hunt by *Blackwood's Magazine* five years later, which speedily threatened to drag down Hazlitt in their train. Secondly, nothing could better introduce an aspect of Hazlitt's political journalism which has escaped some of his biographers who, reared in the tradition of a constitutional monarchy uninvolved in politics, have underestimated the threat to liberty that the throne posed at that time, and have consequently seen his assault on the principle of hereditary monarchy as unwarrantably extravagant, eccentric even. The passage was not solely inspired by the Prince Regent's profligacy but also by the prospect of his embracing his father's policy of interference and intrigue. It was published in March 1812, a month after the restrictions of the Regency Bill of 1811 were lifted and he had deserted his old allies the Whigs. Until then, appearing to support them, he had in reality merely used them in his private struggle with the King. When his father was finally declared mad and the reins of State were in his hands he shrewdly turned to the

[43] *DNB* entry.

[44] Belatedly by Howe (18. 463); also by Baker, 193; and Wardle, 136.

[45] Hunt, 235, 247.

Tories, who had served his father's interests well, as his most reliable servants in his new situation.

The reign of George III had been marked by the struggle between the great Revolution families, who clung tenaciously to the ground won in 1688, and the King, determined to recover as much power as his royal predecessors had lost. Upon the average Englishman the conflict obtruded itself only when some human issue was involved, as at the time of the tyrannical Royal Marriage Act, whose aim 'was not to assure the succession to the Crown but to extend the authority of the individual who wore it'.[46] But there were others in England who could not be described as average Englishmen and who, although private persons rather than politicians, humble citizens rather than members of the great Whig dynasties, were yet compelled to keep as close an eye on public events. The Dissenters, like the Catholics, statutorily excluded from taking part in the affairs of their country by the Test and Corporation Acts (softened though these oppressive laws were by custom and by the expedient of annual indemnity, they still remained on the book), saw the King, who obstinately opposed any relaxation of these disfranchising enactments, as their natural enemy. The Unitarians, the dissenting sect furthest removed from the Thirty-nine Articles of the Established Church, and consequently the most obvious target for persecution, rapidly acquired the legal vigilance and acumen which self-preservation begets in unpopular minorities, so that by the close of the eighteenth century their activities had become not less political than religious. Weak because scattered, self-willed, and incompletely amenable to the disciplines of solidarity, they were, for the same reasons, strong to endure oppression; a fact which throws incidental light upon the ostracism which Hazlitt has often been accused of having deliberately courted and perversely cherished rather than unjustly incurred. In the Dissenters' case, not only was there a danger of attack from without but also an enemy within. The immanent principle of Nonconformity was anarchic, and the stability, the existence even, of the congregations was constantly threatened by the emergence of schisms. Hazlitt's father, who doggedly opposed the danger from without, unconsciously illustrated the insidious threat from within when he said he would sooner die in a ditch than submit to human authority in matters of faith,[47] thereby ignoring the dubious validity of his own pastorate if this principle were to become, as it must, an imperative for his own flock.[48] Whatever may

---

[46] G. O. Trevelyan, *The Early History of Charles James Fox* (1880), 460. [47] Moyne, 51.

[48] The implication is reflected in Burke's unwontedly violent language against certain Dissenters who petitioned for the laws to be relaxed for themselves alone and not for *other* Dissenters: 'arrangez-vous canaille!' (*The Writings and Speeches of Edmund Burke*, ed. P. Langford (Oxford, 1981), ii. 384, Speech on the Toleration Bill).

have been the validity of his pastorate its precariousness was only too
evident. Margaret tells a weary tale of her father's years-long search for a
pulpit, on both sides of the Atlantic. His pastoral vexations and rebuffs
in America were owing to his doctrinal intransigence; in England and
Ireland also, but with a political aggravation. The Wem years held for
the ageing Revd Hazlitt a bleakness not apparent in his son's reminiscences of him. Towards the end, in 1808, he complained of having been
deserted by his '*quondam* Unitarian friends'. Margaret speaks of the 'evil
destiny' that led them to the 'dismal' town of Wem, of the faults and sins
of its people, of parochial cabals and petty jealousies.[49] We remember
also the disobliging verses about the 'ugly jaws a-straining' over the
rights of man. In his own words, he had his share of that evil report
which the orthodox usually circulated amongst their brethren against
Unitarians.[50]

   That evil report was blackest at Wem in the years after the French
Revolution, when the deism and republicanism of the indiscreetly
exulting Unitarians seemed to justify the swing against them, in other
parts of the country, from stealthy persecution to open violence. In July
1791 a Birmingham mob, howling 'Church and King' (which was still
to be the rallying cry of the *Blackwood's* gang thirty years later), burned
down the house of the most eminent of the Unitarians, Dr Priestley, who
fled to America. The violence spread: although none is recorded in
Wem, chapels elsewhere were wrecked and congregations besieged. The
Birmingham outrage inspired young William Hazlitt's first published
words. In his letter of protest to the *Shrewsbury Chronicle* his mind was still
on the religious freedom of the individual, and heresy, bigotry,
persecution were his themes. In the following summer his attention
turned to the question of civil liberty, after he had listened to an
argument between his father and one of the congregation about the Test
and Corporation Acts and the limits of religious toleration in the State. It
was then, he later affirmed, that he first began to think, and the ideas he
then conceived (with immense difficulty) and attempted to work up into
a 'Project for a New Theory of Civil and Criminal Legislation' set their
mark on him for the rest of his life.

   Civil and religious liberty were to be the watchwords of his public
hopes, personal freedom the watchword of his private aspirations, and
the supreme symbol of oppression in whatever guise and from whatever
quarter was to be the Throne.

   During his first spring as reporter the *Morning Chronicle* printed
sensational evidence of the degradation of the Royal family, not this time

----

[49] Moyne, 122, 82, 103.
[50] Ibid. 115.

the Regent but the Princess of Wales, in the depositions, then first made public, relating to her conduct in the year 1806; nor was it spread over the newspapers alone: in one week in March at least ten books or pamphlets appeared on the investigations. Hazlitt may have been interested by the affidavit in the *Chronicle* on 16 March of Thomas Lawrence, RA, the fashionable portraitist, a notorious ladies' man, who was implicated in the accusations against the Princess, but he was not impressed by Samuel Whitbread's Commons speech in her support on the 17th in which the celebrated Whig 'grew pathetic' in an attempt to gain sympathy, — an attempt that failed, he thought, although the report filled five columns of the *Chronicle*.[51]

When it came to marital misfortunes he is more likely to have been touched by events nearer home, by the calamitous story of his colleague John Black. In the month when Hazlitt joined him in the gallery, Black rushed into marriage with a woman he had met only five days earlier at a friend's house. It was a disaster. She despised him, made no effort to conceal her contempt, laughed at his simplicity. Within three months she had run him deep into debt, and when he expostulated she sold his furniture. In this March of 1813 she left him. By now he had at last realized that the whole thing was a put-up job: the woman had been, and in fact was now again, his treacherous friend's mistress, and the marriage was a plot to provide for her. But the story did not end there. Black was both ingenuous and besotted; she twisted him round her little finger, walked out and returned at will. He took it hard. He plunged back into the dissipation that once before, in like circumstances, had dragged him down into the kennels of Edinburgh. He sought crude oblivion in the taverns and gin-shops, neglected his duties, and incurred the anger of Perry, who, however, overlooked the dereliction on learning the cause. In that hard-drinking age a perilous and crippling degree of self-abandonment must have supervened before his editor would have become aware of it. In his desperation Black remembered the special conditions of divorce in his native Scotland; he may have spied a lifeline in two paragraphs in *The Times* of 14 and 16 April on the validity of Scottish divorces in England. He cannot have missed, because it appeared in the *Morning Chronicle*, a report on the 15th of an important decision of the Consistorial Court at Edinburgh laying down that 'adultery committed in Scotland is a legal ground for divorce without distinction as to the country where, or form in which, the marriage was celebrated.' In 1814 he was somehow able to persuade or bribe his wife to accompany him to Scotland to attempt a divorce; but the proofs of domicile they had brought in were deemed insufficient and the project

[51] 17. 10.

failed.[52] It was years before the guileless Black finally shook off this incubus.

[52] Howe wrongly states that Black 'emerged with success' from the undertaking. The *DNB* version, which I follow for this and other details of the episode, is independently supported by the absence of his name from the records of Scottish divorces at Register House, Edinburgh, which are complete for successful actions.

4
Parliamentary Reporter

THERE were other claims on Hazlitt's attention in the spring and early summer of 1813. Soon after the appearance of Coleridge's *Remorse* at Drury Lane came a change in the fortunes of another of his former Lake District friends. Word reached the Lamb circle on 24 March of Wordsworth's appointment as Distributor of Stamps for Westmorland. In 1812 Wordsworth was faced with the same difficulties as Hazlitt. Despite his frugality, parsimony even, he had found it hard to make both ends meet, and so, with no employment, little money coming in from his books, a wife and family of four children to support, he had turned to the dispenser of public patronage in his region, Lord Lonsdale, seeking a sinecure which would bring in money without interfering with his writing. An opening now presented itself. It was Robinson who communicated this news to the Lambs — 'with joy', since he foresaw a consequent improvement in Wordsworth's 'moral feelings', hoping that 'mixing more with the world' he would 'lose those peculiarities which solitude and discontent engender'. Others were less gratified. They saw it as a surrender. It confirmed a suspicion they had held for some time. It was, in their eyes, a dereliction, a reversal of attitude towards the establishment on the part of a writer whose position had for some time been ambiguous, but whose want of sympathy for that establishment had only four years before been so articulate that his friend and patron Sir George Beaumont had occasion to warn his fellow-guests of his 'terrific democratic notions'.[1] The *Morning Chronicle* condemned him in some scathing verses (promptly copied into the *Examiner* and the *Champion*) for yielding to corruption and allowing his silence to be bought by government taskmasters. The poem began:

> When Fortune's golden hook is baited
> How swiftly patriot-zeal relaxes
> In *silent* state see WORDSWORTH seated
> Commissioner of Stamps and Taxes

and went on in three more stanzas to pillory Wordsworth for turning his back on the noble example, formerly invoked in his own verse, of Milton, champion of freedom and civil probity, the high principles that

[1] M. Greaves, *Regency Patron: Sir George Beaumont* (1966), 105.

Wordsworth was now abandoning for the crumbs of Tory patronage.[2]

This was in April. In May came the Royal Academy exhibition, with no fewer than eight canvases by Lawrence. Among these was a portrait of the Marquis of Wellesley, detested member of an unpopular clan, who topped the family's pride and gluttonous ambition with an intolerable superciliousness and a private life disreputable even for those days. This former Governor-General of India and ex-Foreign Secretary, whom the upheaval of the election of 1812 had relegated to the heterogeneous groups of the Opposition, had just made a speech on a burning question, the renewal of the East India Company's charter, a speech which, instead of rising to the heights keenly anticipated from his special knowledge of the case and his reputation as an orator, turned out to be the dampest of squibs. Crabb Robinson, who had come expecting a memorable display (as an apprentice barrister professionally interested in oratory, and perhaps intending also to carry an account of the speech to Lamb, one of the Company's servants), left the Strangers' Gallery in disgust after an excruciating hour and a half, no doubt grimacing sympathetically at Hazlitt as he went.[3] Hazlitt, however, avenged the boredom he had been compelled to suffer by elaborating a two-stage exercise in irony, calculated to ricochet, after demolishing Wellesley, upon another insufferable MP, ministerial this time, John Wilson Croker — a calculation which perhaps sowed the seeds of a dangerous enmity.[4] He first concocted a brief but devastatingly ironical paragraph on the way in which 'the exuberance of [the speaker's] animal spirits, contending with the barrenness of his genius, produce[d] a degree of dull vivacity, of pointed insignificance, and impotent energy, which was without any parallel but itself',[5] and contrived, probably by dropping it at night and unsigned into the *Courier* office letter-box, to get it published in that mouthpiece of the Government — a government Wellesley had recently been harrying. (This has puzzled Hazlitt's biographers, who take Wellesley to have been still among the ministerial ranks and consequently an unlikely target for the satire of a ministerial newspaper.[6]) The second stage was to have it copied in the *Morning Chronicle* the day after it appeared, with this note: 'The Treasury journals complain of the harsh treatment shown to Ministers — let us see how they treat their opponents. If the following does not come from the

[2] 20 Apr. 1813. It has been claimed on, I think, insufficient evidence that the poem is Hazlitt's; it seems more likely to have been Tom Moore's. See J. Kinnaird, 'Hazlitt as Poet', *Studies in Romanticism*, 12 (1973), 426–35; and the writer's 'Regency Newspaper Verse', *KSJ* 27 (1978), 87–107.

[3] Sadler, i. 413.

[4] See a portrait of Croker in the *Examiner*, 9 Jan. 1814, 28, and some verses, 'The Crimp Sergeant' in *MC*, 28 Aug. 1813.

[5] 7. 23.

[6] Maclean, 298; Baker, 193 n.

poetical pen of the Admiralty *Croaker* it is a close imitation of his style:—
Speaking of Lord Wellesley's speech he says . . . '; and the *Chronicle* then
reproduces Hazlitt's jibes at the orator who 'soar[ed] into mediocrity
with adventurous enthusiasm . . . and launch[ed] a common-place with
all the fury of a thunderbolt!'

   Croker, another rabidly ambitious Protestant Irishman and a skilled
debater, was known as The Talking Potato, just as Wellesley was
described by a provoked Prince Regent as 'a Spanish grandee grafted on
an Irish potato'. Secretary of the Admiralty since 1809, he inevitably
and invariably drew attention to the smallness of his origins by the
bigness of his pretensions. These earned him many enemies, but he
made more by his conceit, offensiveness, and irascibility. He had
probably acquired too many really dangerous foes to have either the time
or the touchiness to be scanning the paragraphs of an obscure journalist
for disobliging references to himself. But if Hazlitt was obscure the
Morning Chronicle was not; it was so influential that Croker may have
made it his business to discover the identity of the author of this squib.
For years Hazlitt lost no opportunity to thrust at him, more especially in
the articles he later gathered into the volume of *Political Essays* (1819). By
opening that signed collection with the Wellesley paragraph (for which
he had a particular fondness) and retaining, appended to it, the
disingenuous *Chronicle* note, he threw aside his mask. Here, in 1813, at
the outset of his career, he had already begun to sow the ill-considered
winds which he was to reap in the foul whirlwinds of the *Blackwood's*
persecution of 1818 and the *John Bull* and *Literary Register* reviews of the
Liber Amoris in 1823. It may be that it was Croker himself, as Hazlitt
believed, who finally in 1823 retaliated; by stealth, but devastatingly.

   Hazlitt's satisfaction with the novelty of a regular income continued
unabated, as we learn from Robinson, who met him at the Lambs' on
29 April,[7] but an entry of a week or two earlier in Haydon's diary
suggests a steep decline (perhaps coinciding with the arrival of Sarah
from Winterslow) in the domestic comfort Robinson had been glad to
note in January. Haydon, the angry young man of Romantic art, a
self-confident and enterprising bachelor of twenty-seven, impatient,
intolerant and vain-glorious, had, not long before, been brought socially
into the limelight by the success of his *Death of Dentatus*, and, sponsored
by his patrons Lord Mulgrave and Sir George Beaumont, had dined
with Ministers and ambassadors in the fine houses of the nobility. He
was an unlikely person to enter sympathetically into the muted existence
and plain living of the husband and father eight years his senior whose
opportunity had passed, a failure who had resigned himself to the

[7] Morley, 128.

humble and obscure trade of newspaper reporter. He was by temperament unprepared, on his first visit to the Hazlitts, for their tranquil
indifference to their surroundings. Recently admitted on sufferance into
a new world of wealth and fashion which he deluded himself his genius
had conquered, he could hardly be expected to allow his host's inelegant
style of living to compromise his new-found dignity, although his life in
his lair in Great Marlborough Street was not a whit more polished. His
diary account of the christening party he attended on 9 April 1813,
although exaggerated, cannot but be closer to the facts than the later and
more familiar version in his *Autobiography*. Apart from Haydon the only
other guests present were Mary Lamb and Martin Burney, both of
whom he was evidently meeting for the first time, and whose names he
did not retain.

I dined on Friday last with a man of Genius, William Hazlitt. His child was to
be christened, and I was desired to be there punctually at four. At four I came,
but he was out! his wife ill by the fire, nothing ready, and all wearing the
appearance of neglect & indifference. At last home he came, the cloth began to
cover the table, and then followed a plate with a dozen large, waxen, cold,
clayy, slaty potatoes. Down they were set, and down we sat also: a young
mathematician, who, whenever he spoke, jerked up one side of his mouth, and
closed an eye as if seized with a paralytic affection . . . ; an old Lady of Genius
with torn ruffles; his Wife in an influenza, thin, pale & spitty; and his chubby
child, squalling, obstinate, & half-cleaned. After waiting a little, all looking
forlornly at the potatoes for fear they might be the chief dish, in issued a bit of
overdone beef, burnt, toppling about on seven or eight corners, with a great
bone sticking out like a battering ram; the great difficulty was to make it stand
upright! but the greater to discover a *cuttable* place, for all was jagged, jutting, &
irregular. Like a true Genius he forgot to go for a Parson to christen his child,
till it was so late that every Parson was out or occupied, so his child was not
christened. I soon retired, for tho' beastliness & indifference to the common
comforts of life may amuse for a time, they soon weary & disgust those who
prefer attention and cleanliness.[8]

Haydon's memory was always erratic, and he was seldom hampered by
a pious fidelity to fact. Here it seems that he did not think it worth while
(or thought it unhelpful to the squalid picture he was aiming at) to refer
to the pheasant provided as second course that he mentioned in telling

---

[8] Pope, i. 303. Martin Burney had 'a very unprepossessing physiognomy. His face was warped
with paralysis, which affected one eye and one side of his mouth' (Procter, *Lamb*, 144). He was
given to sudden enthusiasms, and mathematics must have been his current craze. He stood
godfather (with Walter Coulson) when the christening (at which Haydon could not have been
present — he was in Hastings) finally took place on 26 Sept. 1814 (*Memoirs*, i. 213). He was devoted
to Mary Lamb who, as Sarah's closest friend, could not have failed to be present. Haydon applies
the word 'genius' to her as Stoddart on another occasion spoke of Bedlam as 'the receptacle of
irregular genius'. Lamb must have been either ill or detained at the office.

the story to his pupil William Bewick.[9] Again, it is not probable that his host, who had a taste for sherry, would on such an occasion produce nothing to drink. Thirty years later, elaborating the story in his *Autobiography*, Haydon said that Lamb was present, as well as 'all sorts of odd, clever people', and since Hazlitt, in the mean time, had achieved fame he dropped the ironical if not scoffing word *genius*, and had Hazlitt delivering 'disquisitions'.[10] He says also that when he arrived, at 'Milton's house, next door to Bentham's', and found Hazlitt out, he went into the park to look for him. Perhaps he did, and perhaps the park *was* St James's Park; but the house was certainly not Milton's: the rate-books show Milton's old house, 19 York Street, Westminster, as unoccupied on 3 May 1813, when the rates were due; the Hazlitts did not move in until sometime after that date.[11]

Haydon may have had a masterful way with facts but he made no mistake about Hazlitt's devotion to his little boy. He introduces the party, in his *Autobiography*, thus: 'In the midst of Hazlitt's weaknesses, his parental affections were beautiful, He had one boy. He loved him, doated on him.' Everything we hear of Hazlitt's behaviour towards the boy betokens entire affection. Nor is this surprising. His attachment to his own father and to his brother, his desolation at the death of his first-born, and his cordial response years later to an unexpected letter from an unknown distant Irish cousin, indicate in their different ways a family piety, a warmth of feeling, which is plain enough in his essays. Young William was his constant companion from a very early age, going with him to visit the Lambs in the Temple and later in Great Russell Street; John Black in his riverside house at Millbank; and, further afield, the Reynells in Black Lion Lane, Bayswater, so long a trek for a child that he had to be carried piggy-back for the last stage: or John M'Creery in Blackfriars (who was, like Reynell, a printer, and set up the *Memoirs of Holcroft*); and even the Montagus at their elegant house in Bedford Square where Mrs Montagu, seeing him curled up in his father's lap, was struck by their intimacy. Later on, when he was ten or eleven, we find Hazlitt taking him — a hazardous experiment — to the fives-court and the Southampton Arms. Haydon's impression, and Mrs Montagu's, is confirmed by Bewick's account of Hazlitt when he first saw him at York Street in 1816 or 1817: 'Upon hearing a noise at the door and perceiving his only child creeping in upon all fours [Hazlitt] jumped up from his seat, ran to him, and clasping his boy in his arms, hugged, and kissed him and caressed him like some ardent loving mother with her first-born.'[12] About this same time Hazlitt was writing his introductory

[9] Bewick, i. 120.
[10] Haydon, 212–13.
[11] Westminster Public Library, Archives.
[12] Bewick, i. 118.

lecture on the English comic writers in which there are passages clearly deriving from the hours he spent playing with the child. Robinson was not less impressed 'His fondness for his child', he said, 'is a good feature in his character.'[13]

It was inevitable that the child should be spoiled, and Robinson, level-headed as ever, added that he was 'troublesome and forward'. This was in 1821, when young William was nearly ten. Two years earlier Keats had already observed this sauciness and used a harsher word, 'that little Nero', implying not only tyranny over the parents but even imperiousness to guests.[14] The same boldness is later apparent in a quarrel the ten-year-old picked with his father's landlady's twenty-one-year-old daughter; and again when his ill behaviour to his father's second wife finally drove a wedge between her and Hazlitt.

But this was in the future, like the predictable injustice of the son's complaint in later life that his parents, though 'both of the kindliest nature', had 'no more notion of managing children than the man in the moon, as [he] knew to [his] cost'.[15] The injustice is all the greater if we recall the wisdom of the father's essay 'On the Conduct of Life; or, Advice to a Schoolboy'. However, no one would have been readier than Hazlitt to recognize the gulf sundering word and deed, 'the old quarrel between speculation and practice'.[16] In the mean time we may safely assume that the boy's companionship compensated Hazlitt for the discomfort of the York Street house, and for the undemonstrativeness and lack of sensitivity of the mother.

We cannot fairly attribute to Sarah a total lack of feeling, nor anything more than a clumsy unsentimental matter-of-factness. The contrast between her and her husband emerges in their discussion about young William after the divorce: Sarah practical and unsentimental, Hazlitt vague and affectionate. According to Sarah's journal, 'He said he would do very well for the child himself, and that he was allowed to be a very indulgent kind father; some people thought too much so. I said I did not dispute his fondness for him but . . . he never saved . . .and [was not] likely to make much provision for the child.'[17] There was something masculine in her that clashed with Hazlitt's feminine side. We infer from Mary Lamb's letters to her before her marriage that she was something of a tomboy, a Miss Hoyden, with an embarrassing brusqueness, and this, linked to a disagreeable secretiveness, may have come to antagonize and finally alienate Hazlitt, whose tact (Patmore said that, with fewer prejudices than any man he had ever known, he was the last to offend those of others) and openness must have suffered from these opposed

[13] Morley, 265.   [14] Keats, ii. 59.   [15] BL Add. MS 38904, fo. 45.
[16] 1. 182.   [17] Le Gallienne, 255–6; Bonner, 196.

traits in his wife (a somewhat unusual combination, as Mary Lamb remarked) both in his own house and abroad.

This, if we magnify the inelegance, and scale down the word 'Society', may be what Haydon meant when he said, 'As he got into Society, the manner of his wife appeared unpleasant. The poor woman, irritated by neglect, irritated him in return.'[18] But what was more serious was that she could not give him the love he craved, and it was in this deeper and more intimate sense that what his son, speaking of the family character years later, called 'the Hazlitt affectionate susceptibility to tenderness' receded, in the face of disappointment and frustration, before what the son called 'the Hazlitt tendency to irritability and impatience'.[19] If he had hoped she might become the wife he desired, that hope faded long before they came back to town. The marriage was a fatal mistake and some of the blame must attach to her determination at all costs not to remain a spinster, some also to his over-sanguine acquiescence. Years afterwards an unexpected piece of counsel in the 'Advice to a Schoolboy' (written while he was awaiting divorce) cast a curious light on the transaction: 'If you ever marry, I would wish you to marry the woman you like. Do not be guided by the recommendation of friends. Nothing will ever atone for or overcome an original distaste. It will only increase from intimacy.'[20] His disappointment with his marriage can hardly have been lessened by the discovery that Sarah was not a virgin.

Her grandson uncompromisingly described her as 'selfish and unsym- pathizing' (although at the same time, with some inconsistency, as 'of excellent disposition and an affectionate mother').[21] Six years old when she died, he cannot have been speaking from memory; he must have been echoing his father, or more probably his own mother's recollections of her mother-in-law, whom she had known from the beginning of the York Street period when she was the ten-year-old Catherine Reynell. There never was, William Carew proceeded, 'a worse assorted pair', crippled as they were by 'a sheer want of cordial sympathy from the first set-out'.[22] He even went so far as to assert that Hazlitt 'had his individual case and fate in view where he speaks of marriages being brought about sometimes "by repugnance and a sort of fatal fascina- tion" '. These stringent remarks throw a crude light on the relation between husband and wife in the years before their separation in 1819 (indeed almost the only light: there is nothing in Hazlitt's letters; Sarah's have not survived; Robinson is silent; so also are the Lambs; and we have quoted Haydon's only observation — but it, of itself, is almost

[18] Pope, i. 214.   [19] BL Add. MS 38904, fo. 45.   [20] 17. 98.
[21] *Memoirs*, i. 214.   [22] *Ibid*. ii. 12.

enough). Two points are significant here. First, she understood him so ill
that she charged him with hating women to take an interest in his
writings. This suggests in him irritation at the interest *she* took: he
presented inscribed copies of his books to Sarah Walker, and told a
friend he hoped she saw and perhaps approved of his articles. [23]
Secondly, in later years Sarah Hazlitt never lived, as was common with
widows in that age, under the roof of her only son and daughter-in-law,
but preferred the independence of lodgings. And her own grandson
W. C. Hazlitt was not even sure in what year she had died.

The York Street house into which they moved in the summer of 1813
was old and shabby, but Hazlitt characteristically closed his eyes to this,
and saw only its past associations. He was more pleased to think that its
walls had heard the blind poet chanting the long slow surge of *Paradise
Lost* than chilled by their bare and cheerless aspect. When Milton lived
there it had been 'a pretty garden-house', facing St James's Park, but as
the district built up and a street took shape behind it, the back at some
stage became the front, and the entrance to the house was transferred to
the rear. The switch was finally clinched when Bentham, who had
acquired it, annexed most of its front garden to the back garden of his
adjoining but more recent house at the corner of Queen Square Place,
leaving to No. 19 only a walled-in strip of paved yard (wide enough, no
doubt, for a utilitarian washing-line, but for little else). However, he
sufficiently honoured Milton's memory to insert a memorial plaque high
up on the old façade (i.e. at the rear) of his truncated property.[24] A step
led down from the former back porch, in York Street, on to a
red-brick-floored semi-basement encumbered with wooden supporting
posts, beyond which another door opened on to a steep dark staircase
mounting to an obscure landing. The living-room to which it gave access
was gaunt, square, and bleak. The wainscot, painted white, was dull and
discoloured: four years later Hazlitt still claimed with pride that the
room was in the same condition as in Milton's time. The tall windows
looking out on to Bentham's augmented garden were bare of curtains,
and the furniture — a table, three chairs, and a sofa — had the
appearance of having just been set down at random by the remover. The
telling detail in this picture of discomfort and dilapidation was the state
of the fireplace — the whitewashed wall above and on either side of it

[23] Le Gallienne, 256; Bonner, 196; *Letters*, 328.

[24] Hazlitt's biographers call it an awkward house with odd access. It was odd because it had its
back to the street. The expropriation of the garden (which might have been deduced from Bewick,
i. 109) is noted in an account of the demolition of the house in *The Times*, 22 Oct. 1875, as also is
Bentham's posing of the plaque (implausibly ascribed to Hazlitt by his son, grandson, and other
biographers). In *MC*, 19 Aug. 1824, we find: 'In a house looking into Mr. Bentham's garden . . .
lived Milton. The front of it is in York St., and, *without being the ancient one*, is in a very squalid state'
(my italics).

swarmed with pencilled inscriptions, long and short, horizontal, vertical, and diagonal. These scribblings were apparently Hazlitt's memoranda for articles and essays.[25]

The place had other disadvantages. In those days York Street (still just as often called Petty France, although officially and patriotically renamed after the Duke of York during the French Revolution, an irony that must have brought a wry smile to the tenant of No. 19) was nearly on the seedy edge of the no man's land between London and the country. At its eastern end, towards town, a small gateway in Queen Square Place led into Birdcage Walk, which then had no roadway and was simply a parade-ground for the adjoining barracks, sloping down to the wooden fence enclosing St James's Park. The park itself, slashed down the middle from Buckingham House to Horse Guards Parade by a long straight canal, with coarsely gravelled walks and broken benches, was little more than pasturage for sheep and cattle. Westwards, beyond Buckingham House grounds, there stretched as far as the village of Chelsea, fields and market-gardens, the haunt of rogues and vagabonds. And between Pimlico and the Thames lay a succession of osier-swamps almost permanently flooded from the obstruction caused down-river by the waterworks at Old London Bridge.

Immediately to the south and east of York Street swarmed the Rookery, a noisome congeries of narrow courts and crooked alleys spreading down to Old Pye Street, and, in the direction of Westminster Abbey, as far as Great Smith Street, a labyrinthine lurking-place of picklocks, cutpurses, and beggars which was not swept away until 1845, a night-town of which Hazlitt regularly skirted the northern verge in returning after dark along Tothill Street from the House of Commons or from the *Chronicle* office, a journey that on winter nights of high wind, with many of the thinly scattered oil-lit street lamps blown out, must have been sinister indeed for so timid a man as Hazlitt in that footpad age, with the watchmen sparse and half-hearted.[26] A few years before the Hazlitts came, a contemporary writer described the inhabitants as 'of the lowest order', who 'aggravated by their numbers that nuisance which the filthiness of their persons and the narrowness of the avenues already occasioned'.[27] A dozen years after, when the face of London was rapidly changing, a journalist mildly remarked that 'the neighbourhood [had] long wanted a good name', and a little later, when Buckingham

[25] They astonished Patmore (ii. 261), Bewick (i. 118–19), and George Ticknor (i. 293). They are not even mentioned by the squalor-hating Haydon, who was evidently, on 9 Apr. 1813, in a different house.

[26] 'Improvements in Contemplation', *MC*, 5 Sept. 1811, gives (satirically) much information about the London of the day.

[27] J. T. Smith, *Antiquities of Westminster* (1808), text to plate 20, 'Southern Extremity of Thieving Lane'.

House became the royal residence, another protested that 'Tothill Street ought not to be suffered to be the connecting street between the Palace and our venerable Abbey to which it has been long so great a disgrace . . . the villainous district behind that street [is] the western St. Giles's'.[28] We catch an echo of all this in the complaint of the wife of an earlier tenant of this same No. 19, Bentham's protégé James Mill, that the house was unhealthy and the street swarmed with dirty, ragged, and rough-looking children, as well as in W. C. Hazlitt's quaintly evasive reference to his father's childhood playmates in York Street as 'rather a promiscuous circle'.[29] There was condemnation also from the humane William Fielding, the novelist's son, at this time ending his career, as magistrate at Queen Square 'police office' (and well known by sight to Hazlitt who projected on to the son's benign person his admiration for the father's novels:, 'I never passed him, that I did not take off my hat to him in spirit . . . and would willingly have acknowledged my obligations to the father to the son'[30]). The magistrate had occasion to reprimand the chimney-sweeps for neglecting their 'apprentices', the little climbing-boys he noticed 'almost naked and in a most deplorable state of dirt' in the street of a Sunday.[31] Added to this was the proximity of the uncouth military. The secluded Bentham himself objected to being disturbed both at his desk and in his principles by the screams of soldiers being flogged in the barracks.[32] Hazlitt does not tell us that he was affected by these things, although we cannot believe he was indifferent to the last; he did indeed briefly refer to the 'idle rabble' of 'that unruly region', but that was incidentally to his noble tribute to the humanity of the Fieldings.[33]

In general he was able to put up with his surroundings. A born bachelor, he should never have married. Indifferent to comfort, he desired only to be left to his thoughts; furnished lodgings or a house, it was all the same to him; and as to achieving ownership, we may be sure he never dreamt of it. The proof of all this is his remaining so long — six years — in York Street, until he was, in effect, evicted, when its only advantage, nearness to his place of work, lapsed after a mere twelve months from the date of his taking up residence there: for the remaining five years he had dealings only with Fleet Street. Cheerless as No. 19 was, it proved to be Hazlitt's last settled home. After leaving it he was never again a ratepayer: for the final eleven years of his life he moved continually from one set of furnished rooms to another, even during his second marriage, until he died wretchedly in a two-pair back.

The expense at No. 19 must have been more than he could well afford.

[28] *Courier*, 12 Apr. 1827; *MC*, 9 Oct. 1827.

[29] Alexander Bain, *James Mill* (1882), 72; BL Add. MS 37949 fo. 252; *Memoirs*, ii. 18.

[30] 12. 84.   [31] *MC*, 5 Oct. 1816.

[32] Bain (n. 29 above), 143 n.   [33] 12. 84.

The rent alone ran away with nearly a quarter of his salary,[34] and his wife, although fully alive to the advantages of money, was a poor manager. W. C. Hazlitt, again no doubt quoting his mother, described her as one of the least domestic of women, without the slightest turn for good order and economy (a description which lends some support to Haydon's story of squalor). According to him, she now allowed a servant to rule the roost, her housekeeper Mrs Tomlinson, who had two daughters, and what with the junketting in the kitchen and the consumption of meat and drink upstairs 'in the parlour, where the same set came to dinner about three times a week, the household expenses must have been considerable . . . .'.[35]

This is a strange story, and we can only suppose that Mrs Tomlinson, a large, masterful, and unconstrained woman, battened on the lax, unpractical Hazlitts and played the same domineering role in their house as the Lambs' maid Becky did later in theirs.[36] This obscure glimpse of a hidden life hints, like so many other isolated half-explained details, at a mild, compliant Hazlitt (sufficiently unlike the antisocial Hazlitt of the legend) who, even if he did not himself much relish the company of the invaders, did not stop his wife receiving them. If a choice arose between imposing or being imposed upon we may be certain that in his detestation of any kind of tyranny he fought shy of the former, despite his persuasion that the very consciousness of subservience in servants armed them in an unequal contest, that 'any real kindness or condescension only sets them the more against you', and that 'persons of liberal knowledge or sentiments have no kind of chance in this sort of mixed intercourse with these barbarians in civilised life'.[37] W. C. Hazlitt was right. He was no more suited to the role of master than of husband or father. But this does not mean that the disorder of the house, since it was initially Sarah's province, may not have partly contributed to the Hazlitts' unexplained final separation.

With all these expenses to meet, how was he to get extra money? The proroguing of Parliament at the end of July brought him leisure and created in the newspapers a vacuum which, as usual, was filled pell-mell with whatever offered: silly-season letters, facetious sketches, anecdotes of famous men. Much went on in London that summer that must have interested Hazlitt. Not, certainly, the illuminations of 4 July following

---

[34] Bain (n. 29 above), 73 says that Mill's rent was £50 a year; Bentham, in a vaguely worded letter to Francis Place (BL Add. MS 37949, fo. 252), not unwilling to impress Place with his generosity, says it was '£20 or £30'. Mill was his protégé, Hazlitt merely his tenant; he had no reason to be generous to him, nor was he, when the opportunity later arose.

[35] *Memoirs*, i. 215–16.

[36] See a letter from Hunt to Hazlitt in which he compares Mrs Tomlinson to the gross and raffish Queen Caroline, Memoirs, i. 254; and Lucas, *Life*, 609–11.

[37] 8. 308–9.

the news of the battle of Vittoria, or the fireworks and *ombres chinoises* of
the grand victory celebrations across the river in Vauxhall Gardens on
the 20th, the Wellington Fête which went on all night, and by which the
jingoistic Haydon was 'affected extremely' it had, he said, a 'fairyland'
atmosphere of 'chivalrous enchantment'. No, but rather the exhibition
of paintings by Sir Joshua Reynolds (including the *Ugolino*) at the British
Gallery in Pall Mall, across the park, in June and July, and the extracts
from Northcote's recent life of Sir Joshua in the *Morning Chronicle* in
August and September (his temperamental affinity to the old painter was
confirmed by the solemn censuring of this life, which was blamed, as his
still unpublished *Memoirs of the Late Thomas Holcroft* was to be blamed, for
the introduction of 'things which ought not to be recorded').[38] There was
also the arrival of a foreign visitor of great renown at the British Gallery
on 2 July — Mme de Staël, the opponent of Napoleon, who after
escaping from France had got to London, via Austria, Russia, and
Sweden, in June.[39] Drury Lane and Covent Garden closed for the season
in July, but he attended also the summer theatres, not only the fairly
respectable Haymarket but also, if we may judge from his later defence
of these places, the Pantheon in Oxford Street (where *Love in a Village* was
produced on 14 August), the Adelphi in the Strand, famous for its comic
actors, the Lyceum nearby, not yet exclusively devoted to opera, and
probably resumed his visits to Sadler's Wells with Lamb, and sometimes, like Horace Walpole before him, to Astley's Royal Amphitheatre
just across Westminster Bridge. He did not miss the other great
attraction of that summer, the sensational Indian jugglers whose novel
act, with its glittering, stanchless fountain of brass balls, crammed Pall
Mall with eager crowds throughout July and August, and earned them a
command performance at Carlton House before the Prince Regent and
his ministers.[40]

There would have been scope for Hazlitt's pen in any of these things,
but for the time being he did not embark on anything new, but tried
instead to make some money with material he had been keeping in
reserve. Soon after mid-August 1813 he submitted to Perry one or two
extended critical paradoxes to which he gave the ironical title of
'Common Places'. He mentioned in his letter (it is typical of his
diffidence that he did not approach his employer personally) that he had
several other papers of the same kind ready if these were acceptable. The
first to appear was 'On the Love of Life'; another two that appeared in

[38] Pub. by Henry Colburn, (who later published much of Hazlitt's work) 29 June 1813;
Farington, 20 Aug. 1813.

[39] *Drakard's Paper*, 13, 27 June, 4 July; *MC*, 22 June; *Examiner*, 4 July, 418, 426, and
25 July, 827.   [40] *MC*, 17 July 1813.

the *Morning Chronicle* were 'On Classical Education' (the theme that had provoked a controversy in the *Political Register*), and 'On Patriotism'; others, published elsewhere after Hazlitt had left the *Morning Chronicle*, were on the love of 'posthumous fame', on the causes of Methodism, on taste and seeing, 'characters' of writers, painters, and actors.[41] Although Perry's reaction was not unfavourable, his flair seems to have deserted him on this occasion: Mary Russell Mitford said years later that he had not then the slightest suspicion that he had a man of genius in his pay, and not the remotest perception of the merit of his writings.[42] All the same, he printed 'On the Love of Life' on 4 September 1813. This essay, of great significance in the corpus of Hazlitt's work (concerned as it was with hope, a sentiment in which he wryly claimed to be 'deep versed'), occupied an insignificant place in the paper and was evidently included on equal terms with a much longer piece from some other pen, a letter in solemn or mock-solemn defence of Old Moore's Almanack.

The quality of the writing in 'The Love of Life' is plain enough now, but Perry must simply have thought that his parliamentary reporter was a little cranky, and looking down his nose at, the 'damned fellow's damned stuff',[43] thought it at least not much worse, if odder, than the other contributions, so that the success of the piece must have brought him more surprise than gratification. What he could not have perceived, were he ever so sensitive to originality, was the characteristic note fluttering in the tail of this tentative kite launched into the windy skies of journalism. This brief essay corresponds to a profound level already glimpsed in Hazlitt's work to which he would return midway in his career as essayist with 'On the Fear of Death', and again just before its close with 'The Sick Chamber'. It must have been entirely lost on Perry: the cruder colours of an acrimonious controversy were needed to arrest his attention, and presently they flared. In after years Hazlitt gave two accounts of his start in journalism. In the first (in 1828), pointing a lesson from those circumstances, he was evidently thinking of his squib on Wellesley and his surprise at finding that this airy bubble apparently weighed heavier than all the close-reasoned arguments and scrupulous demonstrations he had for so long laboured over and seen ignored:

Finding this method did not answer, I despaired for a time: but some trifle I wrote in the Morning Chronicle, meeting the approbation of the Editor and the town, I resolved to turn over a new leaf — to take the public at its word, to muster all the tropes and figures I could lay hands on, and, though I am a plain man, never to appear abroad but in an embroidered dress. Still, old habits will prevail; and I hardly ever set about a paragraph or a criticism, but there was an

---

[41] *Letters*, 135.   [42] Howe, 146.   [43] Ibid.

undercurrent of thought, or some generic distinction on which the whole turned.[44]

Perhaps he here had in mind also two other articles in the same vein as the Wellesley paragraph, in which he claimed (18 and 20 September) that Southey owed his laureateship to two things. The first was the withdrawal from the contest of another strong candidate, Walter Scott, who retired, not out of charity, as he let it be thought, but from a shrewd assessment of the public feeling of the absurdity of the office;[45] Southey, who needed the salary, was aware of this absurdity, but was assured that the ancient custom of set odes would be allowed to lapse, and although the Prince Regent in the end would not agree to this, Southey was confident that he could write the disesteem down.[46] The second lever was the backing of another member of the same camp, Croker, to whom Southey had dedicated his *Life of Nelson*. Leigh Hunt neatly summed up the office of laureate when he said that it had 'the singular fatality of being impossible to be well-bestowed: if a good poet accepts it the office disgraces him; if a bad one, he disgraces the office'. In the second article ('The Laureat') Hazlitt, in a painterly metaphor pillorying the sad servility of the aspirants, anticipates his theory of the natural and inevitable alliance of poetry and power[47] ('Why what a sort of men these poets are! with what gaiety and alacrity they can lay the brightest colours on the darkest ground!'), and in the first he opens the game of slyly juxtaposing the Lake Poets' youthful revolutionary writings with the loyal addresses of their later *bien-pensant* years, a sport which culminated in 1817 in the disclosure of the skeleton in Southey's cupboard, the republican drama *Wat Tyler*.[48]

There is no mistaking the 'undercurrent of thought' in these paragraphs, but when, in a conversation of 1829 with Northcote, he remembers his début, he does not seem to have had in mind the change in style which he claims set his foot on the ladder, but a later moment which unexpectedly consolidated his foothold and impelled him to act fast to maintain it:

'I found nearly the same thing that you describe [commitment] when I first began to write for the newspapers. I had not till then been in the habit of writing at all, or had been a long time about it; but I perceived that with the necessity, the fluency came. Something I did, *took*; and I was called upon to do a number of things all at once. I was in the middle of the stream, and must sink or swim.[49]

---

[44] 17. 312.   [45] J. Buchan, *Sir Walter Scott* (1932), 108–9.

[46] Southey, *Selections*, iii. 9.   [47] 19. 116.

[48] *MC*, 18 Sept. 1813 (7. 25), where we find this: '. . .and one of his old Sonnets to Liberty must give a peculiar zest to his new Birth-day Odes.' The Hazlitt canon might here be extended to include a paragraph on the Laureate in *MC*, 16 Aug.   [49] 11. 288.

Perry did not call upon him to do a number of things either after the Wellesley piece or the 'Common Places'; it was the controversy on the modern drama in September that fixed his attention and led not only to Hazlitt's employment as dramatic critic but also as political controversialist.

The change in style was neither as abrupt nor as calculated as he made out, and his claim that a flash of illumination sent him abroad in a flowered dress has a certain self-deprecating irony. Nor was the change dictated merely by the widening scope of the subjects treated, after the abstractions of philosophy. This of course had its effect, but both the matter and the manner at once formed part of and were dictated by the organic development of his thought. In the solitary years at Winterslow, in the empty fields and woods and in the quietness of his room, he had leisure to reflect on the curious relation between the two enterprises that preoccupied him. As he walked home from his painting expeditions, and as he sat down at his writing table, the possibility, the necessity, of reconciling the two roles of artist and metaphysician could not have escaped him nor remained indifferent to him. He must have seen that the conclusion of the *Essay on the Principles of Human Action* contradicted its method, although the method was necessary to the conclusion. That conclusion was that the mind tended naturally to sympathize with others, to turn imaginatively outwards, to look abroad into universality. The method — a rigorous, narrow, spare, inward-turned process of reasoning — was indeed necessary to that conclusion, and so far natural, but now that it had served its purpose it required to be surpassed. The conclusion had to be put into practice. His style had to mirror universality, not because it was morally imperative (although that also was true) but because it was natural: it was the direction of life, and the world of logic was only part of life; all around and beyond lay also the world of feeling and the contingent world of the senses. The sister of philosophy was art. In his writing, the pen he had hitherto used was the specialized pen of the student; he must now relinquish it and adopt instead a more familiar instrument capable of reproducing the tone, the plain speech men use to men, the discourse of those who, talking in the inn-parlour or at the social table, sink their personal interest in their general humanity. The ironical paradox is that the mode he thus adopted was ridiculed as eccentric, wilfully idiosyncratic by his contemporaries, who imagined that their own hollow commonplace abstractions were the natural and universal language of men, whereas in reality Hazlitt's familiar style has turned out to be in the mainstream of English prose, and theirs a stagnant backwater.

The second of his two articles on the Laureate (20 September) countered a reply to his first which had promptly appeared in the *Courier* on the evening of the same day.[50] He never says so, but it is clear from the start he deliberately cultivated the strategy of controversy. The Wellesley note — a singular instance of a unilateral, self-generated (although abortive) attempt to start off an exchange — seems to support this. In any case the technique is as old as newspapers. His very first appearance in print, years ago, had been polemical. It has not been pointed out that his schoolboy letter on the burning of Priestley's house, published in the *Shrewsbury Chronicle* on 11 November 1791,[51] was not an isolated protest against intolerance, but merely one letter in a debate lasting from 5 August 1791 to 3 February 1792, eighteenth in a series of over thirty from at least a dozen other participants.[52] In the same way, some twenty years later, his first major contribution to journalism again embroiled him in public debate; and this gravitation towards the exchange of views rather than their unilateral expression was a manoeuvre he never in his writings ceased to initiate, provoke, or become involved in, from first to last, from the *Principles of Human Action* (where he attacked the 'modern philosophy'), *Free Thoughts on Public Affairs*, and the *Reply to Malthus* in 1805–7, via the papers on Reynolds's *Discourses*, on the Duc d'Enghien, on the Elgin marbles, on Owen's *New View of Society*, to the *Conversations of Northcote* in 1830. All such exchanges are evidence of his disinclination to pontificate, of his dislike of the egotistical; his views are commonly defined dramatically, on specific occasions, by opposition to those of others. Ultimately grounded in principle, they are immediately polemical; they manifest his leaning towards the marriage of the general and the particular, of the abstract and the concrete, of the universal and the individual. He saw the clash of argument as a means of escaping from self-communion into the dramatic, and so of acceding to universality, a means of liberation from the sterility of the immured self. This major contribution, which no doubt was the 'something' he did that 'took', was a long letter on 'The Stage' to the *Morning Chronicle* (25 September, signed 'H') in answer to an anonymous two-column article on the degeneracy of the English stage that had appeared in the paper on 17 September.

Although, as we have said, not the first of Hazlitt's polemical sorties, this is the first of his journalistic hand-to-hand literary duels, so it is worth ascertaining how far he was impelled by policy, how far by conviction, and how far by personal feelings. The anonymous analyst of

---

[50] 7. 24–5; 19. 115–17.

[51] See the writer's two notes on this letter, *The Library*, 6/2 (1980), 358, 360.

[52] Including 'R. Swanwick', writing from Wem (*Shrewsbury Chronicle*, 2 Sept. 1791), who was perhaps related to Hazlitt's young friend Joseph Swanwick, but who was, in this debate, in the other camp. See also *Letters*, 113, Hazlitt to his father, 1809, 'I think you had better not enter into a contest with old Swanwick, unless you are quite sure of success'.

degeneracy was almost certainly William Mudford, the accredited dramatic critic of the paper.[53] Hazlitt's appointment as parliamentary reporter ensured that sooner or later he would meet Mudford in the *Chronicle* office or in the gallery, since Mudford would either still be a member of the team or would at least take an occasional hand. Journalists' tasks at that time were neither clearly defined nor stable; the speakers named in Hazlitt's parliamentary notebook[54] show him as still in the gallery long after he made his bow as dramatic critic, and as late as June–July 1814. However, he probably met Mudford almost immediately through John Black. Temperamentally, Hazlitt and Mudford were poles apart and must have taken an immediate dislike to each other. Although Hazlitt was relieved to have secured work on the *Chronicle*, he knew well enough (in his less diffident moments) that he was worth more than a reporter's hire, and although he wryly disdained, indeed was incapable of, the lowly shifts and devices of ambition he could not help envying its prizes. Mudford was almost as assiduous a courtier as Stoddart. At the age of twenty he had been personal secretary to the Duke of Kent, whom he accompanied to Gibraltar. Even earlier, when only seventeen years old, he had badgered a startled John Philip Kemble with a proposal to dedicate a book to him, and four years after, he returned to the charge by sending the same exasperated manager a play, lost to posterity, for production at Covent Garden.[55] In a few years from now we see him rebuffed by the no doubt equally irritated Duke of Wellington whom he had confidently approached for copy for his projected *Account of the Battle of Waterloo*.[56]

By 1812, the year before he met Hazlitt, he had published a dozen books, including a novel, *Nubilia in Search of a Husband* (a shrewdly opportunist answer to the celebrated Hannah More's successful *Coelebs in Search of a Wife*, a stroke which Hazlitt recalled with irony eight years later), and had got himself appointed dramatic critic to the most important daily newspaper of the realm. He was four years younger than Hazlitt; his pretensions were staggering; and he was dull. We spoke of the turgid style of the day: it is seen a few years before this in his periodical *The Contemplatist*, which he inaugurated with a backward glance at the fortunes of the successful *Spectators* and *Connoisseurs* of the previous century and their long-dead editors, in these words:

[53] The *DNB* merely says he was parliamentary reporter; our source here is Hazlitt, 8. 293 n.

[54] The 'Christabel' notebook; see BL R.P. 580(3). He perhaps reported even as late as Feb. 1815: see the writer's 'Howe's Edition of Hazlitt's works: Two Notes', *NQ*, May 1970, 174–5; and Ch. 6, n. 27 below.

[55] BL, cat. of Lefebvre sale, Paris, 2 Mar. 1854 (S.C. 670), lot 513, John Kemble to Mudford, 26 Apr. 1799; and cat. of Laverdet sale, Paris, 3 Dec. 1857 (11900 bbb(12)), lot 550, John Kemble to Mudford, 14 Nov. 1803.

[56] William Jerdan, *Autobiography* (1852), ii. 183; 8. 111.

Yet, can it be doubted that the career of each was begun with equal expectations of success; with equal hopes of receiving adulatory distinction from contemporary gratitude; and with an equally ardent anticipation of delighting and instructing succeeding generations, when their authors would be alike insensible to censure and to praise — to pre-eminence and neglect — to glory and to shame?[57]

Years later Hazlitt remembered him with shattering irony as 'the Contemplative Man'.[58] His peculiar blend of pertness and complacent dullness irritated Hazlitt just as much as Hazlitt's tart alertness must, if perhaps vaguely, have disturbed him.

And now in this autumn of 1813 he made a false move by publishing this over-sanguine article on the decay of modern comedy: he should have stuck to his playgoing notices and kept his own counsel on wider issues. Hazlitt saw his chance. Mudford's article was a patchwork of random reflections, slack hypotheses, and untested theories; and although Hazlitt's reply rested on assumptions perhaps almost as debatable if more homogeneous, the skill with which he found out his rival's weak points and transfixed them with his rapier logic never faltered. His disdain hardly showed, but his criticisms annihilated. Mudford, who had probably intended no more than a casually elegant and informed trifle, was bewildered and annoyed by what he deemed an uncalled-for attack, and also unnerved. Doubtless Black had told him of Hazlitt's little-known but impressive books; of his friendship with the awesome Godwin, and with Coleridge, Wordsworth, Lamb, and so forth; and of his dialectical skill. Of the last Mudford now had alarming evidence; but it was too late to think of drawing back: he had to stand his ground and reply. He did so on 4 October. Hazlitt's second counterblast (running to four columns — nearly a whole page of the four-page newspaper) was even more devastating than the first. Poor Mudford must have been panic-stricken. But why was Hazlitt so angry? It may be that, skilled ironist as he himself was, he read into Mudford's lame reply more of sly insinuation than his stodgy antagonist actually meant. Mudford in his nervousness may simply have been attempting a half-facetious, half-ingratiating disengagement in saying his article had been 'rather intended as a confession of our ignorance and an appeal to the charity of the illuminated than as a dogmatical solution of the difficulty', and in claiming that he had been delighted to find 'that there were still men with warm hearts of flesh and blood beating in their bosoms ready at the first summons to dispel this intellectual gloom'[59] when there had come

[57] *The Contemplatist*, 1 (9 June 1810), 2.
[58] 8. 196.
[59] *MC*, 4 Oct. 1813. The immediate allusion is to Hazlitt's quotation from Burke in his letter. See 4. 13.

forward on 23 September 'a Correspondent powerful in talent and confident in his strength, to whom the Comic and the Tragic muse seem to have been equally lavish of their secrets and whose richness of illustration of a proposition which he considers as almost self-evident sufficiently proves that his powers have not languished in the dense atmosphere of logic and criticism'. This looks very much like a conciliatory salute to the author of the *Principles of Human Action* and the *Reply to Malthus*, but Hazlitt refused to take it at its face value: he knew well enough that Mudford disliked him, with a dislike which was later to deepen when Mudford became editor of the *Courier*, a paper Hazlitt never ceased to attack.[60]

Hazlitt may even have seen a further personal allusion in the italicization of the two words *vanity* and *egotism* in a subsequent paragraph, where Mudford, after quoting from his opponent the sentence, 'Comedy naturally wears itself out [and] in the end leaves nothing worth laughing at', goes on to make this comment:

On reading this we began to ask ourselves if our past existence in the world had been merely a dream, and when we could no longer doubt of its reality, how it happened that in almost every society we had never failed to witness folly in some characteristic mode or other, and that the *vanity* and *egotism* of individuals meandered through many a weary winding and assumed many a disguise; but that, however these individuals might impose on themselves they seldom succeeded in imposing on others . . .

Whether Hazlitt took this to be a shaft aimed at himself or not, his warm heart of flesh and blood was not touched, and he began his reply by probing the sincerity of Mudford's disclaimer:

Sir, I believe it seldom happens that we confess ourselves to be in the dark on any subject, till we are pretty well persuaded that no one else is able to dispel the gloom in which we are involved. Convinced, that where our own sagacity has failed, all further search must be vain, we . . . are very little obliged to anyone, who comes to disturb our intellectual repose.[61]

Why, we ask again, was he so angry? There may of course have been some wrangle in the *Chronicle* office which has left no trace, unless faintly in Hazlitt's later disparagement of 'the man made of fleecy hosiery', and Mudford's equally disobliging assertion after Hazlitt's death: 'I knew the man well, and, besides having the greatest contempt for his turgid nothings and bombastic paradoxes (which passed for fine writing with some) I thoroughly disliked his cold artificial character, and his

---

[60] *CBEL* iv. 1792 dates his editorship at '1827?'. Griggs, iv. 813 n. and 842, shows him installed at least as early as 1818, so that it was certainly he who was the 'hack-writer' Hazlitt luridly 'illustrated' in 1820 (19. 214–15).

[61] *MC*, 15 Oct. 1813 (20. 1–12).

malignant disposition.'[62] The trouble was, no doubt, Mudford's
manner. Hazlitt treated honest ignorance and simplicity with respect,
but he did not suffer pretentious fools gladly. When he himself had
nothing useful to say he held his tongue, and indeed often did so from
sheer diffidence even when he had; he could not see why he should be
expected to accord to others, who impertinently laid claim to compet-
ence, the indulgence he scorned to ask for himself, who did not. (He
nevertheless recognized this inflexibility, this sacrifice of charity to
unswerving justice, as being sometimes a fault, and this accounts for his
rooted admiration for John Hunt, who reconciled these virtues by being
as indulgent to others as he was exacting to himself.[63]) It was especially
irritating to see amateurish mediocrity commanding not only indulgence
but professional success. The man thus shamelessly complacent of his
own ineptitude was dramatic critic to the great *Morning Chronicle*! Hazlitt
lashed out. And the owner and editor of the paper was forced to take
notice.

Not that Perry had the slightest feeling of loyalty towards his dramatic
critic. He would have turned off Mudford with as little compunction as
he later turned off Hazlitt. But he had great loyalty to his paper. And, as
an editor of long experience, he knew the value of controversy. He
evidently did not replace Mudford yet, but gave Hazlitt a share of the
theatres. The season at Covent Garden had begun on 6 September, and
at Drury Lane on the 11th, and there was room for two critics. He seems
not to have lost a moment in putting his unexpected discovery through
his paces. Hazlitt's first letter on modern comedy, as we saw, appeared
on 25 September; three days later a notice of Stephen Kemble as Falstaff
in *Henry IV* at Drury Lane is almost as surely his as the previous
theatrical notice of 24 September, Miss Stephens's début as Mandane in
Artaxerxes, is not. The gulf between Hazlitt and Mudford yawns even
wider in the contrast between this last notice and Hazlitt's notice of the
same actress in the same play a few weeks later, where flabby
conventional generalizations give way to sharp concrete criticism (the
duplication of notices seems to imply that Perry was putting Hazlitt to
the test).[64] It was eight years since Leigh Hunt had inaugurated a new
era in dramatic criticism, with honest and impartial, though not
impersonal, judgements instead of the customary, meaningless perfunctory puffs paid for by the management, but his example had been
little heeded.[65] Criticism in the dailies had remained on the same

---

[62] 8. 196; Geoffrey Oldcastle [Mudford], 'The Late S. T. Coleridge', *The Canterbury Magazine*, 1
(Sept. 1834), 127.

[63] Hunt, 233, and the Dedication of *Political Essays*.   [64] 18 Oct. 1813 (5. 192–3).

[65] Hunt, 154–5; see also his 'Rules for Theatrical Critics' in *Critical Essays on the Performers of the
London Theatre* (1807), and Landré, ii. 99–131.

conventional level of 'measured praise' as before, and Hazlitt's advent was a second and even more invigorating breath of fresh air. We have a witness to its novel tang in young Thomas Noon Talfourd, a law student newly arrived from the provinces whose eyes were opened by Hazlitt's startlingly new interpretation of *The Beggar's Opera*. But it, in its turn, blew fitfully to begin with.[66]

Hazlitt's arrival did not immediately banish the old style from the *Chronicle* any more than it immediately banished Mudford. Either Perry was cautious and slow to convince, or Mudford difficult to dislodge. The canon of Hazlitt's works gives three theatrical notices in October as his; of the nine the paper provided in November, only one, *Antony and Cleopatra*, is certainly his; five are clearly another's, and must be Mudford's;[67] and in December, when there were accounts of five plays, not one seems to be by Hazlitt.[68] In fact, years later when he said he had been 'put on duty' in place of Mudford in the 'theatrical department' of the *Chronicle* 'just before' Kean made his bow on the London stage on 26 January 1814, he evidently meant an interval measured in days, or at most, weeks, and not in months, so that it was not in September 1813, as has been supposed, that he was put in sole charge of dramatic notices, but soon after mid-January 1814. Then it became his official duty to frequent that garish night-world he had known so long as a private playgoer, but which from now on and for nearly eight years he would have to attend regularly, often several times a week, taking his place on the benches of the pit and trying to shut out from his concentration on the play the rowdy, restless audience of swaggering soldiers and sailors in the gallery, of tradesmen, shopkeepers, genteel clerks, and prentice lawyers in the pit, and the whining orange-sellers ('Choice fruit and a bill of the play!'). However noisy, it must have been a welcome change from the dim-lit arena of St Stephen's Chapel, of boring speeches and snoring Members. His nights were also in another sense less demanding than those at the House: the early evening journey up to the *Chronicle* office was leisurely, and separated by some hours of rapt attention in the pit from the short hurried midnight walk, after the curtains had closed, back along Bridges Street and Catherine Street to No. 143, Strand, where he rapidly wrote out the comments he had mentally elaborated during the performance.

His dramatic notices were only one aspect of his activities and interests that autumn. Perry accepted another 'Common Place' — 'On Patriotism', 5 January 1814 — and then no more, probably finding the others

[66] *Literary Remains*, i, p. cxiv n.
[67] The remaining three, however, may well be Hazlitt's: *The Invisible Bridegroom* (11 Nov.), *The Devil's Bridge* (15th), *Romeo and Juliet* (23rd).
[68] I do not believe *Love in a Village*, given to him by Howe (18. 194), is his.

not close enough to the topics of the day; but he made full use of the
versatility they implied in their author, and during the next six months
printed from Hazlitt's pen not only theatrical criticism but also book
reviews, political commentary, and art criticism. This eruption of
energy is not to be alone explained by the need to get money, nor yet by
the novelty of his situation, although his writing had indeed hitherto
been a solitary occupation, scholarly, laborious, and, as far as public
recognition went, ineffective; whereas he now found that he was able to
compose as he walked through the streets or sat watching the play, and
could write at a cleared corner of the office desk, and still command the
attention of the public. He was intoxicated by the consciousness of an
immediately responsive audience, which is one of the charms of the
journalistic novitiate, of what he called 'the honeymoon of authorship'.
If he submitted an essay on modern comedy, he was forthwith invited to
supply dramatic criticism for the paper; if he wrote a paragraph on the
Poet Laureate, the *Chronicle's* political rival the *Courier* was stung into
replying the very same evening; and a favourable notice of Miss
Stephens in *The Beggar's Opera* brought that actress scurrying to the office
to express personally her gratitude (although it is typical of his
diffidence, of his bashfulness, that he held back, and Perry did not
trouble to call him in to meet her but took all the credit to himself).[69] His
letter on 'The Stage' not only whipped up controversy in his own paper
but seems to have had indirect repercussions in *Drakard's Paper* (just then
rising into notice), where four numbers in November and December ran
an 'Attempt to Explain the Causes of the Decline of British Comedy'.[70]

And yet there was something more still in this burst of energy. Miss
Stephens remained a favourite, but the delight he continued to take in
her singing in later life was evidently coloured and heightened by private
associations. It looks very much as though there lies concealed in this
period and behind these associations a poignant love-affair which
remained ruefully present to his memory for years, but of which his
friends, and perhaps even his wife, were unaware. Ten years later, in
1822, when he was in Scotland seeking a divorce as a solution to the
wretchedness and uncertainty of his enslavement to Sarah Walker, there
was a brief moment when he once more felt his brow cooled by the
'breezes blown from the spice-islands of Youth and Hope — those twin
realities of this phantom world',[71] he looked wistfully back on these days
of eagerness and joy in the last months of 1813 and saw them as marking,
together with the fall of Napoleon a few months later, the end of his
private and public hopes. Speaking of the account he wrote for Perry of

[69] 8. 293.
[70] 14 Nov., 358; 21 Nov., 366; 5 Dec., 382; 12 Dec. 1813, 390.
[71] Coleridge's phrase (*Table-Talk and Omniana* (Oxford, 1917), 313).

Miss Stephens's début in *The Beggar's Opera* he tells of a delightful autumn walk back to town from Addlestone after a visit to his parents (his father, now ageing and deaf, had retired there from Wem a few months earlier[72]) and he quotes the beautiful duet 'Hope thou nurse of young desire' from *Love in a Village*, that fresh and innocent pastoral in which he saw Miss Stephens soon after *The Beggar's Opera*; all these things, the morning's walk, the sky-reflecting river and its banks, the golden October sunlight on the brown leaves, the operas, the singers, and the songs, for ever after remained inseparable in his memory from this period of his life:

I shall not easily forget bringing [Perry] my account of her first appearance in the *Beggar's Opera*. I have reason to remember that article: it was almost the last I ever wrote with any pleasure to myself. I had been down on a visit to my friends near Chertsey, and, on my return, had stopped at an inn at Kingston-upon-Thames, where I had got the *Beggar's Opera*, and had read it overnight. The next day I walked cheerfully to town. It was a fine sunny morning, in the end of autumn, and as I repeated the beautiful song, 'Life knows no return of spring,' I meditated my next day's criticism, trying to do all the justice I could to so inviting a subject. I was not a little proud of it by anticipation. I had just then begun to stammer out my sentiments on paper, and was in a kind of honey-moon of authorship. But soon after, my final hopes of happiness, and of human liberty, were blighted nearly at the same time; and since then I have had no pleasure in any thing:—

'And Love himself can flatter me no more.'

It was not so ten years since (ten short years since, — Ah! how fast those years run that hurry us away from our last fond dream of bliss!) when I loitered along thy green retreats, oh! Twickenham, and conned over (with enthusiastic delight) the chequered view, which one of thy favourites drew of human life! I deposited my account of the play at the *Morning Chronicle* office in the afternoon, and went to see Miss Stephens as Polly. Those were happy times, in which she first came out in this character, in Mandane, where she sang the delicious air, 'If o'er the cruel tyrant, Love,' (so as it can never be sung again), in *Love in a Village*, where the scene opened with her and Miss Matthews in a painted garden of roses and honeysuckles, and 'Hope, thou nurse of young Desire,' thrilled from two sweet voices in turn. Oh! may my ears sometimes still drink the same sweet sounds, embalmed with the spirit of youth, and health, and joy, but in the thoughts of an instant, but in a dream of fancy, and I shall hardly need to complain.[73]

---

[72] He is shown in Chertsey Poor-Rate Books, P2/3/6, 1812–16, as occupying a house rated at 15*s.* a year, from between 5 June and 22 Sept. 1813 to sometime after 2 Feb. 1815; on 3 May 1815 the house was empty. No reason is apparent for his choice of Addlestone. There seem to have been Unitarians in Chertsey: one Mr Joseph Every contributed to the Unitarian Fund (BL 4224. aaa. 11); but they were not well known (see *Christian Reformer*, 5 (Aug. 1838), 512).

[73] 8. 292–3; 18. 342–3. This is the place to correct a damaging statement in Wardle, 139, that Hazlitt criticized Miss Stephens's acting in *The Beggar's Opera* without seeing it, 'having decided in

Patmore never knew Hazlitt out of love (and the pathos with which he apostrophizes himself, exaggerated though it may be, shows the cause of such susceptibility: 'Oh! thou dumb heart, lonely, sad, shut up in the prison-house of this rude form, that hast never found a fellow but for an instant.'[74]) His passion for Sarah Walker must have been by far the most violent of these affairs, since so many of his friends got to know of it; and yet at its height he looks regretfully back to this, of all his unknown earlier loves (and not to the Sally Shepherd affair which would have remained hidden if it had not been for his wife). Only this most muted of echoes of it survives, but secret though it appears to have been, it must have moved him a great deal. His walk from Chertsey to Kingston and then on to Westminster was in a period of unwonted health and joy, and his memories of those autumnal days, illumined not only by the mild rays of the declining sun but by the vividness of resurgent hope (that 'painted vapour', that 'glow-worm fire'), are tender with all the nostalgia of lost happiness.

And yet there are signs that not even this brief heaven was unclouded: the old doubts and self-distrust recur in a paper published at the end of this month, where the familiar lack of conviction, the murmuring suspicion of the uselessness of activity, the fatal rift between contemplation and reality, between the ideal and the actual, loom through a fog of weariness and indifference. The theme of 'Baron Grimm and the Edinburgh Reviewers' (which he later thought sufficiently true and sound to include under the new title of 'On the Literary Character' in the *Round Table*) is that same incurable flaw of inadaptation which constantly obtruded itself on his consciousness: he here holds that the same listless staleness which, whether abruptly or gradually, dispirits and dismays the active man after long years of experience of life cripples the bookish man before he has even set out. In him,

the common indifference produced by the distraction of successive amusements, is superseded by a general indifference to surrounding objects, to real persons and things, occasioned by the disparity between the world of our imagination and that without us. . . . Ideas assume the place of realities, and realities sink into nothing. Actual events and objects produce little or no effect on the mind, when it has been long accustomed to draw its strongest interest from constant contemplation. . . . but [the insensibility of Authors, etc.] is less than it appears to be. . . . a disappointment in love that 'heaves no sigh and

---

advance that "her acting throughout was simple . . ." '. Hazlitt, of course, handed in his account of the *play* at the *Chronicle* office in the afternoon, and of the *performance* (the last paragraph of his notice) after the theatres closed.

[74] 9. 127 (*Liber Amoris*); see also 17. 107 ('My First Acquaintance with Poets'): '. . . my heart, shut up in the prison-house of this rude clay, has never found, nor will it ever find, a heart to speak to.'

sheds no tear', may penetrate to the heart, and remain fixed there ever after. . . . The blow is felt only by reflection, the rebound is fatal. Our feelings become more ideal; the impression of the moment is less violent, but the effect is more general and permanent.[75]

An echo here, no doubt, of the solitudes of Winterslow, but the feeling goes further back still, is indeed fundamental in his temperament and in his attitude to life. And this crippling vice of inhibition still occupied his thoughts two years later, in 1815, when he was reviewing Sismondi's *De la littérature du midi de l'Europe* for Francis Jeffrey. There he found a version of the story of Petrarch and Laura which he immediately re-interpreted in terms of his own experience. Harking back to the words he had used in 'Baron Grimm and the Edinburgh Reviewers', he saw Petrarch's love as operating 'in the way in which the passions very commonly operate on minds accustomed to draw their strongest interests from constant contemplation'. He construed their celebrated first meeting at Avignon in a way that spoke even more plainly of what he, William Hazlitt, was going through in London at that moment in 1815. Sismondi, he says, was wrong to wish (as he professed to do) that there had been greater and longer-lasting intimacy between Petrarch and Laura:

The whole is in better keeping as it is. The love of a man like Petrarch would have been less in character, if it had been less ideal. For the purposes of inspiration, a single interview was quite sufficient. The smile which sank into his heart the first time he ever beheld her, played round her lips ever after; the look with which her eyes first met his, never passed away. The image of his mistress still haunted his mind, and was recalled by every object in nature. Even death could not dissolve the fine illusion: for that which exists in the imagination is alone imperishable. As our feelings become more ideal, the impression of the moment indeed becomes less violent; but the effect is more general and permanent. The blow is felt only by reflection; it is the rebound that is fatal.[76]

One of the greatest French writers of the age happened to see this striking passage in the *Edinburgh Review* and instantly recognized its truth, its originality, and its exact application to his own character and history. The impression was so profound that after noting it marginally in a book he was reading, as supplying the key to his own life (as the 'explication anglaise' of his character), he went on to quote it in a note to one of his own books as representing the '*Biography of the A[uthor]*', i.e. of himself,

---

[75] 4. 133–5.

[76] 16. 45. An early example of how tenacious he was (repeating them from essay to essay) of formulas he had carefully elaborated to define his own experience, and also of the care he bestowed on his cadences.

and finally to use it as epigraph for a third.[77] This is the earliest point in the remarkable parallel that the works of Hazlitt and of Stendhal were more or less involuntarily to form together. It is no less remarkable that Stendhal later seizes upon and adopts word for word for his own purposes another anonymous passage by Hazlitt, just as applicable to his own life as the lines on Petrarch, and that came home to him with equal conviction. It is the description of *Don Quixote*, in an earlier *Edinburgh* review written in December 1814, where Hazlitt says, 'The whole work breathes that air of romance, — that aspiration after imaginary good, — that longing after something more than we possess, that in all places, and in all conditions of life,

> — ''still prompts the eternal sigh,
> For which we wish to live, or dare to die!'' '[78]

These are anticipations, of a year, of two years, from autumn 1813, the point we have reached in our story, but it is necessary at this stage to establish the lurking presence and the frequent surfacing of this Romantic, or post-Romantic, at any rate peculiarly modern note of hollowness and arid indifference (as it was later heard in *L'Éducation sentimentale*, in *Á la recherche du temps perdu*, and, its apogee, in *L'Étranger*). To give it added stress we will quote a passage written some seven years later still, during the worst emotional tempest of Hazlitt's life, where he again makes a link between inadaptation arising out of dangerous idealizations ('tendre ses filets trop haut', as Stendhal put it, in judging his own case) and unsuccessful love-affairs, the kind of discomfiture he met with in the Lakes, and the ultimate outright rejection. A haunting line from the confession of the woeful lover Chrysostom in *Don Quixote* kept running in his head at this time: 'He loved and was abhorred; he adored, and was scorned; he courted a savage; he solicited a statue; he pursued the wind; he called aloud to the desert.'[79] So that when he wrote

---

[77] *Molière, Shakespeare, la comédie et le rire*, ed. H. Martineau (Paris, 1930), 59; *Histoire de la peinture en Italie* (Paris, 1817), ii. 198–5, where the anonymous author of this 'explication anglaise' is not identified; *Rome, Naples et Florence en 1817* (Paris, 1817), title-page, where it is attributed to the '*Mémoires d'Holcroft*'. R. Vigneron, 'Stendhal et Hazlitt', *Modern Philology*, 35 (1938), 375–414, points out the 'curieuse coincidence', but does not attempt to explain it. The explanation, I think, is this. Stendhal discovered the *Edinburgh Review* (an event of capital importance in his life: see Ch. 7, n. 7, below) in Milan in Sept. 1816, at a time when he saw much of Jeffrey's colleague Brougham, then visiting Italy. He must have asked him who wrote the Sismondi review, and Brougham evidently replied (since Hazlitt had not yet published *The Round Table*, nor *Characters of Shakespear's Plays*) 'the author of *Memoirs of Thomas Holcroft*' (pub. 20 Apr. 1816 by Longman's, London agent of the *Edinburgh*). Stendhal, a year later, having forgotten Hazlitt's name, or perhaps not having been told it, but remembering the title *Memoirs of Holcroft*, ascribed the quotation to that volume.

[78] *De l'amour*, note to bk. 1, ch. 3; 16. 9; this borrowing also is noted in Vigneron's splendid article, (n. 77 above) which in imaginative and rigorous scholarship was far in advance of its time.

[79] 20. 50. See also 8. 97, 236; 17. 99; 20. 239.

a letter of advice to his young son to save him from making the same mistakes as he had, he said:

There is one almost certain drawback on a course of scholastic study, that it unfits men for active life. . . . Do not fancy every woman you see the heroine of a romance, a Sophia Western, a Clarissa, or a Julia; and yourself the potential hero of it, Tom Jones, Lovelace, or St. Preux. Avoid this error as you would shrink back from a precipice. All your fine sentiments and romantic notions will (of themselves) make no more impression on one of these delicate creatures, than on a piece of marble. Their soft bosoms are steel to your amorous refinements, if you have no other pretensions. It is not what you think of them that determines their choice, but what they think of you. Endeavour, if you would escape lingering torments and the gnawing of the worm that dies not, to find out this, and to abide by the issue.[80]

The theme of these passages, however earnest and effective its expression (striking even, for the period), is familiar. But Hazlitt proceeds in his paper on Grimm to further remarks that stand apart in their frankness from the hypocritical discretion with which most of his contemporaries veil the licentiousness of the age. The paradox he analyses is plainly grounded in his own experience and casts a curious reflection on the ambiguous autumn happiness of 1813. The *Beggar's Opera* passage enshrines an idyllic mood; the concluding page of 'Baron Grimm' hardly depicts an enraptured lover, but hints ominously at a return to the cruder consolations of his thwarted youth:

. . . excessive refinement tends to produce equal grossness. The tenuity of our intellectual desires leaves a void in the mind which requires to be filled up by coarser gratification, and that of the senses is always at hand. They alone always retain their strength. There is not a greater mistake than the common supposition, that intellectual pleasures are capable of endless repetition, and physical ones not so. The one, indeed, may be spread out over a greater surface, they may be dwelt upon and kept in mind at will, and for that very reason they wear out, and pall by comparison, and require perpetual variety. Whereas the physical gratification only occupies us at the moment, is, as it were, absorbed in itself, and forgotten as soon as it is over, and when it returns is *as good as new*.[81]

We may guess from what remorseful reflections upon casual encounters in the darkness of St James's Park or the Privy Gardens this apologia sprang. His wife was well aware of his commerce with the women of the town: 'he had a taste that way', as she drily remarked, dismissing the matter with a casualness that takes half the sting out of it.[82] We have

---

[80] 17. 93, 98.

[81] 4. 136.

[82] Le Gallienne, 258 (cf. 256); Bonner, 197 (cf. 196).

heard Patmore on the girls he knew, whose beat was in Horse Guards
Parade (on his homeward path of nights).[83] Haydon speaks of his
confessing 'his weaknesses and follies' and of his 'fierce passions and
appetites'.[84] And he himself admitted wryly, 'I have been, among other
follies, a hard liver'. Nor was it easy in Regency London to avoid a fall
from grace. The opportunities were probably even greater than in
Boswell's day. *The Times* in 1816 claimed as a well-known fact that in no
other capital in the world were there prostitutes more clamorously
aggressive than the thousands that nightly infested the streets.[85] And
Hazlitt's work took him nightly into theatre-land, the most notorious
haunt of all.

[83] Patmore, ii. 276–7.
[84] Pope, ii. 64, 65.
[85] 4 Sept. 1816. In 1818 the Police Committee reported, in 3 parishes of 9,924 houses and
59,050 inhabitants, 2,000 common prostitutes and 360 brothels (*The Times*, 24 Aug.).

5
Political Controversy and Art Criticism

WHATEVER obscure defeats may have thwarted Hazlitt's hopes of private happiness that winter, it is plain that the possibility he glimpsed of an imminent peace with Napoleon rapidly darkened into anxiety with the progress of events. After news of the French reverses in the battles of Leipzig reached London in late October 1813 (soon after his walk from Chertsey), the speech from the throne in the new Parliament and the applauded declaration of Lord Liverpool the Prime Minister on 4 November[1] indicated that England meant to extort nothing from her enemy that she herself would have felt unable to accord, and that she was prepared for an honourable and reasonable peace. This attitude was confirmed at that time in the Allies' Frankfurt proposals, which Napoleon was unwise enough to neglect and which were later withdrawn when Castlereagh arrived in Europe in January 1814. On 12 November these overtures were attacked in the latest of a series of 'Letters' on public events which one 'Vetus' (Edward Sterling) had been contributing since 1812 to *The Times*. Sterling was an Irish Protestant lawyer whose pugnacious spirit had led him to join first the Volunteers against the Irish rebels and then the Militia, a role he combined with those of gentleman-farmer and political publicist. A partisan of the Wellesley faction and a dogged enemy of the Liverpool administration, his consciously robust style arrested attention, but the bias and violence of his views deflected and weakened their effect. His earlier letters, gathered into a volume in 1812, had achieved a second edition in January 1813 and already by May sufficient prominence to elicit a series of 'Answers to Vetus' by one 'Sempronius'. This latest letter of Vetus, 12 November, thirty-third in the series, applied to Napoleon lines from *Paradise Lost* (on the fall of Satan) which would have caught the eye of Hazlitt, who had whole passages of Milton by heart. The policy Sterling himself recommended was as immoderate as his tone. After criticizing Liverpool's speech and his moves towards composition with the enemy he proceeded, in a pastiche of Burke and Sheridan, to ask who the enemy of the English really was. It was stated (he went on) to be the French nation. And what was that? A nonentity. They had nothing to do with the French nation, but with Napoleon Bonaparte; and with him they

---

[1] *Parl. Deb.* 27 (1813), cols. 5, 22. The speech remained in Hazlitt's memory; he quoted it in 1828 in his *Life of Napoleon* (15. 156). By then he had persuaded himself that the British had already by 4 Nov. 1813 decided to restore the Bourbons.

could do nothing: 'his honour is incurable ambition: his rights are those of universal conquest: his law is force: his peace is perfidy: [he is] restless, turbulent, insatiable, cruel and remorseless — a breaker of oaths — a contemner of treaties — an encroacher upon the natural and social freedoms of all men.' What, then, he continued, should be the terms held out? None. 'We have but one security. Let us keep all we conquer. Let our Allies keep all they can acquire. The principle was admitted by Napoleon himself . . .'

Hazlitt took up the challenge right away. The following Sunday, 19 November, he launched the first furious riposte in a duel with Sterling that was to last into the new year, or rather in an encounter with two adversaries, Sterling, and Sterling's colleague John Stoddart. Stoddart had been writing leaders for *The Times* since October 1812, and his influence in the paper had increased with the death, in November 1813, of the founder John Walter I, and the accession of his son.[2] He now turned on Hazlitt with a crude abuse and a laborious irony that were to be for some years characteristic of *Times* leaders, until at length the new owner was obliged to muzzle him. Not that the tone of Hazlitt's own attack on Sterling was mild. In the fives-court his shirt would get wringing wet with exertion; he worked himself up into a frenzy which ended, if the game went against him, in an agony of self-vilification, but if he won, in equally extravagant triumph. And he often wielded his pen as he did his racket. If we set aside the abuse in his reply (not that his abuse was ever gratuitously crude; in the annals of the lesser genre of political philippic Hazlitt's performances rank high) and proceed to extract from it what is positive, that is, the general position from which he criticizes Vetus's rejection of negotiation, we find that, instinct with common sense and political realism, it is permanently valid:

We will also venture to lay down a maxim, which is — That from the moment that one party declares and acts upon the avowed principle that peace can never be made with an enemy, it renders war on the part of that enemy a matter of necessary self-defence, and holds out a plea for every excess of ambition or revenge. If we are to limit our hostility to others only with their destruction, we impose the adoption of the same principle on them as their only means of safety. There is no alternative.[3]

In his *Times* leader the next day Stoddart quoted his brother-in-law's article with the comment, 'We copy the following paragraph from a paper of yesterday as a proof that the keepers of St. Luke or Bedlam do not keep that watch and ward over their patients which they ought to do: at least they allow them the use of pen and ink: such ravings can certainly

[2] *The Times*, 19 Dec. 1816; Anon., *The History of The Times* (1935–52,), i. 157.
[3] 7. 33.

only have proceeded from one of those receptacles of irregular genius.' He ends in the same jocular vein, 'We understand that soon after the desertion of Smorgonne several of the political idolizers of "Napoleon" exhibited symptoms of a wandering intellect which rendered the occasional use of a straight-waistcoat necessary; and that the great battle of Leipsic bereft them of the last glimmerings of sense.' It is no doubt a solecism to read the moral sensibilities of one age into those of another, and there is perhaps less callousness than we imagine in these words from one who was both the son of an unfortunate madwoman and a friend of Charles and Mary Lamb.

Hazlitt took public events hard. We shall see how heavily he drank in the summer of 1815 to drown the sickening disappointment of Waterloo and the restoration of the absolute monarchies in Europe. We are not therefore surprised to find that the antagonism that had accumulated over the years between him and his brother-in-law now erupted, and a breach was wrenched open that never again closed.

For the moment he concentrated on Vetus. On 2 December, replying to a letter of 23 November, he initiated a series of 'Illustrations of Vetus' (the only good things he ever did, in Godwin's view),[4] in which he laid bare with brilliant and devastating ease the sophistry of Sterling's arguments, while at the same time setting out his own conception of the just basis of a treaty with the enemy. Where Vetus had said that the first policy of a wise people was to make clear to rival nations that to attack them was not only to invite certain defeat but also to suffer irrecoverable losses, Hazlitt's reply was as provoking as it was unanswerable: 'It is not in the nature of things that the losses of rival States should be irrecoverable. Vetus would do better to decree at once that the possessions of nations are *unassailable* as well as *irrecoverable*, which would prevent war altogether.'[5]

His declared aim was limited. It was to supply the 'voluminous effusions of Vetus' with 'marginal notes'[6] based on Vetus's fallacious arguments, but they were more profoundly based on the principle of benevolence. 'Looking abroad into universality' did not only mean sympathizing imaginatively with neighbours but also with nations. Exclusiveness was as distasteful, as baneful, among peoples as among individuals. The abiding part of the 'Illustrations' appears in paragraphs like the following (it is marred only by the final adjective):

'Think, there's livers out of England. What's England in the world's map? In a great pool a swan's nest.' Now this 'swan's nest' is indeed to us more than all the world besides — to cherish, to protect, to love, and honour it. But if we expect it to be so to the rest of the world — if we do not allow them to cultivate

4 8. 285.   5 7. 42.   6 7. 40.

their own affections, to improve their own advantages, to respect their own
rights, to maintain their own independence — if in the blindness of our
ignorance, our pride, and our presumption, we think of setting up our partial
and local attachments as the law of nature and nations — if we practise, or so
much as tolerate in theory that 'exclusive patriotism' which is inconsistent with
the common privileges of humanity, and attempt to dictate our individual
caprices, as paramount and binding obligations on those, to whose exaction of
the same claims from us we should return only loud scorn, indignation and
defiance — if we are ever so lost to reason, as Vetus would have us, who
supposes that we cannot serve our country truly and faithfully but by making
others the vassals of her avarice or insolence; we shall then indeed richly
deserve, if we do not meet with, the natural punishment of such disgraceful and
drivelling hypocrisy.[7]

This series continued until 5 January. But he had noted the leader of 20
November, and finding that he could deal with Vetus with one hand tied
behind his back he took on Stoddart as well; indeed there was no great
difference between his two antagonists: the arguments were the same,
and Vetus was only a little more pompous, more nobly and arrogantly
declamatory than Stoddart. So on the 27th he cut out Stoddart's leader
of that day postulating an imminent (and wholly improbable) defection
of the French armies of the Adour and the Rhine to the Bourbon cause
and used it on 6 December as his text for a bitterly ironical broadside.
One sentence in particular stuck in his gizzard: 'What if the army
opposed to [Wellington] should resolve to avenge the cause of humanity,
and to exchange the bloody and brutal tyranny of a Bonaparte for the
mild and paternal sway of a Bourbon?'[8] The small but determined band
of English partisans of Legitimacy had not yet shown their hand, and did
not with any confidence until the new year. It was generally assumed
that at the Frankfurt negotiations the peace to be made with France was
to be made with Napoleon. Liverpool and even Castlereagh were
persuaded that it was not yet time to think of deposing him, and that in
any event, on the principle of national independence, the choice of a
successor was a matter for the French.[9] But Carlton House, as would
soon appear, had other plans. Meanwhile Hazlitt's apprehensions were
aroused by this straw cast into the wind by his brother-in-law. It was an
inviting irony that it should come from that former revolutionary
'Citizen' Stoddart, but Hazlitt was so shocked by this first tentative hint
at a reinstatement by force of arms of a dynasty expelled by the French
people twenty years before that he neglected the opportunity to enlist the

---

[7] 7. 50.

[8] 7. 29.

[9] As late as 29 Jan. 1814 Castlereagh told the Allies that Britain would treat with Napoleon as *de
facto* ruler of France as long as the French themselves accepted him as such (C. D. Yonge, *The Life
and Administration of Lord Liverpool* (1868), i. 500; C. J. Bartlett, *Castlereagh* (1966), 124).

reader's sense of incongruity and made instead what he deemed a stronger appeal to self-interest:

Why, if the French wish to shake off the galling yoke of a military Usurper, we say, let them do it in God's name. Let them, whenever they please, imitate us in our recal of the Stuarts; and, whenever they please, in our banishment of them thirty years afterwards. But let them not, in the name of honour or of manhood, receive the royal boon of liberty at the point of the bayonet. It would be setting a bad precedent — it would be breaking in upon a great principle — it would be making a gap in the general feeling of national independence.[10]

The violence of his political writing at this time was not solely owing to his idolatry of Napoleon. He was well aware of the arbitrariness of some of his hero's past actions, but thought them mitigated by the beleaguered Emperor's need to defend himself against endless coalitions.[11] Just as he shrewdly noted that Vetus's 'chief objection to . . . Bonaparte . . . is not evidently, to his being a robber, but because he is at the head of a different gang',[12] so his own chief objection (among many) to the removal of Napoleon was not Napoleon's irreproachability, but the risk of his being replaced by Louis XVIII or by that other mild and paternal Bourbon, Louis's brother the Comte d'Artois, later the detested Charles X who in 1830 provoked a second revolution.

Crabb Robinson met Hazlitt at Lamb's in mid-December and fell into a political wrangle; he found him overbearing and rude, a sure indication of how passionately at that moment the usually equable and fair-minded Hazlitt mistrusted events:

He mixes passion and ill-humour and personal feelings in his judgments on public events and characters more than any man I know, and this infinitely detracts from the value of his opinions, which, possessing as he does rare talents, would be otherwise very valuable. He always vindicates Bonaparte not because he is insensible to his enormous crimes, but out of spite to the Tories of this country and the friends of the war of 1792.[13]

Four days later, on 20 December, Hazlitt listened to Sir James Mackintosh make his maiden speech. The Dutch had risen against the French on 15 November and recalled the Stadtholder of their republic, the Prince of Orange, from exile in England, but he had perfidiously assumed the title of Prince of the Netherlands, true to the aspiration of

---

[10] 7. 30–1.

[11] He cannot but have been disappointed at Napoleon's failure to champion Poland. He attempted a consciously lame excuse of this in 1828 (14. 292).

[12] 7. 44. This of course remained the accepted French view of Britain's policy: according to Jacques Bainville the terms of the Treaty of Paris laid bare the real object of the British, which was conquest (*Histoire de France* (Paris, 1925), 328–9).

[13] Morley, 133.

his house (early encouraged by England and Prussia) to establish a
hereditary monarchy. Hazlitt must have seen a clear portent in this, and
he warmed to Mackintosh when he found him echoing his own recent
reaction to Stoddart's leader, in regretting that 'the ancient Dutch
Republic should be subverted . . . by a few gentlemen (the inhabitants of
two towns only) and in the presence of a foreign force'.[14] He had always
liked Mackintosh despite his apostasy. The admiring and yet disapproving analysis he later gave of Mackintosh's style of oratory explains much
that seems excessive and violent in his own political statements. Hazlitt,
however much a man of principle, was a realist and never allowed his
keen sense of occasion and circumstance to be weakened by the
abstractions of theory. He saw very clearly that political speeches, like
political journalism, must persuade. Their function is not that of the
political philosopher or the historian. And he blamed Mackintosh for
mistaking his aim when in addressing the House he conscientiously
explored all the ramifications of a question and, in his care for truth, met
a thousand prospective objections and elaborated a thousand wire-
drawn reservations. In the mean time those he might have persuaded
had either gone to sleep or gone home. Hazlitt well knew that if he made
the same miscalculation and reverted to his old 'logical way of writing'
his readers would likewise yawn and lay aside the *Morning Chronicle*, and
whatever influence he might hope to exert would be forfeited.

By this time Stoddart had certainly told Sterling who his attacker was,
depicting him naturally enough in his own former sansculotte image. On
24 December, in letter XXXVII, Sterling makes the anonymous author
of the 'Illustrations of Vetus' out to be a disciple of Godwin (which he
could hardly have discerned from the 'Illustrations' themselves), and,
referring to him as 'a drawling hypocritical projector whom no natural
affection can move, nor individual happiness enlighten, . . . [a
worshipper] of the strumpet goddess Reason', attributes to him the aim
of *reasoning* mankind out of all the human sensibilities and 'into a new
assortment of sensibilities, on a larger and nobler scale'. Inspired
perhaps by some allusion of Stoddart's to Hazlitt's dissipation and
neglect of his wife, he claimed that among the institutions to be swept
aside was marriage (here Tartuffe reared his irrelevant and indignant
head as grotesquely as five years later in *Blackwood's Magazine* when the
same accusation was levelled at Leigh Hunt). Marriage was to go
because it was a bar to 'those unconfined embraces' which would usher
in the reign of universal love among the vast family of mankind in the
universal republic. Here were all the familiar bugbears of the anti-
Godwinians. But the main target of Vetus's attack was that poisonous

---

[14] 11. 96–7. See the writer's 'Howe's Edition of Hazlitt's Works: Two Notes', *NQ*, May 1970,
174–5.

vegetable, the doctrine of benevolence, which, in advocating negotiation with the French and upholding the rights of the enemy, was 'threaten[ing] the desolation of the moral world'.

Although impatient of this digression and no doubt irritated by the allusions to his private life, Hazlitt was perceptibly less hostile towards Vetus in his sixth and last 'Illustration' than he had been in his first, and, in his valediction, professed himself ready to believe that Vetus 'had talents and acquirements which might be made useful to the public'. The reason was that in the interval Sterling had unexpectedly turned out to be, in one essential regard, of his own way of thinking. We saw that on 27 November Stoddart hinted for the first time at restoring the Bourbons. Vetus reacted immediately, and on 8 December, taking the opposite side, protested against 'the "soft nonsense" whispered throughout the London coteries in favour of their return'. He flatly declared, 'I deny that the Bourbons are more legitimate kings of France in the year 1813 than the Stuarts within twenty years of our Revolution were legal Sovereigns of England.' Hazlitt blinked, and revised his opinion; he began to see Vetus as a potential if wrong-headed ally. He thereupon characteristically overlooked his attacks on his reputation (just as, years later, and long after the monstrous calumnies of the *Blackwood's* gang, finding some unexpected praise of Napoleon in their magazine, he burst out, 'That's great, that's fine; I forgive them anything they said about *me*'). He held out his hand to Sterling: 'The *tempora mollia fandi* do not belong to Vetus any more than to ourselves. He is, like us, but an uncouth courtier, a rough, sturdy, independent politician who thinks and speaks for himself.'

But in the *Times* office, unfortunately for Sterling, it was not his views but the Bourbon-loving leader-writer's that prevailed. Stoddart at this stage seems to have had sole possession of John Walter's ear. Sterling was dropped. And the beginning of 1814 ushered in the succession of egregious leading articles which in the next few years were to earn Stoddart the title 'Dr Slop', and *The Times* the bad name of serving as mouthpiece to a political writer of notorious bigotry and malignant prejudice. The fame of the paper did not begin its spectacular rise until John Walter's patience was spent and Dr Slop in turn was dismissed in 1816.

These restoration rumours were first and most violently condemned by the *Champion*, formerly *Drakard's Paper*. Originally founded as the *Stamford News* by John Drakard it had been devoted from the start, unlike most other provincial papers, to the discussion of political events and the spreading of liberal views. When Drakard was sent to Lincoln Gaol in 1811 (where Finnerty also was imprisoned) for protesting against the savagery of army floggings, a fund was subscribed for his relief by

'Friends of the Liberty of the Press'. The *Examiner* copied the incriminat-
ing article and was in turn threatened with a libel action. It was no doubt
the publicity arising from this imprisonment that brought the paper, on
10 January 1813, from Stamford to the ampler arena of London, under
the editorship of a superb journalist, the thirty-year-old Aberdonian
John Scott (who like Leigh Hunt had been a clerk in the War Office, a
singular but not entirely irrelevant training for the two foremost liberal
editors of the day).[15]

Scott traced these rumours to their source in Carlton House. The
Prince Regent, he said on 2 January 1814 in his 'political essay' (as he
termed his front-page leaders), would never consent to receive Napoleon
into the exclusive club of Europe's sovereigns: 'Such companionship the
nice honour of Carlton House is not inclined to brook, — and, we
understand, there is not a bumper drained there but to the downfall of
Buonaparte, nor a hiccup vented but as an ejaculatory prayer for the
restoration of the Bourbons.' These hiccups were articulated (continued
Scott) in the leading articles of the *Courier*, which was not so much an
organ of the Ministry as the organ of Carlton House (an important
distinction). He went on to state his own view, which was much like
Hazlitt's, that the policy of peacemakers (and the English ought to be
peacemakers, not warmongers) should be guided by a proper respect for
the independence of France as a nation.[16] Scott's allusions to the Prince
Regent, if rather more sober and less satirical, were just as disparaging
as Leigh Hunt's in 1812.

On 11 January Stoddart swung *The Times* more determinedly
(whatever John Walter may have thought) and more explicitly into line
with the *Courier*, wildly asserting that 'we wish to see a Bourbon prince
on the throne of France; we wish to see the old times revived; we wish to
have to fight (if we are to fight) with Gentlemen and Men of Honour, not
with vagabonds', a fervent prayer that Hazlitt ironically applauded in
his article of 21 January 'On the *Courier* and *Times* Newspapers', where
he blandly assumed the *Times* declaration to be prudently inspired more
by the hope of having a less dangerous enemy to face than by mere
niceness of honour.

Stoddart and Sterling were not Hazlitt's only targets in these agitated
months. Part of his attention (and a good deal of the attention of the
Champion and the *Examiner*) was turned towards the Poet Laureate, who
had just brought out his first official offering, the *Carmen Triumphale*,
celebrating the successes of the Allies and the prospect of a restoration in

[15] *Examiner*, 28 May 1811, 321–2; 16 Feb. 1812, 105 ('his paper has been established by [his
prosecution]'). This was the *Stamford News*; Drakard thereupon launched on 10 Jan. 1813 a
London weekly, *Drakard's Paper*, renamed *The Champion* on 2 Jan. 1814.
[16] *Champion*, 2 Jan. 1814, 409.

France. It was issued shorn of its more violently anti-French stanzas on the advice of Croker and Rickman, who foresaw an embarrassment for the official poet of the British throne if France should once more become a friendly power. Hazlitt probably heard of this emasculation from Rickman (he was at his house on 30 December), or else from Lamb.[17] He reviewed the poem on 8 January with a mildness and restraint all the more notable for the anxieties which were pressing upon him:

It is romantic without interest, and tame without elegance. It is exactly such an ode as we expected Mr Southey to compose on this occasion. We say this from our respect for the talents and character of this eminent writer. He is the last man whom we should expect to see graceful in fetters . . .[18]

He had probably seen that the *Champion* of 2 January had taken his hint of 18 September when he had quoted, against Southey, Southey's own early work, *Joan of Arc*, the 'subversive' drama of 1796. The *Champion* returned to the attack (more clumsily than he would no doubt have liked — he could not have approved the reference to 'Bob') on the 13th in a review of a volume of verse by suggesting that the author might perhaps have become Poet Laureate if it were not that 'he had not, like Bob, the "pure weakness" to be tempted by "a hundred pounds a year" ', an attack kept up on the 30th by an allusion in a skit, *Botanical Presents*, to the Laureate's 'rather blighted' bays. Nor did the *Examiner* lag behind: on the 16th it ended a parody of the *Carmen Triumphale* with a double-barrelled line that blasted both servility and venality in one discharge — 'Glory to Kings, thy song! A hundred pounds, thy payment!'.

True to his precept of looking abroad, Hazlitt continued to occupy himself with other things as well as politics, engrossing though these then were. Parliament was in recess and his duties as dramatic critic did not begin until the third week in January, so he had leisure. Fortunately so, since the life of London was eclipsed and the people confined to their houses, from Christmas into the new year, by an impenetrable fog, of a density such as had not been seen since the middle of the eighteenth century, and he probably took advantage of his enforced inactivity to spend long hours at his writing-table.[19] Mme de Staël's *De l'Allemagne*, long banned by the Imperial censor in France, had been brought out by John Murray in November and in translation a few weeks later. Hazlitt could hardly have been unaware of the loudly trumpeted arrival in London the previous June of this celebrated refugee. Received with alacrity by Lady Bessborough and by Lady Lansdowne whose guests clambered on to chairs to get a sight of her, she was invited both to

---

[17] Morley, 134.

[18] 7. 25.

[19] *GM* 84 (Jan. 1814), 87, and daily papers.

Carlton House and to Holland House, and was fêted as much at Oxford as in London. She acquired a reputation for accessibility and met not only royalty and the nobility, but such as Mackintosh (a special case: he had been a fellow-student at Edinburgh University in the 1780s with her lover Benjamin Constant), Southey, Coleridge, Crabb Robinson (whom she had known at Weimar), Godwin, and even William Ayrton — the last not until very late in her stay and probably on account of his musical reputation.[20] Even if Ayrton could have arranged for his friend Hazlitt to meet the woman the *Edinburgh Review* called 'the most eminent literary female of the age', and *Drakard's Paper*, less respectfully, this 'sort of Dr Johnson in petticoats', we cannot suppose Hazlitt would have been very enthusiastic.[21] There was a prospect of discomfort. He loved the personal style in conversation, talk warmed by good humour, spontaneity, and sympathy, and disliked the prevailing and pretentious mode of set debate, of critical and analytical disputation, in which Mme de Staël excelled, outmanœuvring and outshining most of her interlocutors.[22] He relished argument and discussions but not when they assumed all the panoply of a joust, as, from contemporary accounts they did with her:

I am to witness the first meeting between Mme de Staël and Mackintosh today . . . She is mighty fond of general questions, which the English wits are apt to *shirk* in conversation, such as 'whether or not the English constitution is applicable to all the nations in Europe?', 'whether or not it could exist in a country where Christianity was not established?' We had both these questions yesterday . . . but Sheridan and Canning parried them with jokes. Sir James, no doubt, is ready with opinions and arguments.[23]

But that was no reason to ignore the book that had attracted almost as much attention as its author. He had already given a cursory account of it in the *Morning Chronicle* on 13 November, but the reading had carried him back to the themes of his own lectures on philosophy, and it seems also that it encouraged him to persevere in his old project of a major work on 'metaphysics'. The result was four papers on 'Mme de Staël's Account of German Philosophy and Literature' published between 3 February and 8 April in the *Morning Chronicle*, where they stood out from the rest of the matter, war correspondence, political reports, London and provincial news, with an abstruseness that must have made many a reader stare. Only Mme de Staël's popularity can explain Perry's acceptance of these eight closely printed columns.

[20] *Courier* advts., 19 Nov. and (trans.) 10 Dec. 1813; *MC*, 22 June 1813; *Examiner*, 4 July 1813, 418, 426; 26 Dec. 1813, 827; BL Add. MS 51952, at 23 June, 3 Oct. 1813; Southey, *Selections*, ii. 332; Morley, 132; Godwin Diary, 5 Oct., 28 Nov. 1813; BL Add. MS 60372, at 17 Apr. 1814.
[21] *Drakard's Paper*, 27 June 1813, 200 (quoting *ER*); 1 Aug. 1813, 238.
[22] 8. 202; 4. 12; 6. 153.
[23] S. H. Romilly, *Letters to Ivy* (1905), 212, Earl of Dudley to Mrs Dugald Stewart, *c.*July 1813.

Hazlitt's ostensible aim was to consider Kant as a critic of empiricism.[24] After a perfunctory gesture in the direction of Mme de Staël's account of the Kantian system he dealt with that system (inadequately represented as it was in Willich's translation and commentary) exclusively as 'an elaborate antithesis or contradiction to the modern philosophy'.[25] He made nothing — could scarcely at the time have made anything — of what has since been seen as Kant's historical role of mediator between rationalism and empiricism, and confined himself to Kant's relation to the sensationalist philosophers. But the importance of these papers is that Hazlitt is again seen, in the endlessly recurring debate between freedom and determinism, arguing, from however mistaken a ground, the claims of the creative role of the spirit against the dead hand of the mechanists.[26]

To affirm the primacy of the creative over the mechanical in the human mind was also the tacit business of two more contributions at this time to the *Morning Chronicle*, the 'Fragments on Art: Why the Arts are not Progressive'. They dealt with the fallacy of improvement in art, which Hazlitt saw as deriving from a false analogy with the sciences: 'What is mechanical, reducible to rule, or capable of demonstration, is progressive, and admits of gradual improvement: what is not mechanical or definite, but depends on genius, taste, and feeling, very soon becomes stationary or retrograde, and loses more than it gains by transfusion.'[27] Here his theory of liberty is linked with the theory of decadence adumbrated in his letters on modern comedy the previous autumn, which is in turn linked with the pastoral ideal. A long footnote quotes an account of Claude Lorrain, whom he is reluctant to relegate (as his theory requires) to the inferior order of genius found in latecomers in art. This account, borrowed from his friend Northcote's *Memoirs of Sir Joshua Reynolds*, echoes his own passion for those air-drawn landscapes that descend by subtle gradations (defined by diminishing, rapt, lurking figures, rustic or heroic, bent upon their secret task or quest) from the tree-framed sward of the foreground down and away

[24] 20. 12–36.   [25] 20. 20.

[26] *De l'Allemagne* was not the only work of Mme de Staël's he was prompted to read by her visit. 'On the Character of Rousseau' (*Examiner*, 14 Apr. 1816) derived from her *Lettres sur Jean-Jacques Rousseau* (1788), repub. in French in London by Colburn, 1814. Hazlitt's page-refs. are to this edn., and the proof that he read it, and set down his impressions, now, soon after publication, lies in the singular circumstance of his quoting this ostensibly future essay of 1816 in a notice of Miss O'Neill at Covent Garden in 1814 (*MC*, 16 Oct.; see 5. 404, 199 n.).

[27] 4. 161. The original articles of 11 and 15 Jan. 1814 were repr. in 18. 5–10, with an erroneous reading: Howe missed a correction printed in *MC*, Mon., 17 Jan. which is significant of Hazlitt's precision and of his jealousy of his reputation as a writer. It reads, 'Erratum: in "Fragments on Art" in Saturday's paper, the first sentence, for "naturally much the same" read "naturally limited" or "naturally the same at different periods." The expression as it stands conveys a meaning totally different to what we intended.'

through the near distance, past copse and rock and bridge and winding river to the far-away floor of the valley, and then further still across the hazy plains, glimmering from league to league with hints of walled cities and embattled castles, to the remote horizon of the unsubstantial blue hills.

Just as the publication of the two letters 'The Stage' and 'On Modern Comedy' immediately resulted in his being entrusted with a share of the *Chronicle*'s theatrical notices, these articles on painting prompted Perry to send him without delay to the British Gallery, Pall Mall, to report on the annual exhibition of the British Institution. His first published criticism of the artists of his day appeared on 5 February. He found the contemporary scene dismally unexciting, and was particularly out of sympathy with the pretensions of the grand style:

But in the higher . . . style of art . . . this country has not a single painter to boast, who has made even a faint approach to the excellence of the great Italian painters. We have indeed a good number of . . . large canvasses covered with stiff figures arranged in decent order, with the characters and story correctly expressed by uplifted eyes or hands, according to old receipt-books for the passions . . . But we still want a Prometheus to give life to the cumbrous mass, to throw an intellectual light over the opaque image, to embody the inmost refinements of thought to the outward eye, to lay bare the very soul of passion. That picture is of little comparative value, which can be completely *translated* into another language, of which the description in a common catalogue is as good, and conveys all that is expressed by the picture itself; for it is the excellence of every art to give what can be given by no other, in the same degree.[28]

This last discriminating insight would have been lost on his friend Haydon, who confounded literature with painting (and Shakespeare with Benjamin Robert Haydon) in a Romantic swirling mist of hybrid emotions.[29] Neither his *Macbeth* nor his *Dentatus* was mentioned here, not even negatively as exceptions to the strictures. Of course Hazlitt was writing about an exhibition in which Haydon was not represented, but the painter must still have been displeased not to be hailed as that providential Prometheus by the new critic who had known his work for nearly two years. There is no reference to this 1814 exhibition — one of the two main artistic events of the season — in Haydon's diary, but that diary, although voluminous, was irregularly kept (only two entries for this February), and although his bad eyes at the time made reading difficult we cannot be sure that he missed the notice, nor, consequently,

[28] 18. 10–11.

[29] A confusion pardonable in an age when nearly all painters apparently also wrote: Blake, Shee, Westall, Ward, Northcote, Hoppner, Allston, and (although they did not publish) Lawrence and Fuseli.

that his future resentments against Hazlitt did not derive from as far back as this first piece of art criticism. To this excitable votarist, who as Elizabeth Barrett later remarked 'saw maniacally in all men the assassins of his fame',[30] the whole paragraph (together with the later sentence, 'But we should be at a loss to point out . . . any English painter who, in heroic and classical composition, has risen to the height of his subject') would have been an affront, a vexatious act of unfriendliness. We do not know exactly when their differences began, but in a few years Haydon, noting a conversation with Hazlitt on contemporary painting, made out his deplorable opinions to be a corollary of his failure as an artist.[31] Haydon was hoping for a dishonest complaisance that would be quite out of character. Hazlitt would no more spare a friend than he would spare himself. *Amicus Plato, sed magis amica veritas*. When the *Judgement of Solomon* which Haydon was then struggling to complete (it was generally agreed to be his best picture) was exhibited in the summer he praised it, but with echoes of unspoken reservations, and his criticisms of detail were unrelenting.

Posterity has largely confirmed his judgement of Haydon, but would be likely to demur to his motion that the 'English face' was unsuitable for historical paintings. This he had already advanced in his essay on modern comedy ('We have no historical pictures because we have no faces proper for them') and had repeated, no doubt, in conversation with an American artist he had met in Paris in 1802, Washington Allston. He had warmly praised Allston's *Dead man restored to Life by touching the Bones of Elisha* but deprecated what he saw as the bad influence of French painting in its subservience to 'receipt-books', and its neglect of the natural. 'The faces are in the school of Le Brun's heads — theoretical diagrams of the passions — not natural and profound expressions of them.'[32] His acquaintance with Allston was perhaps renewed through John Hazlitt, whose transatlantic apprenticeship and republican sympathies drew him to American artists, but Allston also knew Coleridge whom he had met in Rome in 1805, as well as Wordsworth, Sir George Beaumont, and others of their circle. Hazlitt once asked him where he found models for his heads: he had never seen their like in the streets of London. Allston replied that he painted them from imagination, and in the incredulous stare Hazlitt turned on him he thought he discerned the unspoken comment, 'You are the greatest liar I ever met'.[33] He was probably wrong: what provoked the stare was the offence to Hazlitt's ideal of truth to nature rather than the strain on his credulity. This

[30] *The Letters of Robert Browning and Elizabeth Barrett Browning 1845–1846*, ed. E. Kintner (Cambridge, Mass., 1969), ii. 865.

[31] 6 May 1817 (Pope, ii. 110; cf. Haydon, 211–12).

[32] 18. 13.   [33] J. B. Flagg, *The Life and Letters of Washington Allston* (1893), 123.

traditional ideal, never departed from, is echoed in a saying of Goethe's Werther which he later remembered with approval: 'Nature is inexhaustible, and alone forms the greatest masters. Say what you will of rules, they alter the true features, and the natural expression.'[34] He quoted this in 1820, but it was evidently a passage that had long settled into his mind: he had read *Werther* in the early 1790s, at a time when he was constantly in and out of his brother's studio.

We may take leave of his first venture in art criticism with something more arresting than what he says about models, and that is his comments on a great contemporary whose genius was not then acknowledged. A paragraph in this first 'salon' on Turner's *Apuleia in search of Apuleius* gives early evidence of the pluralism that was personal to Hazlitt and marked his criticism in all fields: he never felt himself obliged to force his judgements into a mould of unity, and saw no reason to suppress apparent contradictions or inconsistencies to gain a spurious homogeneity.[35] He recognized the true nature of Turner's genius and saw the essence of the painter's originality even at this early ambiguous stage and deflected as it still was by the alien conventions of Claude Lorrain:

[His pictures] give pleasure only by the excess of power triumphing over the barrenness of the subject. The artist delights to go back to the first chaos of the world, or to that state when the waters were merely separated from the dry land, and no creeping thing nor herb bearing fruit was seen upon the face of the land . . . The utter want of a capacity to draw a distinct outline [combined] with the force, the depth, the fulness, and precision of this artists eye for colour, is truly astonishing.

He was not of course alone at that time in deprecating Turner's want of this capacity,[36] but he was one of the few who were prepared, not to concede, but to take positive pleasure in, and at the same time to define, his extraordinary eye for colour, and who foresaw in it the principle of his future spectacular development.

---

[34] 'On the Pleasure of Painting', *LM*, Dec. 1820; 8. 6 n.

[35] 18. 14.

[36] As early as 1803, when *Calais Pier* was exhibited, Hoppner, Fuseli, and Sir George Beaumont made the same criticism.

6

Dramatic Critic

WEDNESDAY, 26 January 1814 saw a début which has remained celebrated in the annals of the English Theatre. The weather was appallingly cold. The dense fog of the last days of the year was followed by a heavy snowfall which marked the beginning of the Great Frost. The snow lay on the ground throughout the month of January, right through February, and on into March; indeed it was not until the 20th of that month that a thaw showed any sign of breaking through. The oldest inhabitants could scarcely remember the kennels so choked, the pavements piled so high; the accumulation of snow surmounting the buildings was such that the roofs of ancient houses had to be cleared to prevent collapse; the taps of public stand-pipes were left half-open for fear of burst mains although this turned the streets into skating-canals. On this Wednesday, 26 January the wind veered to the south-east and brought a momentary hint of thaw, but by dusk it was as bitter as ever, and London stayed dumb, like an occupied city, under the snow-lit darkness, listening to the ice-floes on the Thames at ebb-tide exploding like cannon against the bridge piers.

To Drury Lane theatre, out of the thousand frost-bound streets, there came only a hardy sprinkling of an audience to shiver on the draughty benches of the pit, blowing on their nails and watching their breath whiten the air, until a small demonic figure (Edmund Kain, or Kean, the papers called him), a skinny, undersized, down-at-heel, provincial nobody, made his entrance as Shylock in the third scene of *The Merchant of Venice* and electrified them, first into puzzlement, and then into rapt attention, self-forgetfulness, and the warmth of enthusiasm.

Hazlitt, in that audience, was immediately on the alert, and by the time Kean launched into his denunciation of Antonio, 'You called me unbeliever, cut-throat dog, And spit upon my jewish gaberdine', he knew two things: first, that a man of genius had burst upon the town, and secondly, that this genius had in a matter of minutes destroyed for ever the Kemble religion in which he like all his contemporaries had been brought up, and which had never until now been questioned. Yet after a regretful backward glance at his memories of the fifty-six-year-old veteran he turned, open as he always was to the untrammelled, free-ranging spirit of life and the primacy of nature, to welcome this new phenomenon, the 'first gleam of genius breaking athwart the gloom of the stage'.[1] His notice in next day's *Chronicle* was the first and the firmest

[1] 5. 175.

recognition of the actor's genius. The *Champion* said nothing for two Sundays, and the *Examiner* not until the Sunday after that; and even then, although they were both favourable, both were very cautious. By this time the town had made up its own mind: despite the weather, as Jane Austen reported to her sister in Hampshire, the rage for seeing Kean was so great that one had to be satisfied with what seats one could get.[2] The *Champion* made a great show of independence, asserting that Kean had been much overpraised, whereas Hazlitt had been so enthusiastic that it was later rumoured that he had been given a bribe of £1,500 by the management of Drury Lane to write up Kean, a baseless accusation he had no difficulty in refuting.[3] No one has pointed out that what is significant about this rumour and what makes it so evidently a retrospective invention is not the implied corruption nor the size of the sum involved, but the absurd assumption that Hazlitt, who had written no more than a handful of brief notices, had indeed only recently been formally confirmed in this department by Perry, was already of such weight and influence as to be worth suborning; whereas it was on the contrary his criticisms of Kean that created his own public reputation.

By the time the *Examiner* notice appeared the idolators and detractors had formed into two camps, and their violent opposition was marked by strong political overtones. It was not only that Kean's patrons were the Whig committee of Drury Lane; he also symbolized a threat to the Tory establishment in his revolutionary style of acting, a threat to which he gave voice in his off-stage ranting against the social order. Hazlitt as usual held aloof from both camps and went his own way. Not even the first glow of pleasure blinded him to Kean's faults. He saw exaggeration in his pauses and miscalculation in his reliance upon facial expression unsupported by gesture. What succeeded on a small provincial stage was ineffective in a large London theatre. But he also realized that he would show himself even greater in parts better suited to his manner than Shylock. He soon saw and preferred him in the more difficult role of Richard III, and ultimately came to consider that his masterpiece was Othello (a part Hazlitt had pondered deeply and with which on grounds of temperament he was peculiarly qualified to sympathize).[4] He was no more likely to gloss over Kean's vulnerable points than to revoke his admiration for Kemble's merits. To dub Kemble classical and Kean romantic is to oversimplify but it does suggest their totally opposed styles, the gentlemanly dignity and formal grace of the one, and the plebeian vehemence and inspired inelegance of the other.

---

[2] Letter of 2 Mar. 1814.

[3] L'Estrange, ii. 47 n. Kean's success saved the theatre from foundering.

[4] He quoted *Othello* oftener than any other play except *Macbeth*, and, of course, *Hamlet*.

What effect Kean's success had on the popularity of Kemble is hard to guage, but within three years he retired and we find Hazlitt condemning the cant '. . . that Mr Kemble has quite fallen off of late — that he is not what he was; he may [continued Hazlitt] have fallen off in the opinion of some jealous admirers, because he is no longer in exclusive possession of the Stage: but in himself he has not fallen off a jot.'[5]

On the same day as his first notice of Kean there appeared also in the *Morning Chronicle* his refutation of the current claim that both French royalists and French republicans were at one in their hatred and fear of Napoleon, and that this justified the movement to recall the Bourbons which had been fitfully rumoured to be on foot since the first days of January. Observing in passing that 'the lucubrations of a celebrated writer' (Sterling) had been 'withheld from the public' (i.e. he had been sacked from his paper) for opposing the mooted restoration, he goes on to refer to another 'writer in *The Times*' (Stoddart) 'who has lately raved very vehemently about the ''restoration of the Bourbons, and the good old times, and gentlemen and men of honour'' [and who] gave us not long ago a definition of a Republican and a Jacobin, which he nicely distinguished'. This distinction Hazlitt found incomprehensible,[6] and it gave him an opportunity to propound his own definition of a Jacobin. It is one which reveals the peculiarly abstract, intensely personal, and almost poetic nature of his political attitude.

He who has seen the evening star set over a poor man's cottage, or has connected the feeling of hope with the heart of man, and who, although he may have lost the feeling, has never ceased to reverence it — he, Sir, with submission, and without a nickname, is the *true Jacobin*.[7]

The strikingly unexpected evocation of 'the evening star . . . over a poor man's cottage' (in a definition as incomprehensible to Stoddart as his was to Hazlitt, and therefore duly satirized in the next day's *Times*) seems, as we learn later, to derive by rooted association from some kind of personal illumination that came to him in December 1805 and which compounded, in hope and joy, his early aspirations as a painter, his reflections on benevolence, and the arrival at Wem of the news of the battle of Austerlitz. No more is said to explain this mysterious epiphany at this time, in 1814, but in later years it seems to have become an important image in his recollection, bound up with his dead hopes and the memory of his dead father.

The visual element in this 'pastoral definition' (as he called it in 1817,

[5] 5. 375.

[6] It is given in 7. 386–7.

[7] 7. 370. The passage was accidentally omitted from the newspaper on the 27th and supplied in correction on the 28th.

recollecting with some irritation Stoddart's jeering failure to understand it) is repeated in 1815 when he ends 'Mind and Motive' with the same coda as the *Reply to Malthus* in which he had envied with studied ambiguity those 'to whom the guiding-star of their youth still shines from afar, and into whom the spirit of the world has not entered. They have not been "hurt by the archers", nor has the iron entered their souls.'[8] This last and unexpected page of the *Reply* may well have been written just after the experience of December 1805. At any rate it seems to have been inspired by the unworldly idealism of his father's life, and to be obscurely linked with childhood memories of the story of the Nativity, the Magi, and the angels' message to the shepherds, 'Peace on earth and good will toward men'.[9] In 1820 the episode recurs to his mind once more, in speaking of a portrait of his father, and it is here that he gives it a date.

I think, but am not sure, that I finished this portrait (or another afterwards) on the same day that the news of the battle of Austerlitz came; I walked out in the afternoon, and, as I returned, saw the evening star set over a poor man's cottage with other thoughts and feelings than I shall ever have again. Oh for the revolution of the great Platonic year, that those times might come over again! I could sleep out the three hundred and sixty-five thousand intervening years very contentedly! — The picture is left: the table, the chair, the window where I learned to construe Livy, the chapel where my father preached, remain where they were; but he himself is gone to rest, full of years, of faith, of hope, and charity.[10]

His last 'Illustration of Vetus' had appeared on 5 January, and was followed by 'On the *Courier* and *Times* newspapers', but after that he withdrew from the arena, keeping his own counsel, working at his literary notices and art criticism; and recording the triumphs of Edmund Kean, but evidently missing little of politics that appeared in the press at this critical period.

No doubt he had always read the *Examiner*, but he must now have studied its columns with special care. (If he could have looked into the seeds of time he would have paused a moment over an item unconnected with the European situation — a lengthy account on 3 April of the iniquities of the Vice-Governor of the West Indian island of Grenada based on an affidavit sworn before one Henry Bridgwater).[11] But it was politics that engrossed his attention. As well as the *Examiner* he must have studied the *Champion*, but he himself maintained during the early months of 1814 an almost total silence. The exception was an article on

---

[8] 7. 151; 20. 52–3
[9] The expression 'hurt by the archers' is Cowper's, who applied it to Christ (*The Task*, 3. 113).
[10] 8. 13.   [11] *Examiner*, front-page leader, 'Vice-Governor Ainslie'.

the *lex talionis* principle on 26 February, setting forth some very advanced views: he saw that in international politics the principle of revenge is self-defeating and so far irrational, and saw it with a clarity that was beyond the scope of even so intelligent and humane a man as Southey. Southey was then saying much the same as has been repeated in different forms during every successive war in modern times. He was saying (and compounding the ineptitude by an equally illiberal comparison) that the French 'ought to be the Jews of Europe — a people politically excommunicated and never to be forgiven, and above all never to be trusted'.[12] In addition Hazlitt half-glimpsed the economic fallacy of war-reparations.

His silence gave him leisure to follow John Scott's sober, thorough, and liberal discussion of the developing situation, in front-page leaders. The first of the new year was on 'The War-mongers at Carlton House', and the carefully elaborated titles of the succeeding 'essays' give a fair idea of their tenor. They embraced all the problems of the impending peace, stating them either in the clear, popular form of specific questions, or, rather cumbrously, as a single-sentence abstract. His admirable articles, in their clarity, urbanity, breadth of reference, common sense, and nobility of principle, as well as in their notable modernity of tone, bear comparison with the best political journalism of the late nineteenth or the present century. On 13 and 20 February appeared a two-part article entitled 'Can the Allies with Propriety refuse to make Peace with Buonaparte?' The subject announced for 8 March was, 'How the Future Tranquillity of Europe is likely to be affected by the Duration of the Present French Government', but in the mean time the growing prospect of a restoration and the mounting hatred of Napoleon impelled Scott to substitute a comment on 'The War-newspapers and the Character of Buonaparte'.

Although Hazlitt watched events in silence he watched with anxiety. Thus he did not miss in a corner of *The Times* on 18 January the following monstrous proclamation by Blücher (dated 1 January) to the inhabitants of the left bank of the Rhine: 'All connection with the French Empire must cease from the moment of the entrance of the Allied troops. Whoever infringes this order will render himself guilty of treason against the Allied Powers; he will be carried before a Military Tribunal and condemned to death.'[13] On 17 March his political allegiance and his devotion to art fused in a burst of mingled anguish and fury over a *Times* leader eagerly looking forward to the sacking of the Louvre by the

---

[12] BL Add. MS 30927, fo. 221, letter of 26 Apr. 1814. (Not in pub. version, Southey, *Selections*, ii. 350-2.)

[13] Howe (7. 39,372) is wrong in thinking that Hazlitt's reference to Blücher's 'manifesto' is to his declaration of 3 January to the troops, which was in no way sinister.

Cossacks whose atrocities had terrified Europe and revolted mild England. The nostalgic recollection of those gallery walls hung with beauty had not ceased to haunt his dreams, and he quailed at the prospect of their obliteration. Asserting at the same time the claims of common sense and the priority of the imagination, he pointed out that those who wanted to see the Louvre demolished as a reprisal for the burning of Moscow forgot that the one could be rebuilt but the world's masterpieces could never be replaced.

His brother-in-law, now confirmed in the *Times* editorial chair by the dismissal of his colleague Sterling, blandly continued his reactionary exhortations. His leader of 7 March had proclaimed that 'our rallying cry should be "Europe as it was in 1788!" ', and on 11 March he, the naïve and overweening *ci-devant* Citizen Stoddart, had unblushingly pinned his hopes on 'the return [of France] to those safe and practicable principles of policy which prior to the fatal Revolution had rendered France the envy and pride of the world'. At this same time Southey was writing to Daniel Stuart of the *Courier* in precisely the same sense, wringing his hands over the feebleness and disabilities of parliamentary government and the deplorable 'readiness of our Ministry to make peace against the opinion both of the Prince and the people'.[14] Hazlitt, at last breaking silence, resumed his pen and summed up his views on revolutions, wars of extermination, and restorations in an article that Perry for some reason (perhaps the perplexing uncertainty of events) rejected. Hazlitt carried it to the *Champion*, and Scott published it as 'A Correspondent's Observations on the Motives and Principles of those who are for Maintaining Everlasting War against Revolutionized France'.[15] This was on Sunday, 3 April. The editor expressed disagreement with much of it and promised reasons in the next issue. They never appeared; he was overtaken by events. In his own leader of the same date he had said of a reported royalist insurrection at Bordeaux that if indeed there was popular support for the Bourbons the Allies must bow to it, but that the results must be awaited of the talks with Napoleon's emissary at Châtillon. Hazlitt had said in that same article of the 3rd that an Allied march on Paris was but a splendid reverie not yet accomplished. He was wrong. Neither Scott nor he knew how fast the wheel of fortune had spun. The Prince Regent was one of the few in England who knew that the talks had ended on 18 March, but not even he knew that on the 31st

---

[14] *Letters from the Lake Poets to Daniel Stuart*, ed. M. Stuart (privately printed, 1889), 416, letter of 14 Mar 1814.

[15] The title it is given in *Political Essays*, 'On the Late War' (7. 72), is misleading: the war was not over. It is also worth remarking that Hazlitt in this article recognizes and emphasizes a point always popularly ignored in Britain: I mean the unconscious assumption of moral and political superiority deriving from her position as an island but never ascribed to it.

the Allies had entered Paris. The news burst upon London in a late edition of the *Courier* on Tuesday, 5 April, a complete surprise.

On Sunday 10th Crabb Robinson hurried early to the coffee-house and found Napoleon's rumoured abdication confirmed in the newspapers. On the 12th he remarked on the *Morning Chronicle*'s 'unpatriotic spirit' and observed that among public figures only Cobbett and Sir Richard Phillips (editor of the *Monthly Magazine*) kept aloof from the rejoicing. But he soon encountered another malcontent in Hazlitt. Glum indeed must Hazlitt have been during this week of celebration, and all London ablaze with illuminations. Silver lamps and elaborate transparencies bedecked the public buildings; above the Ionic portico of East India House (his friend Lamb's place of work), always to the fore in the manifestations of loyalty, were the arms of the Bourbons and the name Louis XVIII; all the way from the Guildhall westwards along Cheapside and past St Paul's, along the Strand through Temple Bar to Charing Cross and down Whitehall was one long garland of lights. Everywhere he looked the looming façades glittered with the detestable legends 'Louis XVIII', 'France Restored', or the grammatically as well as politically obnoxious 'Vive les Bourbons', and as he made his way home down Whitehall at night after leaving the *Chronicle* office he could see on the brick front of the Foreign Office on the south side of Downing Street, beyond the joyously illuminated windows of the corner pub, the Cat and Bagpipes, the final insult picked out in the last of the blazing lights along that route — a crown, 'G.R.', 'P.R.', and 'The Triumph of Legitimate Sovereigns'.[16]

He felt acutely the moral isolation he later so effectively defined in his essay on public opinion:

The weight of example presses upon us (whether we feel it or not) like the law of gravitation. He who sustains his opinion by the strength of conviction and evidence alone, unmoved by ridicule, neglect, obloquy, or privation, shows no less resolution than the Hindoo who makes and keeps a vow to hold his right arm in the air till it grows rigid and callous. To have all the world against us is trying to a man's temper and philosophy. It unhinges even our opinion of our own motives and intentions. It is like striking the actual world from under our feet: the void that is left, the death-like pause, the chilling suspense, is fearful.[17]

He had not the heart even to see his friends. The weather was fine, and he probably walked alone a good deal in the fields outside the town. He was on edge. He had recently quarrelled with Lamb for his unwelcome flippancy at this juncture, growling at him that it was because of his 'infinite littleness' that he was incapable of understanding Napoleon's

---

[16] *The Times*, *MC*, and *Courier*, *passim*.
[17] 17. 310–11 (cf. 7. 137–8).

greatness.[18] The chronic levity that constituted Lamb's stubborn pro-
phylaxis against the stresses of his domestic situation, against his own
maimed life, sometimes misfired. The breach between the two (a
paradoxical evidence of Hazlitt's regard both for Bonaparte and for
Lamb) lasted all summer; indeed it was never thereafter completely
healed. The fifteen years during which Lamb later proudly recalled
having stood well with Hazlitt were drawing towards a close; the old ease
and freedom of the companionable evenings never returned after
Europe was 'overspread with that night which saw no dawn' (as Hazlitt
said ten years later in his nostalgic account of one of Lamb's Wednesdays in Mitre Court, 'Of Persons one would wish to have seen'). Lamb
told Robinson of the rift at the close of the month, on 29 April, adding
that Hazlitt 'was confounded by the conduct of Buonaparte and
ashamed to show his face', a strange interpretation of events.[19] Also
present at Lamb's that night was Stoddart. A few days earlier, in a
particularly violent leader, after first referring to the triumphant
departure of Louis XVIII from London to claim his throne, he had
turned to Napoleon, the perjured and bloodthirsty monster who would
be the object of the contempt and scorn of all ages, the vile animal whose
infamous career made him for ever execrable, and whose vile carcass
ought not to be allowed to escape the gibbet.[20] Dr Slop had no notion of
magnanimity or even moderation in victory; it was simply a cue for
redoubled vilification.

Hazlitt, confronted everywhere by the rejoicings of his hero's enemies
and unwilling to encounter the complacency of his own friends, also had
the ill luck at this very moment to run into difficulties with his editor.
Perry, caught in a dilemma, uncertain what attitude to adopt towards
the Bourbons, had little patience left for this intransigent and now
embarrassing partisan of Napoleon. Also, apparently about this time
Perry asked him to answer on the spot some article or speech on the
freedom of the press. He succeeded in concocting a reply, in which,
however, he had no great confidence. Perry glanced at it, and said, 'This
is the most pimping thing I ever read. If you cannot do a thing of this
kind off-hand you won't do for me'.[21] Even for the brusque and
capricious Perry this was somewhat excessive, addressed to someone to
whom at another time he acknowledged himself indebted for having
done more for his paper than any other man on it.[22] It accords however
with Miss Mitford's view of their relationship: 'He had not the slightest
suspicion that he had a man of genius in his pay — not the most remote
perception of the merit of the writing — nor the slightest companionship

---

[18] Morley, 142.   [19] Ibid.   [20] *The Times*, 25 Apr. 1814.
[21] Morley, 154.   [22] Ibid. 153.

with the author.'[23] He is partly excused in having been hard-pressed at the time, domestically as well as professionally; his consumptive wife had been ordered eight months earlier to Portugal, leaving him in charge of a young family (next month, May 1814, she boarded a homeward-bound vessel which was captured by Algerian pirates: she died of her privations, at Bordeaux on 18 February 1815, aged thirty-eight).

Hazlitt, however, was deeply offended at this unexpected affront. But his next contribution after the rejected article got him into even worse trouble: he covered the Royal Academy Exhibition the first week in May, and spoke of Lawrence's portrait of Lord Castlereagh as having 'a smug, smart, upstart, haberdasher look'. For Perry this was the last straw. Not only was the artist a friend of his, he was actually painting *his* portrait at that moment. Futhermore his strained relations with Hazlitt at the time may have prompted him to see in his reporter's words a veiled allusion to his own spell in the drapery trade at Aberdeen (of which Hazlitt was probably ignorant). The king of London journalism was not anxious, in that age of aggressive snobs, to be reminded of his early days as counter-jumper. The result for Hazlitt was a humiliating public reprimand from his own editor: within forty-eight hours Perry inserted a paragraph in the *Chronicle* rapping his art-critic over the knuckles for 'mix[ing] the ebullition of party spirit with his ideas of characteristic resemblance.' Politics, he said, had nothing to do with the fine arts and the portrait of Lord Castlereagh, remarkable for the ease of the attitude, was one of the best in the exhibition.[24] This was not the judgement of another artist, John Flaxman, who agreed with Hazlitt about the portrait. He also told Robinson that they both held Lawrence's social success to be injurious to him as an artist, quoting a saying of Hazlitt's which has an autobiographical ring: 'No good talker will ever labour enough to become a good painter.'[25]

Some weeks after this, Perry told Hazlitt he would have to go, without, apparently, giving any clear reason. He certainly said nothing of the grounds he complained about to Miss Mitford, who thought his behaviour high-handed (if she had not been so fond of him she would probably have called it brutal). 'He hired him', she said, 'as you hire your footman; and turned him off with as little or less ceremony as you would use in discharging the aforesaid worthy personage, for a very masterly critique on Sir Thomas Lawrence, whom Mr P., as visiting

[23] Howe, 146–7

[24] *MC*, 5 May 1814. Lawrence's biographer thought otherwise: 'The portrait, it must be confessed, conveyed no idea of the figure and carriage, and very little of the face of Lord Castlereagh.' D. E. Williams, *Life and Correspondence of Sir Thomas Lawrence* (1831), i. 289 (by a curious error Williams ascribed the notice to Finnerty).

[25] Sadler, i. 429 (7 May 1814). Robinson Diary continues, 'Lawrence's Lord Castlereagh is objected to as mean and haberdasher-like by Hazlitt and Flaxman . . .'

and being painted by him chose to have praised.'[26] Hazlitt seems never to have suspected any connection with the Academy notice. It appeared on 3 May. His last avowed contribution to the *Morning Chronicle* was a theatrical criticism on 27 May, but he continued to report in Parliament until 6 June, and perhaps even until 8 or 13 July.[27] Perry evidently held his tongue until Hazlitt's contract terminated with the end of the session, and then refused to renew it.

The pretext Perry seized upon to allege incompetence (without any specific charge) related to his reporting, which cannot have been other than skilful. What was behind the allegation was probably drink. There is a hint of this in the already quoted reminiscence of a friend after Hazlitt's death ('he preferred his wine with a few friends'), as well as in his grandson's *Memoirs*.[28] The complaint may have been justified at this time, when Hazlitt, disappointed in love and dismayed by Napoleon's fall, probably had recourse to the gin-shops just as he did later on, after Waterloo. The charitable Crabb Robinson, who thought Perry vulgar and overbearing, continues the story (which he had from the mortified Hazlitt the following autumn) thus:

Perry said to [Hazlitt] expressly that he wished [him] to *look out for another situation* — the affronting language Hazlitt could not easily forget — as he was not fit for a reporter. Hazlitt said he thought he could do miscellaneous things. Perry said he would think of it. However, when Hazlitt afterwards went to the office for his salary he was told Mr. Perry wished to speak with him. He went to Perry's room. Perry was not alone, and desired Hazlitt to wait. Hazlitt went to another room. Perry then seeing Hazlitt there, went out of the house. Hazlitt, in consequence, never called again.[29]

Perry's promise to 'think of it' reappears, differently phrased, in a briefer account of the interview in an essay on the reasons for success and failure in life, published by Hazlitt six years later, where yet another explanation of his dismissal is brought forward:

A writer whom I know very well . . . having written upwards of sixty columns of original matter on politics, criticism, belles-lettres, and *virtù* in a respectable Morning Paper, in a single half-year, was, at the end of that period, on applying for a renewal of his engagement, told by the Editor 'he might give in a specimen of what he could do!' One would think sixty columns of the *********** *********** are a sufficient specimen of what a man can do. But while this person was thinking of his next answer to Vetus, or account of Mr. Kean's performance of Hamlet, he had neglected 'to point the toe,' to hold up his head

---

[26] Mitford, Letters, iii. 356. L'Estrange, ii. 48, interpolated the word 'damaging'.

[27] The 'Christabel' notebook (BL R.P. 580(3)), which Hazlitt used in the Gallery, could not otherwise have included, as it does, reports of D. Browne, MP, who spoke only on 6 June, and 8 and 13 July, 1814. The notes do not make clear which of these dates are referred to.

[28] *Memoirs*, i. 195–6.

[29] Morley, 153–4.

higher than usual (having acquired a habit of poring over books when young), and to get a new velvet collar to an old-fashioned great coat . . . This unprofitable servant of the press found no difference in himself before or after he became known to the readers of the ******* *********, and it accordingly made no difference in his appearance or pretensions.[30]

Compared to his earlier bitterness this paragraph in the *London Magazine* is innocuous, but Perry was not at all pleased. The editor of the *London*, writing to Perry soon after, apologized for the reference, and Perry replied with self-betraying hauteur that such personalities '. . . never gave [him] a moment's uneasiness. That in the *London Magazine*, [he] knew, came from a person who knew [him] only by the favours [he] had conferred on him.'[31] Poor Perry! he never dreamt (nor evidently did his employee) that, as time would tell, any favours conferred had been conferred not by him but by Hazlitt. By the year of this rebuke, however, he was old and ill. If he had lived a little longer he might have discovered some satisfaction in Hazlitt's warm portrait of him in a later essay in *Table Talk*, 'On Patronage and Puffing'. Many people disliked him as much as Crabb Robinson did, but Hazlitt penetrated beneath the rough exterior, and saw that the surface brusqueness and the occasional irascibility concealed a lively generosity.

There was one thing however that Hazlitt may not have perceived, and that was the connection between Perry and Holland House. In the *Table-Talk* portrait he is comically repentant of the embarrassment he sometimes caused his editor: 'Poor Perry! what bitter complaints he used to make, that by *running-a-muck* at lords and Scotchmen I should not leave him a place to dine out at! The expression on his face at these moments, as if he should shortly be without a friend in the world, was truly pitiable.'[32] Who one of the Lords was we have already seen; among the 'Scotchmen' were Thomas Campbell and Sir James Mackintosh, in an Illustration of Vetus, 3 January 1814, and at the end of 'Why the Arts are not Progressive', 15 January, where he poked reckless fun at his *bêtes noires*, including, for good measure, as well as the two Scots, Samuel Rogers, Benjamin West, Fanny Burney, Miss Edgeworth, Mme de Staël, and the *Edinburgh Review*.[33] Mackintosh did not complain direct to the editor but within a few days of the second article he mentioned the 'slighting paragraphs' in a letter to John Allen at Holland House, asking

---

[30] *LM*, June 1820, 651–2; 12. 204–5 (*The Plain Speaker*, 1826, where the asterisks were replaced by the name *Morning Chronicle*).

[31] NLS MS 1706, fo. 114, letter of 10 Feb. 1821.   [32] 8. 292.

[33] 7. 62, where Mackintosh's Lincoln's Inn lectures are characterized, like Dr Parr's prose, as 'very tolerable, dull, commonplace declamation, a little bordering on fustian'. The second is given in 18. 10 (Hazlitt omitted the disobliging last sentence when he incorporated it in *The Round Table* (4. 163)).

him to find out if they were intentional on Perry's part, although he rather thought Perry might not have observed their import.[34]

Mary Russell Mitford tells a story of how Hazlitt later, as she supposed, 'got his revenge'. In the winter of 1817–18, when *Characters of Shakespear's Plays* and his lectures on the English poets had brought him into some prominence, Perry thought to take him up again, and invited him to dinner. Miss Mitford was not present; it sounds as though she is repeating an exasperated recital by Perry:

Hazlitt's revenge was exceedingly characteristic. Last winter when his *Characters of Shakespear* and his lectures had brought him into fashion, Mr Perry remembered him as an old acquaintance and asked him to dinner, and a large party to meet him, to hear him talk, and to show him off as the lion of the day. The lion came — smiled and bowed — handed Miss Bently to the dining-room, — asked Miss Perry to take wine — said once 'Yes' and twice 'No' — and never uttered another word the whole evening. The most provoking part of this scene was, that he was gracious and polite past all expression — a perfect pattern of mute elegance — a silent Lord Chesterfield; and his unlucky host had the misfortune to be very thoroughly enraged without anything to complain of. Even Champaign failed to open his lips — Not having been there I admire this piece of malice very much — If I had been present perhaps my opinion might have altered.[35]

His behaviour was certainly characteristic, but Miss Mitford and Perry were wrong in believing it to be actuated by revenge. Such deviousness, which would seem a natural if irritating tactic to the gregarious Perry, acquainted with urbane insolence, would never have occurred to the solitary Hazlitt; if he had wanted revenge he would simply have refused the invitation. He genuinely liked Perry, else he would never have gone near him on this occasion. It was an effort for Hazlitt even to enter a drawing-room. He never shone in strange company; he was uneasy in grand surroundings; he hated to be lionized.[36] If his behaviour was constrained, it was by the need to carry off the situation without discredit, not from the perverse curbing of an ambition to hold the table in thrall. He bore Perry no grudge. We find him on 23 February 1818

---

[34] BL Add. MS 52182, letters of 18 and 19 Jan. 1814 to John Allen (they may be quoted as illustrating the journalistic vendettas of the age): in the first (fo. 53) he says, 'I never shall directly or indirectly do so foolish a thing as deprecate the hostility of a newspaper . . . but for both personal and political reasons the slight of the *Morning Chronicle* surprised me . . . [it] might . . . have come from my enemy in the *Examiner*'; and in the second (fo. 55) he is curious 'to ascertain the writer of the slighting paragraphs — because I think it might lead to a discovery of the critic in the *Examiner* who is, I believe, one of two bad writers on either of whom it would be easy and very proper to inflict condign punishment'.

[35] Mitford, Letters, iii. 356. Howe, who did not see the last sentence, which is not in L'Estrange, believed Miss Mitford to have been present.

[36] See 17. 96.

writing with sufficient friendliness to thank Perry for a favourable notice of his previous week's lecture on Swift, which declared 'We know indeed of no author of the day who excels Mr Hazlitt in critical discernment'.[37] (Coleridge also was lecturing at this time.) The tone of Hazlitt's response (a response he was not obliged to make) may be judged from the only part of the letter available: speaking of his lecture of 24 February, he says, 'That tomorrow will be on Burns, but I am afraid "A very lame and impotent conclusion".'[38] The letter may well have been the occasion of the invitation. In any case it is probable that Perry had already, several months before the lectures and the *Characters*, held out the olive-branch: on 14 November 1816 the *Chronicle* praised the *continuation* of the Holcroft memoirs as 'executed . . . with great feeling and judgment' and 'interspersed with many just and enlightened remarks on Holcroft's principal productions'.[39] The link between the *Chronicle* and Holcroft at the Treason Trials makes it certain that Perry was aware of this notice.

However, this imperfect reconciliation lay in the future. For the moment he was determined to resent the shabby treatment he thought he had received and for which he could see no justification. No one of Hazlitt's susceptibility, with his love of independence and horror of self-interest, would have attempted to hang on. He did not call at the *Chronicle* office again for many years, not until his friend Black became editor, and by then, with his usual capacity to see both sides even when his interest was directly involved, he had probably concluded that his touchiness had magnified Perry's insolence.

Fortunately at that time, when he had quarrelled with his closest friend, was on the brink of losing his job, and, in the wider arena of the European conflict, witnessed the overthrow of his hard-pressed hero, one of the few persons with whom he could talk politics in the assurance of an attentive hearing if not of entire agreement, was also a man in a position to give him work. He had certainly met Leigh Hunt before now, and probably at Lamb's. In a London of just over a million inhabitants, writers and journalists lived very much in each other's pockets, and the absence of a particular name from a writer's biography (especially one as plagued with gaps as Hazlitt's) need not signify an absence of relations. Haydon, like Lamb, had known Hunt since 1808 (he thought him 'the purest and most virtuous character, witty, funny, amusing and

[37] *MC*, 23 Feb. 1818.

[38] Sold at Sotheby's 28 Nov. 1913, it does not figure in *Letters*. See the writer's 'Some New Hazlitt Letters', *NQ*, July–Aug. 1977, 339.

[39] Notice not recorded in Houck. Perry was an early friend of Holcroft's, and there is so much about him in the latter's *Memoirs* that we may not unreasonably suppose he wrote the notice himself.

enlivening — an opinion that like many other opinions he later modified); Haydon himself was by way of being the *Examiner*'s golden boy of British art; and in a diary entry of January 1813 he speaks of daily arguments with Hazlitt, with Leigh Hunt, and with Wilkie — it is unlikely that their visits to Haydon's studio never coincided. Hazlitt's strangeness, in Hunt's description of his arrival at Horsemonger Lane Gaol, was simply his usual shyness: it was his first visit to Hunt, and probably to a prison, and it was Hunt who, delighted with his own singularity and wallpaper, unwittingly gives the impression that they were strangers:

. . . William Hazlitt, who there first did me the honour of a visit, would stand interchanging amenities at the threshold, which I had great difficulty in making him pass. I know not which kept his hat off with the greater pertinacity of deference, I to the diffident cutter-up of Tory dukes and kings, or he to the amazing prisoner and invalid who issued out of a bower of roses.[40]

Hunt's friends flocked to visit him in gaol. 'The Lambs came . . . in all weathers, hail or sunshine, in daylight and in darkness, even in the dreadful frost and snow at the beginning of 1814.'[41] We may assume that Hazlitt called oftener than on the two occasions we know of (the second was when he and Thomas Barnes and Haydon dined at the gaol[42]), and also that it was there that he met John Scott.

The work Hunt gave him on the *Examiner* enabled him, with some help from the *Champion* and the *Edinburgh Review*, to keep afloat for the next three years until the tide carried him into harbour with the publication of his first book of essays and his first work of criticism (not that the Tory journals allowed him to remain there long). It gave some regularity to his income although it was hardly a princely living and no doubt compared ill with what he had earned on the *Chronicle*: the *Examiner*'s circulation was small[43] and the Hunts' finances perilous, but it ensured his survival.

He certainly could not afford the great attraction of the day, which was a trip to France. No sooner had Napoleon fallen than curious English tourists thronged the Channel packets, as at the Peace of Amiens twelve years earlier when Hazlitt had joined them with his sights on the Louvre. Thelwall arrived there towards the end of summer, and also the ubiquitous Crabb Robinson; both of them in time to visit the studio of David, painter of the celebrated *Oath of the Horatii*, just before he was exiled by the Bourbons.[44] What would Hazlitt not now have given to see

[40] Hunt, 246. [41] Ibid. [42] Pope, iii. 640–1.

[43] The relative circulation of these journals and of *The Times* and the *Champion* may be estimated from the following statements of Stamp Duty paid, PRO, A.O.3, 1004 (year 1814): *MC*, £833; *The Times* (already outstripping the *MC*), £1,301 8s. 0d.; *Champion*, £9 9s. 0d.; *Examiner*, £0.15.0. The first two sums are for dailies, and should therefore be divided by six.)

[44] Robinson Diary, 3 Nov. 1814.

once more the objects of his daily veneration at that time, the pieces he had of late so passionately defended against threatened destruction, the *Transfiguration*, Titian's *Man in Black*, the two *St. Jeromes*, the *Marriage Feast at Cana*. He could have made shift to shut his eyes to the marks of the Restoration in the streets on the way to the Louvre, although it would have been difficult not to look for signs of the degradation of the people themselves, the French whom he furiously accused of levity and irresponsibility in deserting Napoleon, remarking with savage hyperbole, 'they are the only people who are vain even of being cuckolded and being conquered'.[45] The two bachelors Haydon and Wilkie set off together in high glee on 25 May; they had money in their pockets (Haydon fleetingly from his recent exhibition of his *Judgement of Solomon*, Wilkie more solidly from thrift); whereas Hazlitt was held fast in town: he had a wife and child to support, rent to find, and no regular income.

He had just criticized Haydon's picture, exhibited from 2 May at the Water Colour Society's rooms, Spring Gardens, and the artist, although pleased, rightly suspected that the critic had not spoken his whole mind. Hazlitt would not have set foot in Haydon's studio (any more than later in Perry's drawing-room) if he had not thought well of him, and he was as anxious to be able to praise his work as he would be, within a few months, to praise Wordsworth's new poem *The Excursion*, but in neither case would he dream of unduly heightening praise or softening blame. He recognized in Haydon the passionate and stubborn nature he had half-subdued in himself (as well as his brother John's wayward obstinacy to choose unsuitable subjects); he respected the courageous if wrong headed integrity that sustained Haydon against the established league of academicians, patrons, and connoisseurs; but he harboured no illusions. Haydon, he said, was born to be bos'n of a man-of-war, and he perceived in the violence of his character a clue to his uneven performance as a painter. Had he got a sight of Haydon's diary he would have recognized a further analogue in the furious torrent of its prose. Brilliantly effective though such ebullience and abruptness were in that chronicle of a headlong life, they were fatal to the tone and proportion of a painting. Haydon was the type of the excessive romantic, erratic, unsteady, ill-governed, He lacked the strong, unvarying, unflagging power of his French Romantic counterpart Delacroix.

So Hazlitt remained behind in London, deprived of his regular salary, having to rely on what he could make as a freelance. There was no opening in dramatic criticism in the *Examiner*. Formerly the prerogative

[45] *Examiner*, 19 June 1814 (4. 30), an accusation unconsciously echoed a century later by a great French writer, who at the fall of France in 1940 thus apostrophized his fellow-countrymen: 'O incurablement léger peuple de France!' (André Gide, *Pages de journal* (Paris, 1944), at 21 mai 1940).

of the editor, it was taken over, while he was in gaol, by Thomas Barnes, a contemporary of his at Christ's Hospital, a Cambridge graduate and student of the Inner Temple, who was then moving over from law to journalism. Already dramatic critic and parliamentary reporter on *The Times*, the energetic Barnes not only attended the theatres for the *Examiner* but also contributed a series of 'Parliamentary Criticisms', as well as furnishing the *Champion* with a series of very competent 'Literary Portraits' over the signature 'Strada'.[46] Nor was there any room in art criticism: Robert, the third of the Hunt brothers, looked after the exhibitions. In the *Champion* this field was unoccupied but seemed about to fall to the anonymous author of two articles on the fine arts, of 17 and 24 April. However, Scott, perhaps prompted by Haydon, had read Hazlitt's 'Fragments on Art' and his *Chronicle* notices. On 26 June he published an attack by Hazlitt on Benjamin West's very successful *Christ Rejected* in which the critic's impatience with the President of the Royal Academy's insinuation of self-praise in a descriptive catalogue dropsical with pretentious prose was nicely balanced with indignation that so eminent a painter should produce a picture so wooden, mechanical, and utterly devoid of gusto (the intense feeling Hazlitt deemed essential to great art). The following week Crabb Robinson, visiting Flaxman, West's colleague, Professor of Sculpture at the Academy, an admirer of Hazlitt whose wife (and perhaps himself also) had attended the philosophy lectures of 1812,[47] read out to him this 'bitter and severe but most excellent performance' and we sense Flaxman's uneasiness at this plain speaking when we read that he was 'constrained to admit the high talent of the criticism though he was unaffectedly pained by its severity.'[48] Eager as always to impart his delight in Hazlitt's prose, Robinson some months later read it to Mrs Barbauld.[49] He constantly read Hazlitt's articles aloud to anyone who would listen.

The enigma of genius preoccupied Hazlitt more and more. He was always obedient to that essentially Romantic instinct defined a hundred years later by Rémy de Gourmont as the fundamental urge in the man of genius to convert his personal impressions into universal laws, and about this time he reflected this view in a striking formula: 'Every man, in reasoning on the faculties of human nature, describes the process of his own mind.'[50] He saw in this very preoccupation with genius the peculiar bent of the age he lived in, the bent towards criticism rather than creation. Was genius originality, or feeling, or imagination, or energy? He pondered specifically the relation between the artificial and the

[46] D. Hudson, *Thomas Barnes of The Times* (1944), 16, 18.   [47] Morley, 63.
[48] Ibid. 146.   [49] Ibid. 152.
[50] 18. 53. Also at this time he said, 'It has been received as a maxim that no painter can succeed in giving an expression which is totally foreign to his own character' (18. 86).

natural, and found himself inclining more than ever to justify his own failure as a painter by the theory of decadence: the great artists were those who came early on the scene, in the first ages; spontaneous, natural, comprehensive, they disregarded their narrow ego, and looked abroad to the life about them; they hardly knew they were artists, so utterly were they men. By the beginning of the nineteenth century there was no longer much left that the creative mind had not already encompassed, and the spirit of the age was predominantly critical. He saw the decline of creativity as balanced by a growth of artificiality, and of this he saw a symptom in the increasingly self-conscious aspiration to fame. He had turned over the idea in his mind for years. It furnished forth his first paper for the *Examiner*, 'On Posthumous Fame — Whether Shakespeare was Influenced by a Love of it', which, although (like some later essays) it probably originated in a discussion at Lamb's 'evenings', was almost certainly rooted in a consciousness of his own ambitions, both past and present. In an article in the *Champion* later that summer, 'Fine arts. Whether they are Promoted by Academies and Public Institutions', the theme is again tied to his own failure, and tied yet more closely. 'The arts have in general', he said, 'risen rapidly from their first obscure dawn to their meridian height and greatest lustre, and have no sooner reached this proud eminence than they have as rapidly hastened to decay and desolation.' Academies are powerless to halt or hasten their growth or their decline, and the formal study of models cannot nourish and often stultifies the individual talent. And he proceeds to a mercilessly ironical dissection of the career of the failed artist, that is, of his own decline (as he saw it) from the Orléans Gallery revelation of the spring of 1799 to the paper he is now writing for the *Champion* in 1814:

. . . from having his imagination habitually raised to an over-strained standard of refinement, by the sight of the most exquisite examples in art, he becomes impatient and dissatisfied with his own attempts, determines to reach the same perfection all at once, or throws down his pencil in despair. Thus the young enthusiast, whose genius and energy were to rival the great Masters of antiquity, or create a new aera in the art itself, baffled in his first sanguine expectations, reposes in indolence on what others have done: wonders how such perfection could have been achieved, — grows familiar with the minutest peculiarities of the different schools, — flutters between the splendour of Rubens and the grace of Raphael, finds it easier to copy pictures than to paint them, and easier to *see* than to copy them, takes infinite pains to gain admission to all the great collections, lounges from one auction room to another, and writes newspaper criticisms on the Fine Arts.[51]

During this bad period when he was most in need of the company of

---

[51] 18. 42.

the estranged Lamb, he often visited Thomas Alsager, a businessman almost exactly his contemporary who owned a bleaching factory, but whose interests were literature and music (he did much for the reputation of Beethoven in England). A financial wizard, he became the first, and very influential, City Correspondent of *The Times*. He was sociable, and like Lamb, fond of whist parties; his warmth and enthusiasm must have been very agreeable to Hazlitt. His house in Great Suffolk Street, Southwark, was a stone's throw from the County Gaol in Horsemonger Lane, and Hazlitt probably met him on a visit to Hunt, to whom Alsager had recently been introduced by Barnes, and to whom he proved a good friend; it was he who alleviated the chill of Hunt's arrival in his prison cell by sending over his first dinner.

Lamb himself at this time was enjoying the celebrations of the first summer of peace, smoking, and drinking bottled stout with the returned soldiers at the hastily erected booths in Hyde Park, watching mimic sea-fights on the Serpentine (the 'Naumachia' derided by the *Morning Chronicle*)[52] and firework displays among the scores of towers, alcoves, and drinking-stalls which had invaded the park by the authority of the Prince Regent, while the grass, on the other hand, had retreated and then disappeared ('not a vestige or hint of [its] ever having grown there', said the volatile Lamb[53]), and the open spaces in the semi-urbanized half-ruined park had become a sandy desert. Others, including we may be sure Hazlitt, were less tolerant than Lamb. The *Examiner* stigmatized the park as 'the headquarters of drunkenness and obscenity'; to *The Times* it was 'a convention of all the dancing-dogs, gingerbread stalls, slack-rope vaulters and fire-eaters in the kingdom'; and the *Champion* complained more soberly of these 'expensive fooleries'.[54] Other branches of the Grand Jubilee, or 'The Regent's Fair', as Hazlitt called it, were opened in Green Park and also in St James's Park beneath his windows. Thousands of sightseers invaded the three parks from an early hour on 1 August, the official day of celebration, and that night the thronged walks of St James's Park were illuminated by a blaze of coloured lights and Chinese lanterns, and finally, after midnight, by a firework display that glared through the great curtainless windows of his living-room. There was no escape from the public rejoicing. It seemed a symbolic climax to this feverish agitation when the Chinese Pagoda erected on a bridge over the canal accidentally caught fire at one in the morning and fell in flames into the water.[55]

In the preceding weeks the town was kept on the tiptoe of excitement by the arrival of the Emperor Alexander of Russia and the King of

[52] 19, 22 July.   [53] Marrs, iii. 95–7.
[54] *The Times*, 2 Aug. 1814; the *Examiner* and the *Champion*, 7 Aug. 1814, 504 and 249–50, respectively.
                                                            [55] *MC*, 2, 3 Aug. 1814.

Prussia at the Pulteney Hotel in Piccadilly. The autocratic Emperor, Tsar of all the enslaved Russias, prided himself with no sense of contradiction upon his liberal ideas; the freedom of his behaviour must have startled, as much as it puzzled, Carlton House (where there was also resentment at the contempt he barely concealed for the gorgeously uniformed, fat, and stay-at-home Regent). He summoned Erskine to his hotel and complimented him on his speech in Hardy's defence, which he said he had read at the time of the Treason Trials.[56] He also called on Jeremy Bentham and presented him with a gold snuff-box. Hazlitt, despite the Emperor's role in the Alliance that had overthrown his idol, was at first not disinclined to give credence to the genuineness of his pretensions to liberalism (he was clutching at straws). He cannot have ignored the convulsion York Street was thrown into by the Tsar's visit, by the startling incursion of the royal pageantry of Cossack outriders, gilded coach, and liveried footmen. He may actually have witnessed the arrival of the autocrat who was later to play so considerable a part in his *Life of Napoleon*, although he would not have gone out of his way to see it: we cannot imagine him accepting, like Miss Mitford and his friend Ayrton's wife,[57] the hospitality of Perry — even if he and the editor had still been on good terms — to view from the *Chronicle* office-windows the royal procession (the Emperor Alexander, his sister the Grand Duchess Catherine, the King of Prussia, and the Regent — who was hissed) making its way along the crowded Strand to the Guildhall on Saturday, 18 June.

We know he was still on good terms with John Black. Among those fêted at this time was the Russian general Platoff, more popular with the London mob than even his master. This enables us to fix at a date not long after these occurrences a visit (doubtless one of many) that Hazlitt paid to Black's Thames-side house at Millbank when his small son fell into the water and was fished out by his host's enormous dog, topically named Platoff.[58]

It is, as we have seen, uncertain when his engagement as reporter terminated, but we know that he attended debates in June, when Castlereagh, back from Paris, presenting the Peace Treaty to the House, defended against the scandalized assaults of Wilberforce, Horner, and others the shameful clause empowering the French to continue the slave trade. That defence inspired his *Examiner* paper of 10 July[59] where, in another fencing display of verbal dexterity, he punctured with a dozen flashing thrusts its inconsistencies and absurdities: he imagines Talley-

[56] Broughton, i. 144.
[57] Mitford, Letters, ii. 231; BL Add. MS 60372, Mrs Ayrton's diary.
[58] *Memoirs*, i.213 (young William was born 26 Sept. 1811).
[59] Howe, mistakenly, has 3 July.

rand (Prince Maurice) giving instructions to his parrot (Castlereagh) in a devastatingly ironical catechism of which the beautifully balanced item 5 reads, 'That the French Government simply wished to begin the Slave Trade again as the easiest way of leaving it off, that so they might combine the experiment of its gradual restoration with that of its gradual abolition, and, by giving the people an interest in it, more effectually wean their affections from it.' A severer and more direct item was No. 8, 'That we are not to teach the French people religion and morality at the point of the sword, though this is what we have been professing to teach them for the last two and twenty years.' He concludes with another thrust at Croker, advising Castlereagh, who in his remoteness had always reminded him of a fine 'taffeta lining to a court dress', to 'leave the buckram of office to his friend the secretary of the Admiralty'.[60]

Attentive as he was to the 'grand gaol-delivery of princes and potentates' on the Continent, to art, and to the theatre, something else occupied his mind that summer. A love-affair, perhaps the same that had stirred his heart with hope and joy the previous autumn, was turning out badly. The circumstances are obscure. All we know is the little that is hinted in shadowy nooks and corners of his work. Earlier, as we have seen, and evidently in a moment of lucid depression, he spoke (in words his friend Black might have echoed) of bookish men as being helpless against the wiles of women. And later, looking back on the defeat of Napoleon, he mourned that period of his life as one branded by the final overthrow of his public and private hopes. But at the present moment, when it seems that his love-affair is collapsing in ruin, he is in an agony of wretchedness, and speaks more harshly than ever, striking out (in unconscious conformity to Gourmont's law) brusque and broad generalizations from his individual case. Of all Shakespeare's plays *Othello* was one of the three most often in his thoughts, and at this time it could hardly ever have been absent from them. He had seen it again at Drury Lane in early July. On the 24th and on 7 August he contributed to the *Examiner* a two-part paper on Kean's interpretation of Iago, and to this he appended a savage note on Desdemona which, said his editor, startled some readers:

If Desdemona really 'saw her husband's visage in his mind', or fell in love with the abstract idea of 'his virtues and his valiant parts,' she was the only woman on record, either before or since, who ever did so. Shakespeare's want of penetration in supposing that those are the sort of things that gain the affections, might perhaps have drawn a smile from the ladies, if honest Iago had not checked it by suggesting a different explanation. It should seem by this,

---

[60] Scott was so impressed that he reprinted the article as a pendant to a piece on the slave trade in the *Champion*, 18 Sept. 1814.

as if the rankness and gross impropriety of the personal connection, the difference in age, features, colour, constitution, instead of being the obstacle, had been the motive of the refinement of her choice, and had, by beginning at the wrong end, subdued her to the amiable qualities of her lord. Iago is indeed a most learned and irrefragable doctor on the subject of love, which he defines to be 'merely a lust of the blood, and a permission of the will.' The idea that love has its source in moral or intellectual excellence, in good nature or good sense, or has any connection with sentiment or refinement of any kind, is one of those preposterous and wilful errors, which ought to be extirpated for the sake of those few persons who alone are likely to suffer by it, whose romantic generosity and delicacy ought not to be sacrificed to the baseness of their nature, but who, treading secure the flowery path, marked out for them by poets and moralists, the licensed artificers of fraud and lies, are dashed to pieces down the precipice, and perish without help.[61]

This looks like a reply to a long letter of two and a half columns in the *Examiner* of 15 May sent in by one P. G. P[atmore], an assiduous playgoer whom we remember from the O.P. riots and who was now apparently attempting to break into journalism. He had already in March got one and a half columns of detailed interpretation of Kean's Hamlet accepted, and he now turned to *Othello*.[62] He praised Desdemona: 'Shakespeare delighted to honour the female character, and how exquisitely has he done so in Desdemona [who] loves the Moor in spite of his personal defects . . . *she* saw his beauty in his goodness, *she* saw Othello's visage in his mind".' The quotation is the point of departure of Hazlitt's supererogatory note. Perhaps 'P.G.P.' took up the challenge, but if so, Hunt did not let him be heard, judging perhaps that there were already enough gladiators in the field: there were further handles to controversy both in the note and in the body of the article, one of which was seized by Hunt himself in vindication of courtly love, and another by Barnes (who, as the regular dramatic critic, may have been piqued by this intrusion on his province) in a defence of his own interpretation of Kean's Iago.[63]

The article's main interest, however, is in this note, and the dim light it casts on his unhappy love-affair. A further fitful glimmer is shed six months later in an essay where he tells the story of the sixteenth-century Italian philologist who went out of his mind when his books and papers were accidentally burnt:

Almost every one may here read the story of his own life. There is scarcely a moment in which we are not in some degree guilty of the same kind of

---

[61] 20. 401 (Howe has, in mistake, 'treading securely').

[62] *Examiner*, 15 May 1814, 317–19; 27 Mar. 1814, 205–6. P.G.P. 's two letters have hitherto escaped notice.

[63] Ibid., 14 Aug. 1814, 525–6; 4 Sept. 1814, 571–2.

absurdity, which was here carried to such a singular excess. We waste our regrets on what cannot be recalled, or fix our desires on what we know cannot be attained. Every hour is the slave of the last; and we are seldom masters either of our thoughts or of our actions. We are the creatures of imagination, passion, and self-will, more than of reason or even of self-interest. Rousseau, in his *Emilius*, proposed to educate a perfectly reasonable man, who was to have passions and affections like other men, but with an absolute control over them. He was to love and to be wise. This is a contradiction in terms.[64]

It is impossible to plumb the depths of self-disgust hinted at in the Desdemona note but it is easy to guess that it was inspired by some harrowing scene that had recently filled him with shame. As embittered by his own absurd 'romantic generosity and delicacy' as he was appalled by the 'baseness of his nature', he spilled it all out — a catharsis — in the columns of the *Examiner*, much as he was later to do in the *New Monthly* when he was in love with Sarah Walker.[65]

Cursing his folly, and no doubt resolving amendment while at the same time conscious of the futility of the resolve, he turned his attention in an unlucky moment to a subject which offered full scope to his disinterestedness. Hearing that Wordsworth's long-awaited *Excursion* was about to appear and that the Lambs had received a copy, he thought of a way to review this very expensive volume for the *Examiner* by sending Martin Burney to borrow it. Wordsworth claimed to have so poor an opinion of the *Examiner* that he had no wish for a review in it, and he was fortunate that someone on that paper was prepared to go to such embarrassing lengths to lay hold of it. Of the two poet-heroes of Hazlitt's youth he was persuaded that Wordsworth owed to Coleridge an immense and unacknowledged debt,[66] but he saw little chance of Coleridge's ever fulfilling the promise that had gleamed about his path during those radiant months of 1798 in Shropshire, Llangollen, and Somerset, and he continued to hope instead for some kind of achievement from the other, whose work at least was continuing to appear. It is sad that he died without seeing (and without writing about) the *Prelude*. He was perhaps over-eager to admire the *Excursion* from his early recollection of Wordsworth reciting parts of it in the Lakes in 1803, so that his disappointment on reading it now in its entirety was bitter. Crabb Robinson reports, sceptically, Lamb's statement that he cried 'because . . . he could not praise it as it deserved',[67] and yet it was just the kind of frustration, complicated by a clash of loyalties, that would bring tears of vexation to the eyes of this passionate but discriminating man; he

---

[64] Ibid., 26 Feb. 1815; 20. 43.

[65] Feb. 1822, where, by a seeming fatality of association, he also repeats the story of the Italian philologist (8. 237–8); see also S. C. Wilcox, *Hazlitt in the Workshop* (Baltimore, 1943), 17–18.

[66] Morley, 179 n.

[67] Ibid. 202.

detested Wordsworth's politics but did not see that as a reason to detest the poetry of the *Excursion*, which indeed he remembered as admirable. All his life he quoted Wordsworth more than any other contemporary poet; Wordsworth figures more largely than any living writer in his anthology *Select British Poets* of 1824; and he was confident that Wordsworth was the one most likely to survive. His discernment has not met with the credit it deserves. From the serene heights of Wordsworth's present reputation, bathed in the rays of the *Prelude*, it is hard to imagine how ill-esteemed he was in those days, and how, even when his politics might be acceptable, his poems invited derision; and how, in any case, his egotism effectively turned the scales against his popularity. De Quincey in the 1856 edition of the *Confessions of an English Opium Eater* apologized for not having in the earlier 1822 edition named the 'great modern poet' he had had occasion to quote, because at that time it might have provoked an explosion of 'vulgar malice'.[68] His most recent poems, in 1807, had met with almost unanimous disparagement, even contempt. Jeffrey in the already influential *Edinburgh Review* had poured scorn on what he later termed Wordsworth's 'system', 'the lowliness of tone which wavered so prettily . . . between silliness and pathos', and had singled out the 'Ode on the Intimations of Immortality', one of the pinnacles of Romantic literature, as 'the most illegible and unintelligible part of the publication'.[69] Hazlitt's admiration for this poem was profound, and its imagery so insinuated itself into the texture of his thought that he was hardly conscious of quoting it.[70]

Leigh Hunt concurred with Hazlitt's estimate, declaring Wordsworth in this same month of August 1814 to be 'the greatest poet of the present time',[71] but while he recognized his 'real genius' he also felt 'indignation at his puerile abuse of it'. Hazlitt saw deeper. He was not offended by occasional, incidental puerilities. The liberal, imaginative, sympathetic principle of criticism to which he adhered took a broad view that spurned the niggling pedantry of such objections, but it was this very principle that was bound to encounter a stop in the *Excursion*. He was disappointed in the self-centred exclusive spirit of the poem, which instead of looking abroad into humanity turned its back on it. The flaw stood revealed in Wordsworth's pretension to dramatic form when he had no sense of dramatic interest to sustain it. Hazlitt's own inclination, manifested consciously or unconsciously in the literary controversies he contrived to stir up, was, as we have seen, towards the dramatic form as being the most faithful reflection on the essential spirit of life. The lyrical

[68] World's Classics edn. (Oxford, 1902), 257 n.
[69] *ER* 11, no. 21 (1807–8), 216, 217, 218; 24, no. 47(1814), 1; 11, no. 21 (1807–8), 227.
[70] See E. W. Bratton, 'Unidentified Wordsworthian Echoes in Hazlitt', *NQ*, Jan. l968, 25–7.
[71] *Examiner*, 14 Aug. 1814, 525.

mode, central in the romanticism of the age, corresponded to a vital part of life, but, however beautiful and alluring, it was incomplete because one-sided. He felt that the lyric poet's implicit claim to be the sole mouthpiece of life was a denial of life. A wider sense of humanity, no less than the prompting of modesty, impelled him to give his allegiance to the more natural (and more classical) epochs of the past when men lived not exclusively to themselves but also and at the same time with their fellow-men, lived in society as well as in solitude (in London as well as in Grasmere or Winterslow), notably the Elizabethans and the Augustans. If the poet really was a man talking to men, then he ought to be able, and ought to wish, to record some other response than a self-repeating echo in the mountains.

Despite his disappointment, his account of the poem constitutes a juster appreciation than Jeffrey's notorious 'This will never do' review, or the anonymous writer's in the *British Critic*, or even the poet James Montgomery's in the *Eclectic Review*.[72] Neither his conviction of Wordsworth's genius nor his awareness of Wordsworth's irritable resentment of criticism could induce him to abandon or soften legitimate asperities, nor could his dislike of Wordsworth's unyielding egotism or his dismay at the inroads it had made into his style persuade him to deny the poem its due measure of praise. Nor did he allow the disquieting political implications of Wordsworth's appointment to the stamp distributorship to influence the spirit in which he read the Solitary's account of the French Revolution, although it deepened the discouragement, the pessimism, of his own reflections, which form a peroration to the second part of the review (buttressed — how inevitably — with an unattributed quotation from the 'Ode on the Intimations of Immortality'):

. . . nor can we indulge with him in the fond conclusion afterwards hinted at, that one day *our* triumph, the triumph of virtue and liberty, may be complete. . . . It is a consummation which cannot happen till the nature of things is changed, . . . though we cannot weave over again the airy, unsubstantial dream, which reason and experience have dispelled ——

> 'What though the radiance, which was once so bright,
> Be now for ever taken from our sight,
> Though nothing can bring back the hour
> Of glory in the grass, of splendour in the flower':—

yet we will never cease, nor be prevented from returning on the wings of imagination to that bright dream of our youth; that glad dawn of the day-star of liberty; that spring-time of the world, in which the hopes and expectations of the human race seemed opening in the same gay career with our own; . . . when, to the retired and contemplative student, the prospects of human happiness and glory were seen ascending, like the steps of Jacob's ladder, in

[72] Robinson praised the *Examiner* notice, Morley, 151.

bright and never-ending succession. The dawn of that day was suddenly over-cast; that season of hope is past; it is fled with the other dreams of our youth, which we cannot recal, but has left behind it traces, which are not to be effaced by birth-day odes, or the chaunting of *Te Deums* in all the churches of Christendom. To those hopes eternal regrets are due; to those who maliciously and wilfully blasted them, in the fear that they might be accomplished, we feel no less what we owe — hatred and scorn as lasting.[73]

He was long acquainted with Wordsworth's susceptibility to criticism but probably thought that even he could hardly resent the attributing to Shakespeare and Milton of powers superior to his own.[74] He was wrong. He began the third part of his review with a distinction between the poetry of imagination and the poetry of sentiment, in a passage which he suppressed when reprinting the criticism in the *Round Table*. The greatest poets, he said — Chaucer, Spenser, Shakespeare, and Milton — possessed both kinds, and in the highest degree. Others, like Young and Cowley, turned to fancy, to the neglect of feeling; Wordsworth, on the other hand, 'whose powers of feeling were of the highest order [was] certainly deficient in fanciful invention'. He could hardly have said worse. Wordsworth, that autumn, was preparing a collective edition of his poems, in which for the first time he adopted the classification that has since given his editors so much trouble: Poems of the Fancy, Poems of the Imagination, Poems Founded on the Affections, and so forth. I do not mean that the distinctions were new in Wordsworth's thoughts.[75] I am merely suggesting that Hazlitt's review may have finally impelled him to make them explicit and give them prominence. At the same time, in the 'Essay, Supplementary to the Preface' which he added to the edition and which has embarrassed his biographers, he compared the imagination that informed his poetry with Shakespeare's imagination, and Milton's, going on to say, in a furious paragraph that might as well have named Hazlitt and Jeffrey outright:

. . . Justified by a recollection of the insults which the Ignorant, the Incapable and the Presumptuous have heaped upon these and my other writings, I may be permitted to anticipate the judgment of posterity upon myself, [and] I shall declare . . . that I have given, in these unfavourable times, evidence of exertions of this faculty upon its worthiest objects . . . which have the same ennobling tendency as the productions of men, in this kind, worthy to be holden in undying remembrance.[76]

[73] 19. 17–18.

[74] And so did Robinson (Morley, 166), De Quincey and Mrs Clarkson (ibid. 195–6), and Keats (i. 237).

[75] They are hinted at in Lamb's droll letter of 15 Feb. 1801 to Manning, and so is the obsession with Shakespeare and Milton.

[76] *Poems* (1815), i, p.xxxi. Carlyle noticed that Wordsworth's opinion of Shakespeare and Milton was strongly influenced by Wordsworth's opinion of himself (Carlyle, 360).

Hindsight encourages us to a sympathetic view of this 'Essay', but it shocked Wordsworth's contemporaries and has given discomfort to some of his most devoted admirers. Hazlitt must have had this 'Essay' in mind, as well as his own theory of genius and the degree to which genius is conscious of its own powers, when years later, writing about egotism, he asked what one would think of 'a poet who should publish to the world, or give a broad hint in private, that he conceived himself fully on a par with Homer or Milton or Shakespear?'.[77] In justice we should recall that Wordsworth himself (and Hazlitt could not know this) was uneasy about the preface and the supplementary essay. He consulted Lamb who (somewhat disingenuously) reassured him. Crabb Robinson, a more scrupulous counsellor, warned in vain that 'it betrayed anger', and that he thought it better left unpublished.[78]

Hazlitt's review was in the event unfortunate, not only because it strangely offended Wordsworth, but because it almost certainly prompted Wordsworth to repeat to a man who happened to be with him when he read it, and who later became Hazlitt's enemy, the story of Hazlitt's discomfiture at Keswick in 1803. This was John Wilson, who in 1814 was living at Elleray, above Windermere. No points are to be noted in Wilson's account of the incident.[79] At first Wordsworth was pleased by the review, and it was not until he came to the signature of the reviewer, from whom he could not think of accepting praise, that he became irritated. We are not told why. Whether it was because Hazlitt had omitted to acknowledge the 'great and unmerited obligations' Wordsworth assured Wilson he had conferred upon him, or because he had failed to merit them, is not clear.

In retailing the story in 1818, Hazlitt rejects the imputation of ingratitude. So far from being ungrateful, he said, he had seized the first opportunity of publicly praising Wordsworth's poetry. His account of Wordsworth's reaction to his review was meant to show that nothing could abate Wordsworth's animosity and his determination to believe in Hazlitt's ingratitude and ill will, and perhaps also to convey indirectly to Wordsworth that at last, in 1818, he had realized that it was Wordsworth who had maligned him to John Wilson and others immediately after the *Excursion* review. In 1814, with his usual ingenuousness, he never dreamt that Wordsworth might use that old story to injure him, nor evidently in how sinister a light it could be placed. And indeed the episode had not appeared in any such light either to Wordsworth or Coleridge when it happened in November 1803. In March 1804, just four months later, we find Wordsworth still writing to Hazlitt in the most

---

[77] 12. 158.

[78] Morley, 166.

[79] As given by Hazlitt in *A Reply to Z*, not pub. until 1923 (9. 6).

relaxed and friendly terms (he talks of 'our wish to promote your interest', and concludes, 'I am very affectionately yours').[80] Clearly something occurred between 1804 and 1814, and probably before 1808, to change Wordsworth's attitude to Hazlitt, but what it was remains obscure.[81] It is certain that when Wordsworth launched the Keswick story (as soon as the *Excursion* notice appeared) it was in order to ruin Hazlitt's credit, but certain also that his true motive was not the sole destruction of Hazlitt's reputation but the indirect defence of his own poetry. He naturally and very properly conceived his first duty to be towards his work. His conviction of the sublimity of his verse so overwhelmed him that he could not believe there was anyone properly capable of appreciating it at its true worth, least of all the youngster his weak and erratic friend Coleridge had brought to see him at Alfoxden years before, when *Lyrical Ballads* first came out.

Hazlitt resented having his friendly offices spurned, and his impression of Wordsworth's character (although never of his poetry) thereafter hardened into dislike. 'His egotism', he said three years later, 'is in some respects a madness; for he scorns even the admiration of himself, thinking it a presumption in any one to suppose that he has taste or sense enough to understand him.'[82] Wordsworth's inordinate self-esteem was well known, and was bound to repel so nervous a person as Hazlitt. In the poet's estimation no good man could be blind to the nobility of his verse, even if it was too deep for him; consequently anyone who criticized it must be bad. His condemnation of Hazlitt to John Scott, when indirectly telling him the Keswick story in 1816, was ingenuously based on that simple premise: 'His sensations', he said, 'are too corrupt to allow him to understand my Poetry.'[83] But this outright avowal was exceptional. He could not really persuade himself that because Hazlitt could not understand his poetry he must be a bad man. So up surged, pat, the memory of something unpleasant at Keswick, which, warped by his guilt-ridden and perhaps incest-fearing attitude towards sexual appetite, had grown swollen and discoloured, and now became a formidable and avowable surrogate for the immediate efficient cause (whatever more remote cause may have come into operation around

---

[80] Wordsworth, i.446–7. Southey is supposed to have helped in Hazlitt's escape, and yet there is no reference to this dramatic incident in his published letter to Duppa of 14 Dec. 1803 (Southey, *Corr.* ii. 237–8), nor in unpublished letters of 19 Nov., 5, 15, 17, and 31 Dec. to his brother and to close friends, long and gossipy though they are (BL Add. MS 47890). See also Southey, *New Letters*, i. 335–47.

[81] Douglas Grant (letter in *TLS*, 19 Sept. 1968, 1062) thought Wordsworth was influenced by his brother Christopher whom Hazlitt believed to be author of a review of the *Essay on the Principles of Human Action* insinuating that he was immoral.

[82] 5. 163 (lecture of 3 Mar. 1818). He had said almost the same thing on 24 Feb.: 'He is repelled and driven back into himself, not less by the worth than by the faults of others.' (5. 132).

[83] *TLS*, 27 Dec. 1941, 660.

1806–7) of his dislike of Hazlitt, viz. Hazlitt's supposed incapacity to admire his work and consequent disservice to it. Lamb also reviewed the *Excursion*, and Robinson reported Hazlitt as remarking; 'if Lamb . . . had found but one fault with Wordsworth he would never have forgiven him.' With this even the loyal Robinson was forced to agree; 'And some truth there is in the extravagant statement.'[84]

Wordsworth's attempt to influence Lamb against Hazlitt failed,[85] but with Robinson and John Scott, who unfortunately were more important to Hazlitt's career, he had some success. But his greatest success, unintentional, unforeseen, long-delayed, and probably deplored by him when it emerged into print, was with someone who had never met Hazlitt — John Wilson. The story Wilson heard in 1814 disappeared underground like a train of gunpowder laid by sappers, and did not explode beneath the victim's feet until four years later, but with devastating effect, in the columns of *Blackwood's Magazine* for August 1818.

[84] Morley, 166.
[85] Marrs, iii. 125.

7
The End of Public Hopes

HAZLITT in the mean time knew nothing of this gossip, but the concealed
train of ill omen it started off was balanced and compensated in the last
months of 1814 by a similarly hidden, but this time auspicious, sequence
of events working for his good, which also led towards Edinburgh. His
articles on the inertness of institutions in encouraging the fine arts,
completed in the *Champion* on the same day as the last part of his *Excursion*
review appeared in the *Examiner*, had caught the eye of the energetic,
high-spirited, and intelligent wife of Sir James Mackintosh. She and her
husband knew the editor of the *Champion*, so she immediately wrote to
Scott to enquire about the identity of his contributor. She must have
been intrigued to find that the object of her admiring curiosity was a
journalist whom her husband had suspected earlier in the year of cocking
a snook at him, but relations between her and her husband were
strained, with accusations flying from side to side, and she may have felt
a mischievous satisfaction at being in a position to give the screw another
turn.

At all events the pleasure the articles procured her was in no way
diminished. She corresponded with, perhaps even met, Hazlitt, and
either he hinted at an ambition to contribute to the *Edinburgh Review*, or
more probably she suggested the idea to him. She offered to recommend
him to Jeffrey, who was known to her (an unpublished letter of her
husband's suggests that she may even at some time have submitted
contributions to the *Review* herself).[1] Mackintosh was in Paris, and so
by now was John Scott. Not knowing Scott's address, Catherine
Mackintosh wrote to him care of her husband, who, having no relish
either for the curtness of her tone (further evidence of her 'contempt and
scorn') or for the role of messenger-boy assigned him, was unscrupulous
enough to open the letter, and replied testily on 16 November:

Finding that Mr Scott had returned to England I ventured to open your letter
to him supposing that it might contain something of political and literary news
with which you do not indulge me. I was not disappointed — but I was rather
amused at the effrontery of Mr Hazlitt either the writer or the writer's brother
who abused me in the *Examiner* and who now employs you to procure him
influence and income in the *Edn. Rev.* This would more suit my placability or

---

[1] BL Add. MS 52443, letter of 7 Dec. 1818. She may be the unidentified reviewer of 'Religion
and Character of the Hindus' in Feb. 1818.

forgetfulness or indifference or meanness (call it by what name you please) than
your anger or pride or dignity.[2]

Mackintosh is here more indignant at his wife's 'indifference or
meanness' than he is at Hazlitt's abuse, and he might have been less
resentful even of that if he had recognized in Scott's contributor the
author of the anonymous works that had given him so much pleasure in
India. We do not in fact know what particular abuse he had in mind. He
says it was something in the *Examiner*, but whatever that was it could not
have been Hazlitt's. If he was thinking of the *Morning Chronicle* and 'Why
the Arts are not Progressive', then it was, we remember, a very muted
affair, cloaked in mild and obscure irony, glancing cursorily not at
Mackintosh alone, but at a whole catalogue of (it was obliquely implied)
tedious contemporaries, including however — another gaffe — the
Edinburgh Review.

The next day, 17 November, Crabb Robinson met Hazlitt at Lamb's
(the breach with Lamb was evidently healed) and learned that he had
been 'in a very flattering manner enrolled in the corps' of *Edinburgh*
reviewers, having followed up Lady Mackintosh's recommendation by
sending some of his work to Scotland. We may take it that he prudently
omitted 'Why the Arts are not progressive' and 'Mr Malthus and
the Edinburgh Reviewers'. Jeffrey had certainly seen the second in
Cobbett's *Political Register* in November 1810,[3] but he had either
forgotten these shafts or now brushed them aside in the prospect of
acquiring such a valuable recruit, just as Hazlitt's annoyance at the
Edinburgh's having ignored his *Reply to Malthus* was dissipated in the
brilliant attraction exercised by the most celebrated journal of the time.
Lady Mackintosh's recommendation itself must have carried much
weight with Jeffrey, emanating as it did from the wife of a member of the
Holland House circle, but 'his attitude to Hazlitt was not solely
influenced by the Mackintoshes, it also coincided with his own judgement, and above all with his own temperament. His active, liberal,
forward-looking spirit had a warmth and generosity, and above all an
energy, that quickly overcame and obliterated all minor differences and
squabbles. The phrase 'in a very flattering manner' implies that he
extended his hand to his new contributor no less cordially than
promptly.

'Enrolled in the corps' may mean that Hazlitt had already by
mid-November been set to work. Reviewers were not salaried; and he
cannot have been less painstaking nor taken less long over his first
commission than over his second, to which we know he devoted three

----

[2] BL Add. MS 52441, letter begun 12 Nov. 1814.
[3] See 7. 408–10.

and a half months.[4] True, in the second he had to deal with a four-volume work in French,[5] the reading of which alone was a considerable task even though he had kept up the language since his sojourn in Paris, but neither was the first narrow in scope: he was unwilling to stay confined to its ostensible object, Fanny Burney's latest novel *The Wanderer*, and ranged over the whole field of European fiction. It is certain that he was anxious to please his new editor. If his references to the *Review* before his appointment were unflattering, he seems quickly to have conceived both admiration and respect for Jeffrey once he came to deal directly with him, so that when he spontaneously entitles him ten years later 'the prince of critics and the king of men' his sincerity is beyond doubt.[6] Even without his praise of Jeffrey in *The Spirit of the Age* the ironic final line of his poem 'The Damned Author's Address to his Reviewers' shows that he thought him an impressive writer. There was no irony in his saying that to be an *Edinburgh* reviewer was the highest rank in contemporary letters, an assessment supported by the expression 'influence and income' used by Mackintosh. It was the only British journal of European stature, as is apparent from its momentous impact on Stendhal.[7]

Not that in literary criticism it was particularly sensitive: Jeffrey's irreversibly Augustan taste did not incline him to sympathize with the new wave of Romantic poets. The *Review*'s greatest virtue was its comprehensiveness: it took the whole business and desires of civilized man — the *honnête homme* — for its province, and its scrutiny was as liberal and humane as it was comprehensive. The bird-like briskness and great capacities of 'little Frank Jeffrey' himself maintained a rare equipoise, in him as editor, between a theoretical self-conceit and a practical humility. The *Edinburgh Review* alone mattered, and he and his contributors were its servants. To him, his own fame and the fame of his authors, or indeed, according to Hazlitt, the merit of the works selected for review, were of little account compared to the maintenance of the *Review*'s authority. He was also a lawyer, trained to look at a question from all angles, skilled in pleading and expert in dialectic. Hazlitt's claim that he looked on writers with the eye of a Justice of the Peace sentencing a poacher, was more than a mere *jeu d'esprit*.

This supercilious patronage of the arts and especially of literature was

[4] 'Standard Novels and Romances', *ER* 48 (Feb. 1815, pub. London mid-Apr.).

[5] The first English trans. of Sismondi's *De la littérature du midi de l'Europe* (Paris, 1813) did not appear until 1823; we may assume that Hazlitt made his own (very faithful) translations in his review.

[6] 9. 126.

[7] See Stendhal's letter of 28 Sept. 1816 to Louis Crozet, *Correspondance* (Pléiade edn., Paris, 1962–8), i. 818. There is no record of *Blackwood's* having influenced any European writer of comparable stature.

common to members of all the professions, not to mention people of rank. To be *merely* a writer in those days was to be rather disreputable. Of the principal contributors to the *Edinburgh Review*, Jeffrey was a barrister, Sydney Smith a clergyman, Brougham a lawyer and politician. The want of a profession could only be excused on grounds of noble birth (as Byron and Shelley) or of having gone to one or other of the English universities (as Wordsworth, Coleridge, Southey, De Quincey). The rest — Leigh Hunt, Keats, Hazlitt, and the like — were accounted mere hacks and hangers-on. The distinction was a very real one and largely underlay the violent animosity later shown by the two Edinburgh lawyers and University men, Wilson and Lockhart, towards Hazlitt and Hunt. Not even Jeffrey was entirely unbiased. He was engaging, frank, and relatively free from prejudice and snobbery, but there is a striking difference in his tone towards Hazlitt and towards Moore, both of whom became contributors in the same year (when Jeffrey returned from America to find the *Review* languishing from the neglect of his associates). True, Moore had reputation and Hazlitt was obscure, but we cannot fail to discern the prestige of the relationship enjoyed by Moore (who once said 'poets and authors are the pest of my life') with Lords Holland, Byron, Lansdowne and indeed the whole Holland House circle, influencing Jeffrey's cautious approach to him via Rogers and his mock-humble letters when contact was established. The extraordinary tone of these letters cannot be explained as reparation for the long-since forgotten duel. 'Tell me too, that you will come for a fortnight to Edinburgh early next winter, and see our primitive society here. It is but thirty hours travelling, and will at least be something to laugh at in London, and to describe at Mayfield . . . I can insure you of being very much admired, and you must bear and excuse anything that may be asinine in our courtesies.'[8] He would never have addressed Hazlitt so (nor indeed would Hazlitt ever have addressed anyone at all so). His tone with Hazlitt, though friendly enough, inclined to the patronizing, and on one occasion at least was downright insulting, in a letter beginning, 'I am sorry you ascribe so much importance to the omission of your little paper on Dr Reid's book'.[9]

Not that his editorial shrewdness was in the least affected by all this. Distinctions of rank and birth meant nothing when he sat down at his desk. Then he valiantly laid about him with his blue pencil, and went on to fill up the resultant gaping holes with sentences, with whole paragraphs, with pages even, of his own. No wonder that years later when asked to identify his contributions he claimed articles that have since been proved to belong to others. His obsessive devotion to his

---

[8] Moore, ii. 42, 54. [9] Howe, 230.

journal, coupled with his supreme self-confidence, hyperbolically embalmed in Sydney Smith's neat pastiche of how he would review the
solar systems ('bad light — planets too distant — pestered with comets
— feeble contrivance — could make a better with great ease'[10]), ensured
that he often did make a better article with great ease, and his cavalier
way with contributors' manuscripts has involved this area of the Hazlitt
canon in inextricable confusion.

Strange that the scope offered to Hazlitt by the *Edinburgh* should have
been narrower than what he already found in the *Champion* and the
Examiner, but so it was. The literary-political nature of the journal and its
formal limitation to book reviews meant that the editor's leanings
combined with the hazards of book-publication to deny Hazlitt, for six
years after his excellent series in the *Champion*, the opportunity to write
on the subject so close to his heart which had got him into the *Review*, and
his contributions to the *Edinburgh* during all that time related to
literature; indeed he was never able to write directly about pictures in
that journal.

However, he continued his articles for Scott. In his account of the
inconsistencies between Sir Joshua Reynolds's practice and the principles of his *Discourses* we discern behind the protest of the romantic
against the classic a hint of the disappointment that blames ill success on
inadequate guidance. He accuses the *Discourses* of systematically fos-
tering these weaknesses which he says foreigners detected in British
artists: they are in fact those he recognized in himself. We do not need to
suppose that it was John Hazlitt, a pupil it is said of Sir Joshua's, who
recommended the *Discourses* to his brother: anyone in the neighbourhood
of Rathbone Place, the artists' quarter of the day, would have done so;
they were the Bible of the apprentice painter at that time, although their
inconsistencies were no very great secret either.[11] He uses their
ambiguity (their 'half-principles') to explain their inconsistencies. What
he defines as 'errors' — the primacy of the sedulous imitation of the best
models; the cult of the ideal; the suppression of the particular, of the
individual, in the pursuit of the general; the neglect of nature — are the
characteristic marks of neo-classicism.

He was right.[12] But the real inconsistency was that Reynolds's hand,
which held the brush and imparted an evident and individual truth to the
canvas, sweepingly contradicted itself when it took up the pen and made
its official pronouncements, that 'expression detracts from beauty, and

---

[10] G. Bullett, *Sydney Smith* (1951), 272.

[11] See Pope, i. 100 (Nov. 1809).

[12] He could not then know that he had William Blake on his side in this: see G. Keynes, *Poetry
and Prose of William Blake* (1927), 770–812, and esp. 778, 'I always considered True Art & True
Artists to be particularly Insulted and Degraded by the Reputation of these Discourses.'

that the whole beauty and grandeur of art consists . . . in being able to get above all singular forms, local customs, particularities and details of every kind'. This was a charter for a return to the uniform banalities of the early eighteenth century, and although these precepts were quietly set aside by Reynolds himself when in front of his easel, and by his contemporary Gainsborough (despite what Hazlitt called his want of 'that vigour of intellect which perceives the beauty of truth'[13]) the effect of the teachings and authority of Sir Joshua, thrown into the scale on the inevitably imposing side of 'the ideal', was disastrous. Confirmed by the promptings of ambition and the urgency of making a livelihood, the need to 'tamper with nature and give what is called a flattering likeness' became more and more evident to portrait painters, favoured as it was by the immense popularity of portraiture towards the end of the century, when the tradition of Hoare and Hudson entered into a new phase with the productions of Hoppner and Romney (as in the latter's *Lady Hamilton*), and reached it's gorgeous, genteel apogee in the apotheoses of Sir Thomas Lawrence, the Tom Moore of painting, where sartorial splendour mocks moral seediness and eyes and lips glisten with a synthetic sentiment that has little to do with human feeling.[14] We might find difficulty in understanding Hazlitt's contentiousness if we did not know how artificial portraiture had become. An anecdote of Northcote's shows well how things were: Hoppner told him that he 'used to make as beautiful a face as he could, then give it a likeness to the sitter, working down from this beautiful state until the bystanders should cry out, "Oh I see a likeness coming!", whereupon he then stopped, and never ventured to make it more like'.[15] Now, if that was the principle on which Hoppner worked — a great name in his day — what must have been the stultifying influence of the hundreds of mediocre portraits by unknown artists, long since sunk into well-merited oblivion, that covered every inch of the walls of the Royal Academy and the British Institution from floor to ceiling every year?

Hazlitt's reflections on art became more concrete and immediate in the accounts he gave this year of these exhibitions, as well as of Lucien Buonaparte's collection in the New Gallery, Pall Mall.[16] He also reported in the *Champion* at this time a one-man exhibition in Pall Mall by Wilkie.[17] What he said of it, though favourable enough, upset

[13] 18. 36.

[14] Hazlitt's phrase 'a smart haberdasher look' is echoed in Northcote's description of Lawrence as 'a sort of man-milliner painter' (quoted by David Piper, *The English Face* (1957), 232; see also on the same page the comparison of the two Dukes of Wellington, by Lawrence, and by Goya).

[15] *Conversations of James Northcote R.A., with James Ward*, ed. E. Fletcher (1901), 190.

[16] *Champion*, 22 Jan. 1815; the collection was not opened to the public until 6 Feb. (*MC*, 31 Jan.), and for some reason was not sold until much later, 14–16 May 1816, Great Room, 29 St James's St. Howe failed to locate a copy of the cat.; there is one in V. & A.

[17] *Champion*, 5 Mar. 1815 (18. 96–100).

Haydon, who swore to retaliate, despite Wilkie's dissociating himself with ostentatious calmness for any such move.[18] Haydon's irritation with Wilkie over this matter is a token of the support he demanded of his friends in his own squabbles. If he was to be martyred he required that they also should immolate themselves as an affirmation of loyalty. Wilkie demurred not from timidity but from good sense. The whole of Haydon's subsequent behaviour to Hazlitt is explained by the refusal to allow independence to his friends shown in this incident. In the end he decided against answering Hazlitt, but a subsequent letter of Hunt's on Wilkie to the editor of the *Champion* (an unnoticed anticipation of his 'Round Table' of September 1816 'On Washerwomen') led Scott to counter the criticisms of his fellow-countryman by both Hazlitt and Hunt in the *Champion* of 14 May. Scott's reply was applauded by Sir George Beaumont, according to Robinson, reporting a conversation with Haydon and Scott (whom Robinson calls 'a little swarthy man with rather an unpleasant expression') at Wordsworth's London lodging on 18 June.[19] On 25 May Wilkie, who had cannily bided his time, wrote to thank Scott for his attentions.[20]

In mid-February Hazlitt was so gratified by a cordial letter from Jeffrey regarding his *Wanderer* article that he took a brief jaunt into the country to celebrate the successful launching of his *Edinburgh* reviewing career, and at the same time acquired a copy of the next work he was to undertake, Sismondi's *De la littérature du midi de l'Europe*. Here was a happy change! In the previous weeks his uneasy relations with another of his editors, Scott, cannot have been improved by his drinking, nor by his splenetic rashness (to use his own phrase). The unhappiness of the shadowy love-affair of the autumn and winter had been exacerbated by resentment at Napoleon's removal from the European scene. Robinson mentions an uncomfortable argument at Alsager's in early February when Hazlitt, in his cups according to the diarist, insisted that 'there was at no time so great danger from the recent and unestablished tyranny of Buonaparte as from that of ancient governments'.[21] He took a savage pleasure in the errors, the iniquities of the liberated or restored monarchs of Europe (he was in the gallery for the debate of 21 February on the annexation of Genoa by the King of Sardinia which shocked liberal opinion quite as much as the imposition of a Prince of Orange upon the Dutch Republic had done in December 1813), and this

---

[18] See Pope, ii. 213, 11 Jan 1819, 'When Hazlitt attacked Wilkie in his lectures . . . my first impulse was to answer him', but Haydon cannot have meant the lecture of a fortnight earlier (29 Dec.) in the series on the English comic writers, which incorporated this article of Mar. 1815, but the article itself, published at a time when Haydon and Wilkie were constantly in each other's company. See also Pope, iii. 131, 3 Aug. 1826.

[19] Robinson Diary.

[20] NLS MS 3278, fo. 76.

[21] Morley, 161–2, 5 Feb. 1815.

schadenfreude was no doubt inflamed or turned to plain anger by the Corn Bill riots of early March, in the repression of which a harmless passer-by, one Edward Vyse, was shot dead. The chief witness against the military in this affair was William Hone, the Fleet Street bookseller who later became embroiled with the authorities on his own account, and who also became Hazlitt's friend and publisher.[22] Even Hazlitt's bellicose brother-in-law of *The Times* quailed during the riots at the spectacle of troops drawing round his own metropolis, however eager he was for them to encircle the capitals of other countries.[23]

On Friday, 10 March London was startled by the incredible news of Napoleon's escape from Elba and landing at Cannes; to Hazlitt it was as the rumour of the distant unbarring and unbolting of the iron doors of despotism, which he thought had closed upon mankind forever. It has been assumed that when he switched from the *Champion* to the *Examiner* in this month it was because of some kind of disagreement between him and Scott over this momentous intelligence.[24] A breach undoubtedly came about, but it was personal, not political, in origin, and the cause, which has hitherto remained obscure, lies concealed behind an entry Haydon made year's later in his diary. The breach originated in fact in yet another of those damaging assaults upon Hazlitt so frequent in his life and which can only be explained by his dangerous frankness about himself and his unsuspicious guilelessness. According to this entry Scott 'made Hazlitt tipsy, and got out the secrets and weaknesses of his nature, and then assailed him the very next Sunday in an anonymous letter, touching on these very points'.[25] This was evidently not a private letter, and there is no such open letter in the *Champion*, but there is such an attack. Hazlitt's confidences must date from the very weekend of the news from Cannes, for on the following Sunday, 12 March, there appeared in the leading article in the *Champion* a remarkable reply to the previous Sunday's *Examiner* in which Hunt had said of Napoleon's return and the threat to the stability of Louis XVlll's throne, 'As to the wishes which it becomes people to entertain on this occasion, we have only to repeat our old regret, — that there is no party in the business, to whom we can wish success altogether'. It looks as if on that same Sunday Scott discussed the news of the landing and Hunt's leader on it with Hazlitt, and that there followed a long conversation going far beyond

---

[22] *The Times*, 9 Mar.

[23] *The Times*, 8 Mar.

[24] See Howe, 162; Maclean, 320; Baker, 223.

[25] Pope, ii. 315, 9 Mar. 1821, just after Scott's death. In these crowded reminiscences Haydon quotes no dates. Always an inaccurate reporter, although (because?) a brilliant one, he had forgotten that it was not an anonymous letter but an anonymous leader (in the paper of the 'very next Sunday') that alluded to Hazlitt (unless it is his handwriting that is misleading: I have not seen the manuscript).

political views to a reckless revelation of details of private life. It further
looks as if Scott at the time gave no hint of disapproval, but privately
concluded that Hazlitt was a bad influence on Hunt's politics and a
worse on his morals. He accordingly went on in his reply of the following
Sunday to attempt to drive a wedge between Hunt and Hazlitt by
circuitously implying, while paying tribute to the excellent 'qualities of
head and heart' of the *Examiner*'s editor, that responsibility for the
capricious politics of the paper lay not with Hunt but in the baneful
influence of an unnamed, skulking, misanthropic scowler who lived a life
of headlong, irreclaimable dissipation. The meaning of the allusions
would be accessible only to the few who, like Haydon, knew the three
persons involved, Scott, Hunt, and Hazlitt. They must have been
impenetrable to casual, and perhaps even to regular, readers of the
Champion. But they gave mortal offence to Hazlitt. Here is what those
readers found in the editorial of 19 March (Scott speaks of himself and of
Hunt in the editorial plural):

They confess that their 'wishes are inconsistent' — Now, dissenting as we do,
with something of warm feeling, from this way of putting the matter, we are
bound to oppose it publicly. Inasmuch as it comes recommended by every
quality of head and heart, by every circumstance of life, and ornament of
reputation that can give sanction and force to a sentiment, we, who think it
erroneous, are bound to grapple with it; but we enter the lists in perfect
consistency with undiminished respect for those whom we are to oppose. This,
we confess, we could not do, if we traced to them what we are sure does not
exist in their bosoms, — an unnatural turning with welcome towards the
plagues of the earth, caused by a tormenting consciousness that its delicacies
are contrasted to the coarseness of the individual scowler, — its beauties to his
deformity, — and that its best rewards have been lost to him by a wayward and
unmanly disregard of those observances, which are necessary to render talent
respectable, because they are necessary to the harmony, elegance and safety of
society.

> 'How sour sweet music is
> When time is broke and no proportion kept,
> So is it in the music of men's lives.'
>
> Shakespeare

As gratuitous and muffled an attack as ever puzzled compositor, and
evidently written in a spasm of unreflecting irritation. For all that he was
an excellent editor, Scott was a violent man. Haydon said that he was a
wife-beater, and even the mild Procter significantly conceded that he
was 'a little irritable'. Haydon also said that the 'worst feelings of his
nature [had been] called forth by a father's brutal treatment'; although
he 'had a good heart at bottom, . . . it was so buried in passions that its
native goodness had seldom power to force its way . . . He mentions at

the same time (this was when Scott died) something that suggests that Scott's resentment of former injustices infected his present situation, and that he was jealous of Hazlitt's superior talents, and was consequently all the harsher: 'to this [ill-treatment in childhood] I trace many of his vices and all his bad feelings. The curse of his life was a rankling consciousness of his inferiority to some of his Friends.'[26] As well as aiming this vicious dart at Hazlitt in 1815, Scott's envy was directed at various times to many others, to Byron, Hunt, and Haydon himself.

The shock to Hazlitt may easily be imagined. It was the unexpectedness that took his breath away, no less than the treachery of it. Not that we need give entire credit to the diabolical premeditation imputed to Scott by the ever-dramatic Haydon. Heaven knows it was never necessary to make Hazlitt tipsy to get him to tell the secrets and weaknesses of his nature: he was naturally from a boy so little prone to suspect bad intentions, so devoid of guile, that it was years before he questioned the wisdom of Godwin's teaching on the imperative of sincerity, and even at the end of his life, in discussing the importance of reputation in the last Northcote conversation, he does not accept concealment, nor yet caution, even while tacitly agreeing with Northcote on the danger of setting worldly opinion at defiance. His distress, his resentment on this occasion can be briefly seen by reflection in what Haydon goes on to say of Scott's own behaviour after the publication of the 'anonymous letter': 'Poor Scott, he was his own assassin, living or dead. . . When he met Hazlitt again, his feelings in my room were punishment enough.'[27] It is unclear whether this meeting took place soon or long after the offence: what is certain is that Hazlitt wrote no more for the *Champion* while it continued under Scott's editorship, as it did for another two years. He ceased all communication with him and they were not reconciled until after Scott's return from Italy in 1819. Scott went to Paris after the defeat of Napoleon to gather material for his *Paris Revisited in 1815* and remained there until October.[28] Six months later he made the *Champion* the vehicle of a personal attack on Byron whom he had disliked ever since the poet had ignored him when they met on a visit to Leigh Hunt in Horsemonger Lane.[29]

When Hunt left that prison on 3 February 1815 the nervous dread of the outside world endemic in discharged prisoners, but exacerbated in his case by ill health, forced him at first to take refuge at Alsager's house

[26] Pope ii. 314.   [27] Ibid. 315.

[28] NLS MS 1706, fo. 176, 6 Nov. 1815, Horatio Smith to J. Scott.

[29] Haydon letter quoted in Maggs cat. 367 (1918), item 227: 'At the very time Scott was doing this [publishing Byron's 'Farewell'], as he asserted from principle, he was ill-using his own wife and before or after I *know struck her*, on some dispute about a carpet and its colour. John Scott was the most openly malignant man I ever knew'.

nearby, and then, as his carriage drove up the Strand, to shrink back into a corner and shield his eyes from the ghastly meaningless activity of the street and the crowds streaming spectre-like along the pavements; all this to the distress of his unimaginative friends, who expected to find him immediately relishing as formerly the movement and bustle of existence. 'The whole business of life', he said, 'seemed a hideous impertinence.'[30] He hardly seemed fit to attend the crowded theatres and resume his dramatic criticisms, as he meant, and for all his determination he was often forced in the next months to get Hazlitt to stand in for him. It was not until mid-May that he finally gave in and handed over the regular 'Theatrical Examiner' column to Hazlitt.[31] In the nine weeks that elapsed between mid-March when Hazlitt slammed the door of the *Champion* office behind him and his appointment as dramatic critic in mid-May, he had only five pieces published in London, but Jeffrey's generous payment of £25 for the review of *The Wanderer* made it possible to bridge the gap in regular employment.[32]

Relatively free from money worries, he was able to savour the satisfaction that welled up in him more and more strongly (banishing the black thoughts induced by Scott's malignancy) as he walked out on those invigorating March mornings, with each successive detail from France confirming Napoleon's triumphant progress, a satisfaction that mingled, as he skirted St James's Park on his way to the *Examiner* office in Maiden Lane, with the blissful promise of the return of Spring immanent in the rough winds that wrinkled the surface of the canal. He looked back on it years later with the same reflected joy, and the same sense of loss, that haunted his recollections of Miss Stephens's early appearances in the *Beggar's Opera* and *Love in a Village*, and his prose grew lyrical in the light of former hopes:

And here let us take a brief retrospect of it, brief as was the triumph itself. It was indeed a merry march, the march from Cannes. Those days were jocund and jubilant — full of heart's ease and of *allegresse*. Its footsteps had an audible echo through the earth. Laughed eyes, danced hearts, clapped hands at it. It 'loosened something at the chest'[33]

Nothing could stop what Stoddart mildly referred to in *The Times* as the accursed course of the abhorred monster (henceforth not to be

---

[30] Hunt, 250 (he here briefly anticipates the remarkable first pages of Döblin's *Berlin Alexanderplatz* (Berlin, 1930)).

[31] Not mid-Mar., as has been supposed; see the writer's 'Hazlitt and the "Theatrical Examiner": Two Additions to the Canon', *EA* 38 (1985), 427–33.

[32] Jeffrey's liberality was well known. Hazlitt's review was just over 1¼ sheets, so that in 1814 Jeffrey was paying a new contributor at the rate of £20 a sheet. In 1826 Miss Mitford regarded 12 gns. a sheet as fair payment for magazine articles (L'Estrange, ii. 220).

[33] 15. 229–30.

protected by any law, nor ranked among mankind; nay, who was not to live[34]. The bulletins of the next few weeks — interrupted on the 20th by a rumour of his defeat by Ney, given further impetus on the 22nd[35] by the arrival of the Golfe Juan proclamation of 1 March, the imagery of which must have delighted Hazlitt ('Victory shall march at the charge step; the eagle, with the national colours, shall fly from steeple to steeple, even to the towers of Notre Dame'), and on the 23rd by the hurried departure of *Louis le Désiré* from Paris and Napoleon's entrance close on his heels — culminated on the morning of the 27th in Napoleon's own vivid account of the 'joyous march', copied from the *Moniteur* of the 23rd, a march which seemed, said the *Morning Chronicle* in a phrase poles asunder from the language of *The Times*, 'more like a jaunt of pleasure than the progress of an invader'.

On 15 April Crabb Robinson again got into an argument with Hazlitt at Alsager's and was worsted, but owing, as he says in his careful discriminating way, rather to the dialectical skill of his opponent than the rightness of his views. We are in no doubt about the rage that Hazlitt's admiration for Napoleon aroused among his rabid Church-and-King enemies; Robinson's report of this argument (a valuable statement, evidently scrupulously exact, of Hazlitt's declared position at this time) shows that it also troubled certain among his friends, men of good will, men of liberal views and with a strong bias to justice and humanity:

Hazlitt and myself once felt alike on politics, and now our hopes and fears are directly opposed. Hazlitt retains all his hatred of kings and bad governments, and believing them to be incorrigible, he from a principle of revenge, rejoices that they are punished. I am indignant to find the man who might have been their punisher become their imitator, and even surpassing them all in guilt. Hazlitt is angry with the friends of liberty for weakening their strength by going with the common foe against Buonaparte, by which the old governors are so much assisted, even in their attempts against the general liberty. I am not shaken by this consequence, because I think, after all, that should the governments succeed in the worst projects imputed to them, still the evil will be infinitely less than what would arise from Buonaparte's success. I say destroy him at any rate and take the consequences. Hazlitt says: 'Let the enemy of old tyrannical governments triumph, I am glad, and I do not much care how the new government turns out.' Not that either I am indifferent to the government which the successful kings of Europe may establish or that Hazlitt has lost all love for liberty. But his *hatred*, and my *fears*, predominate and absorb all weaker impressions. This I believe to be the great difference between us.

[34] 24 Mar. A satirical 'Dictionary of Abuse and Scurrility' announced in the *Examiner* 21 May, 328, listed 38 abusive epithets attached to Napoleon by 'the learned Ecclesiastical Doctor, the Editor of *The Times*'. However, Stoddart's proscription here echoed the language of the Allies' declaration of war at Vienna on 13 Mar., pub. in the newspapers on the 28th.

[35] In the *Courier*; not until the 23rd in the *MC*.

The special dilemma facing the friends of liberty at that time, dismayed as they were by the record of the year-old Restoration and by events in other European kingdoms, has faded into impalpability. The startling realities of the refusal to reinstate Poland, the division of Saxony, the enslavement of Italy, the annulling of the Spanish constitution and the imprisonment of the *liberales*, the restoring of the Inquisition, the transfer of Norway and Finland like counters from one power to another, all have become unrecognizable to us through the cobwebbed glass partition of the Victorian era and the maelstrom of the twentieth century, and dimmed still further in France by the superimposition of two later monarchies and four republics (although variants recur in all ages, are recurring even now), but how urgent and harrowing a dilemma it was is evident from this wrangle of a spring night at Alsager's. But for most Whigs it was no dilemma; like Hazlitt they were for Napoleon. George Ticknor, the American lion-hunter, landing at Liverpool in April, discovered a Bonapartist in William Roscoe, and shortly after found that the equally celebrated Dr Parr boasted that he never went to bed without praying for the victory of Napolean.[36] And in that same week Crabb Robinson was offended by a similar vehemence in abuse of the Allies by his friend Anthony Robinson, but he found a compensating pleasure in talking to 'so noble-minded a man as Burrell, a zealous anti-Bonapartist and on high principles',[37] and in ostentatiously declining to see David's exhibition-picture of Napoleon.[38]

The war declared on 23 March hung fire, and the debate in consequence continued. On the 21st the *Morning Chronicle* blamed Napoleon's return on 'the bad faith of the restored Bourbons', and on 5 May Perry printed a long letter by Capel Lofft against the war (two facts which seem to undermine another of Hazlitt's theories about why Perry dismissed him, viz. that the fall of Napoleon made his 'Illustrations' of Vetus 'in the event very unfortunate').[39] On 18 May a public meeting in Palace Yard, Westminster, addressed by Major Cartwright and Sir Francis Burdett, declared that it was not only unjust but impolitic to make war on France. The following Sunday the *Champion*, while questioning whether such meetings expressed 'the public voice' did not entirely dissent from their attitude and wholly rejected Stoddart's violent warmongering: the 'English nation' was 'far from participating in the ravings of the madman who writes in *The Times* in favour of all rottenness' (and upon this 'severe castigation' Sir James Mackintosh hastened to congratulate Scott the very next day[40]). On 13

[36] Ticknor, i. 50.   [37] Sadler, i. 478.
[38] Robinson Diary, 12 June.
[39] Howe, 163, quoting Robinson.
[40] Unpub. letter of 22 May [1815], NLS MS 1706, fo. 89.

June *The Times* in turn stigmatized the *Morning Chronicle* for 'disgrace-fully and insultingly' applying to Napoleon the expressions 'Emperor' and 'Majesty'.

To get back to the argument at Alsager's. Robinson plumed himself on his debating skill, and although he was a fair-minded man (despite the aspersions of many of Hazlitt's biographers), he was after all a barrister and seems to have made it a point of honour, or more exactly an obligatory exercise in professional training, always to exert himself in drawing-room discussion. He could at least console himself for his defeat with the thought that he was sharpening his dialectical armoury. He notes with complacency the congratulations of Serjeant Rough upon his victory over Thelwall in an exchange on much the same theme in 1813.[41] Thelwall should have been a formidable opponent, on a professional showing, presiding as he did over an Institution for the Cure of Impediments and Preparation of Youth for the Pulpit, Bar and Senate in Lincoln's Inn Fields, where he instructed law-students in elocution.[42] Hazlitt is modest enough in his accounts of himself in social gatherings, and we think of him as the unbuttoned but self-effacing talker of 'Of Persons one would Wish to have Seen' and 'On Coffee-house Politi-cians'. We seldom see it asserted that he was also a rigorous and determined disputant, as may be gathered from 'On the Conversation of Authors', possibly because his vanquished opponents (unlike Robinson) would be in no hurry to let their discomfiture be known, but we remember Hunt's word about his 'cannonade reasoning', and his decisive part in the discussion at John Dyer Collier's house of the Coleridge lectures in 1812.

Before long, Wordsworth came to town to see three volumes of verse through the press — the *Poems* (two volumes) for 27 April, and *The White Doe of Rylstone* for 12 May.[43] Robinson, we recall, remonstrated very mildly, about the tone of the preface to the *Poems*. Wordsworth, he reports, denied that there was any expression of anger in it 'though he has nothing but contempt, etc.'[44] for his detractors. It is a pity Robinson did not expand this *etc.*, but we may be sure that Wordsworth was thinking of Hazlitt, among others, even if he did not name him. He had taken the extraordinary course of asking Lamb not to receive Hazlitt at his house when he himself was there, and Lamb, as a rule stubbornly independent, had complied, either because of the previous year's quarrel with Hazlitt, or from fear of losing Wordsworth, or from a shrewd calculation that the unassuming Hazlitt, living in London and

41 Robinson Diary, 15, 19 Apr. 1813.
42 See an advt. in the *Courier*, 20 Apr. 1815.
43 *MC* and *Courier* advts. of those dates, respectively.
44 Morley, 166.

not in the Lake District, was always available. Not that there was much danger. We know from Hazlitt's letter of 13 May, Whit Saturday,[45] to Charles Ollier (who subsequently published *Characters of Shakespear's Plays*), declining an invitation to a concert of Corelli's music because he was 'held fast' by half a dozen of the Italian's countrymen, that he was strenuously engaged on the Sismondi review on that holiday weekend when even Lamb, cruelly overworked as he was, managed at least one excursion into the country (the Mackery End visit).

On 11 June a dramatic notice appeared in the *Examiner* that must decisively have linked Hazlitt in his mind to the fate of his poems. Wordsworth, seemingly, was as much present to Hazlitt's mind as Hazlitt to his. Is it impossible that Hazlitt had heard of Wordsworth's high-handed interdiction of his visits to Inner Temple Lane? This review of a production of *Comus* at Covent Garden hinged upon a distinction between dramatic and didactic poetry, colloquy as opposed to soliloquy, the theatre as opposed to the study, and it harked back to his *Excursion* articles, to the division between poets' private worlds and their public responsibilities, and the way in which the poets he knew were neglecting, even betraying, the cause he felt they should uphold.

The end of the article disclosed his bitterness at the turn of events since Napoleon's arrival in Paris: the joy of those early days had gone sour; the possibility (it was never a hope) that the Allies would treat with Napoleon had rapidly faded;[46] the kings had once again closed ranks to crush the people's hero; the armies were massing in Belgium, and every day the newspapers were filled with the movements of troops. Hazlitt's thoughts went back to that earlier seventeenth century restoration in his own country, turning away from the apostasies of present-day poets to the integrity and stubborn consistency of the blind old Puritan poet who survived into the hostile reign of Charles II: 'He did not retract . . . he was not appointed Poet Laureat to a Court which he had reviled and insulted; he accepted neither place nor pension' (and then, in a startling reference to Wordsworth's poem 'November 1813', which was hardly off the press) 'nor did he write paltry sonnets upon the "Royal fortitude" of the House of Stuart' (and with a final unnecessary turn of the screw hardly consistent with his deep admiration of Wordsworth's finest poetry) 'by which, however, they really lost something'.[47] We will be the less likely to think him unwarrantably callous in this allusion to the

---

[45] *Four Generations*, i. 116, dates this at 4 Oct. 1815; *Letters*, 143, at [May 1815]; the correct date is given in Sotheby cat., 19 Feb. 1963, lot 486.

[46] The *Courier* of 23 Mar. announced Napoleon's entry into Paris, and by the 28th the formal declaration of war against him by England, Austria, Russia, Prussia, Sweden, Spain, and Portugal was known.

[47] 5. 233.

mad George III when we think of Lord Holland's searing verdict, reflecting Whig opinion generally, at the time of the King's death in 1820: 'The lavish and unmerited praises bestowed on the late King makes me sick and almost angry . . . There was nothing good done in his reign that was not done against his will.'[48] But the reckless bitterness injected into Hazlitt's tone by the return of absolute monarchy to Europe is very striking.

If Wordsworth did not discover the allusion for himself it was drawn to his attention 'in a manly way' by Hunt whom he visited on the day the newspaper appeared, and who apologized, claiming it had been printed without his knowledge. Robinson, to whom we owe this information, does not report Wordsworth's reactions.[49] Nor does the less reliable Haydon, who ascribed the meeting, at which he was present, to Monday, 12 June. Wordsworth breakfasted that morning at Haydon's studio in Great Marlborough Street and there met for the first time John Scott. Scott, impressed, did not long delay in imparting to his readers his view of 'the sterling value of Mr Wordsworth's talent', 'a Poet of the first class . . . the greatest poetical genius of the age'.[50] After Scott had gone off to see to his paper, Wordsworth and Haydon made their way along Oxford Street and up the Edgware Road to call on Hunt. Wordsworth at this time was zealous to cultivate editors. Earlier in the year he had sent Hunt a copy of his 1815 *Poems*, and had sounded Lamb (without success) on Scott,[51] and during his present visit to London he called on Daniel Stuart of the *Courier*. Haydon seems to have been unaware of the awkwardness of the *Examiner* note, but he does say that Hunt paid Wordsworth 'the highest compliments' and quotes him as declaring that 'as he grew wiser and got older he found his respect for his powers and enthusiasm for his genius encrease'.[52] Nor do we know whether Hunt, like Haydon, heard at this time the Keswick story, but in a few days Wordsworth, in telling Robinson of his visit to Hunt, was 'led to mention the cause of his coolness towards Hazlitt'.

Haydon mentions only one meeting with Wordsworth during this summer, but years later he recounted an incident that must have taken place now. He and Wordsworth, walking along Pall Mall,[53] turned into

[48] BL Add. MS 51748, Lord Holland to his son, 16 Feb. 1820.

[49] Morley, 169.

[50] *Champion*, 25 June 1815, 206, 'Mr Wordsworth's Poems'.

[51] Marrs. iii. 125.

[52] Pope, i. 451

[53] Ibid. ii. 470, 29 Mar. 1824, which has been taken (Baker, 137) to be the date of the conversation with Wordsworth, who was not then in London. It is plainly a reminiscence and, of all Wordsworth's visits, that of June 1815 alone coincides with the sale of a copy of the *Transfiguration* at Christie's and the telling of the Lakes story to other London friends of Wordsworth's. For the sale, see *MC*, 7 June 1815 and Christie cat., 10 June 1815, lot 74.

Christie's to see a copy of Raphael's *Transfiguration*. This copy by Andrea Sacchi, was then making a great stir; it had been sold the previous Saturday, but was not yet removed.

Haydon, who knew something of painting, agreed with the general view that it was a fine picture, but Wordsworth 'abused [it] through thick and thin'. Wordsworth (who did not much or oft delight to season his fireside with personal talk, but did so now that he was in London) had just been telling his companion about Hazlitt's escapade in the Lakes, and in his account, according to Haydon's son, who said later that it had become a stock anecdote in the painter's family, the fiendish mingled with the ludicrous.[54] This unlikely combination simply means that to Wordsworth it was fiendish, whereas to Haydon (as to Lamb) it was ludicrous.

He was relating to me with great horror Hazlitt's licentious conduct to the girls of the Lake & that no woman could walk after dark, for 'his Satyr & *beastly* appetites.' Some girl called him a black-faced rascal, when Hazlitt enraged pushed her down '& because, Sir,' said Wordsworth, 'she refused to gratify his abominable & devilish propensities, he lifted up her petticoats & *smote* her on *the bottom*'.[55]

The story is not told in the diary for its own sake but to illustrate the superiority of Tom Moore, whom Haydon had just met. Haydon is contrasting Wordsworth's narrowness with Moore's 'naturalness', and all the phrases quoted from Wordsworth are used ironically. The name-calling and the absurd chastisement are evidently authentic; what is implausible is the 'devilish propensities'. The exaggeration in Wordsworth's phrase, and the tangle of prurience and prudishness in the mind of the man who used it, are implied in another incident Haydon ascribed to that day.[56] He says that as they were leaving Christie's they saw a statuary group of *Cupid and Psyche Kissing*, and 'after looking for some time Wordsworth turned round to me with an expression I shall never forget, and said, "The *Dev-ils*" '. In retelling the story in a letter to Miss Mitford, Haydon exclaims, 'There's a mind!' Respectability had made damaging inroads upon Wordsworth's judgements, as his disapproval of De Quincey's 'unfortunate acquaintance', his 'entanglement' with 'the daughter of a statesman' shows.[57]

---

[54] Haydon, *Corr.* i. 110.

[55] Pope, ii. 470 (facsimile, Baker 138).

[56] I say 'ascribed' because the *Cupid and Psyche* is not listed in the Christie cat. It may have arrived too late for the sale, but it seems to be connected with Canova's visit to London about that time (it is evidently a copy of the *Cupid and Psyche Recumbent* he executed in 1793 and which in 1815 was at Compiègne). see E. George, *The Life and Death of Benjamin Robert Haydon* (Oxford, 1948), 138.

[57] Morley, 187, 195.

In the mean time, unconscious or careless (Hunt must surely have mentioned it?) of the impact of his scornful reference, Hazlitt anxiously watched events in Europe. To the reference he had appended an explanatory footnote, reading, 'In the last edition of the works of a modern Poet there is a Sonnet to the king, complimenting him on "his royal fortitude", and (somewhat prematurely) on the triumphs resulting from it'.[58] Not as prematurely as Hazlitt hoped: the triumph was at hand. It was known on the morning of Tuesday, 20 June that the British and French had joined battle. At 11.15 on the night of the 21st Wellington's dispatch reached the War Office and in the early hours of the 22nd the newspaper editors stopped the press to announce what even the hitherto unenthusiastic *Morning Chronicle* roused itself to describe as 'the most brilliant and complete victory ever obtained by the Duke of Wellington, and which will forever exalt the Glory of the British Name'. Glorious tidings indeed to such as Haydon, 'who should have been the boatswain of a man-of-war', said Hazlitt [and who had] no other ideas of glory than those which belong to a naval victory, or to vulgar noise and insolence,[59] who was mad with excitement, and to the greater part of the nation, whirled away in a Mafeking tide of rejoicing; but to Hazlitt a dreadful blow, plunging him into a depression that darkened his existence for weeks. He had either genuinely not believed that Napoleon threatened England, or had been prepared to accept the hazard because it was a threat aimed also at the heads of all the legitimate sovereigns of Europe and so held out the promise of their ruin. Now, it was his hopes that were ruined: the restoration in France was confirmed; and reaction all over Europe was certain.

He had set out in life with the French Revolution, with the first dawn of liberty. That dawn was far in the past; it had merged in the long day of the Napoleonic epic, and now he was forced to acknowledge that the sun of Austerlitz that had illumined that day had finally set. And by a fatal coincidence the love-affair he obscurely hints at seems to have collapsed at exactly the same time. He was doubly possessed, by the lodestar of principle as well as by the trade winds of passion (as other men are swayed by the ground swell of self-interest), and he lamented the loss of the general liberty in the same way as he bewailed the loss of his personal happiness. Unable to shake off his despair, he sank for a time into neglect, into self-abandonment, and the oblivion of drunkenness. He was thirty-seven. He had always been buoyed up by the eager sanguine temperament that prolongs youth indefinitely, but now his hopes were at an end, and his youth eclipsed with them; from that Thursday on he felt himself no longer young, for on that day his hopes fell. Talfourd, who

---

[58] 5. 233 and corresponding n., 409.   [59] 20. 392.

met Hazlitt just at this time (probably through Lamb, since he was living on the next staircase to Lamb in the Temple), was struck by the painful expression of his face, which, though handsome (Hunt also describes him at this period as 'a very good-looking fellow'[60]) and eager, was worn with sickness and thought:

He was staggering under the blow of Waterloo. The reappearance of his imperial idol on the coast of France and his triumphal march to Paris, like a fairy vision, had excited his admiration and sympathy to the utmost pitch; and though in many respects sturdily English in feeling, he could scarcely forgive the valour of the conquerors; and bitterly resented the captivity of the Emperor in St. Helena, which followed it, as if he had sustained a personal wrong.[61]

Haydon was more shocked by his misery than impressed by his passionate resentment:

. . . it is not to be believed how the destruction of Napoleon affected him; he seemed prostrated in mind and body, he walked about unwashed, unshaved, hardly sober by day, and always intoxicated by night, literally, without exaggeration, for weeks; until at length wakening as it were from his stupor, he at once left off all stimulating liquors, and never touched them after.[62]

This is perhaps the place for a brief word about Hazlitt's overdrinking. Haydon probably did not mean that thenceforward he practised rigid abstinence but that he never again touched spirits. 'Advice to a Schoolboy' shows that he knew from experience and feared their addictive poisonousness, but there are instances after 1815 of his taking drink of other kinds. In 1816 Haydon himself gave him a bottle of wine to beguile the tedium of sitting for one of the faces in a crowd; and in 1821 his bill for a stay at Winterslow Hut shows a charge of 12s. for wine, as well as an isolated grog, probably medicinal.[63] In Edinburgh in 1822 we find him with a glass of ale in his hand, but he observes that this was an unwonted celebration, and indeed within a few days we see him taking tea to the other guests' wine at his friend Ritchie's and telling the company how he had 'hurt himself by drinking too freely and had given up strong potations'.[64] Possibly his wretchedness in the bad year of 1823, when he had lost Sarah Walker, induced a relapse, which would explain why his departure for France in 1824 was spoken of but by one who had never actually met him, the self-righteous young Thomas Carlyle — as a loss to 'the ginshops and pawnbrokers'.[65] And the following year, during

---

[60] *Examiner*, 19 Feb. 1815.

[61] Talfourd, *Memorials*, ii. 170.   [62] Haydon, 283.

[63] Pope, ii. 64; Buffalo, at the end of the vol. of letters of Patmore.

[64] See the writer's 'Hazlitt in Edinburgh: An Evening with Mr. Ritchie of the *Scotsman*', *EA* 17 (1964), 113; the expressions are those of George Combe, another guest.

[65] Carlyle, *Letters*, iii. 234.

his stay in Switzerland, he apparently did not deny himself the wine of the country.[66] But he was never again, as far as we know, the worse for drink. But neither did he join Basil Montagu in preaching temperance. He had found that excessive indulgence was bad for him (just like Lamb, who had set down a record of his own case in the bleak pages of his 'Confessions of a Drunkard') so he left off (unlike Lamb, however, who wanted either resolution or luck, and whose life was twisted by stresses infinitely more dreadful). Hazlitt probably did think his escape a matter of luck. It was certainly not something he took any credit for. Alcohol was bad for him, but, he reasoned, not necessarily so for others; and even if it was, he would not interfere: he would be the last to parade his abstinence and cause awkwardness when Booth's Best or Whitbread's Entire was circulating at the Lambs'. We have a pleasing picture of him in after years in convivial company taking up his neighbour's glass of wine to sniff the bouquet, and saying 'By God! that's fine!' before handing it contentedly back. And we can be sure he would never have shown the lack of feeling of the Procters on the Enfield Road years later, when they pretended not to see Lamb whom they encountered half-seas-over.[67]

The banked fires of his irritated disappointment at Waterloo and the banishment to St Helena made him prone to unwonted injustices. An instance occurred in late August in a bad piece on the changes of front of Chateaubriand, the 'prose laureat of the Court of the Tuileries', where he let fall a remark about the Duc d'Enghien, 'whom Bonaparte is accused of having murdered because he was not willing that he, the said royal Duke, should assassinate him'.[68] This indiscretion was bound to arouse protest. It is difficult to appreciate the horror felt even in France at the time of the Duke's execution in 1804; its repercussion in England was such that Haydon, when he visited Paris over a decade later (and a decade of resounding events that convulsed Europe from end to end), went out to Vincennes to see for himself the scene of the crime.[69]

The remark led to a controversy with a reader of the *Examiner*, signing himself *Fair Play*, who, before many exchanges had passed, was objecting quite legitimately that Hazlitt was insisting on *his* proving the Duke's presumptive innocence of plotting against Napoleon's life before he would discuss or even entertain the prospect of Napoleon's responsibility for a crime actually committed. The tone of Hazlitt's four contributions to this protracted controversy, extending from 24 September to 6 December, is often peevish and sometimes violent. It bears evident marks of a consciously insecure position, and four years later Hazlitt significantly omitted them from his collection of *Political Essays*, and never reprinted any of them. In attempting to justify the seizing of the Duc d'Enghien on non-French territory, the forcible return to France, the hasty trials and the summary hole-in-corner execution he was in an impossible position; he knew it; and the knowledge exasperated him. But he was betrayed into it, not only by his hatred of kings ('the said *royal* Duke'), but by his instinctive bias towards fact in preference to theory. Just as, in order to live by his pen, he had been forced to abandon the abstract logical style of his first writings and adopt the concrete strategies of simplification and paradox, quitting the study for the arena; and just as he had ascribed Mackintosh's failures in debate to the disability of a fair-minded overscrupulousness which was disastrous against force and cunning, and just as he had always held that the means employed by patriots in the struggle against embattled power were often (though not always) justified by the end, and that those who 'to do a great right, do a little wrong'[70] are not always to be blamed, so now he saw the justification of Napoleon's act, not in a principle or a theory, but in the realities of Napoleon's dilemma. As he said, in practice, 'political is like military warfare', and 'you have not your choice of ends or means'.[71] But as a result he, like his hero, was here caught in a dirty business. He had to defend an act of violence, and probably a terrible injustice, and almost the only argument he could bring to bear against *Fair Play* was the disadvantage of those who play fair when faced with an unscrupulous foe. In a few years, speaking of Brutus's role in *Julius Caesar*, he put it this way:

. . . the whole design of the conspiritors to liberate their country fails from the generous temper and overweening confidence of Brutus in the goodness of their cause and the assistance of others. Thus it has always been. Those who mean well themselves think well of others, and fall a prey to their security. That humanity and honesty which dispose men to resist injustice and tyranny render them unfit to cope with the cunning and power of those who opposed them.[72]

Violence, then, and probably injustice. He had to persuade himself, first, that Napoleon was right to kill; and then, that he had killed the right man.

On the first point he could not quite bring himself to invoke the *lex talionis* argument and claim justification by such earlier arbitrary acts of the British as the seizure of the Irish rebel Napper Tandy in Hamburg (although he disingenuously circumvents his own scruples here by quoting Cobbett, who did in this case adduce the precedent).[73] No. Finding himself trapped in the everlasting coil, the relentless paradox, of means and ends, he took refuge in the plea of self-defence, or rather, of political expediency. Conscious of the taint of corruption clinging to the

[70] 17. 36.   [71] 17. 40.   [72] 4. 198.   [73] 19. 132

plea as well as the argument, he nevertheless puts the case, such as it is, with admirable force in the best page of the whole correspondence, when he identifies the realities of political action:

The cool, calculating, moderate patriotism which your Correspondent professes, will do very well to keep him on the safe side in opposition — from becoming obnoxious to persons of literal understandings and weak nerves, but it will not prevent him from being made a handle of by those who have the power and the will to go all lengths on the other side of the question, and who will be sure to convert his concessions of speculative and partial right into the means of practical and universal wrong. The cobwebs that entangle him will not stop them. The tide of corruption and oppression will not be stemmed by pretty speeches about purity and morality. The love of freedom is no match for the love of power, because the one is urged on by passion, while the other is in general the cold dictate of the understanding. With this natural disadvantage on the side of liberty, I know what I have to expect from those persons who pique themselves on an extreme scrupulousness in the cause of the people. I find none of this scrupulousness in the friends of despotism: *they* are in earnest, the others are not.[74]

The argument for the use of counterviolence to stem the returning tide of absolutism had indeed been gaining ground since the 'great gaol-delivery of monarchs' in May 1814, and if Hazlitt appears savagely inveterate in these pages a glance at what was happening abroad will afford some vindication. The first acts of Ferdinand VII, restored to the Spanish throne, were to reject the constitution he had agreed to in 1812, re-establish the Inquisition, and imprison the *liberales* who had fought for him; and of Victor Emmanuel I, King of Sardinia, restored to Piedmont, to abolish religious freedom and re-introduce the Church's censorship, trampling underfoot, in company with Ferdinand I of the Two Sicilies, those popular liberties which England during the temporary administration of Lord Bentinck had pledged herself to preserve. Now in the autumn after Waterloo the pace quickened and the news that although the Spanish courts had found nothing against the *liberales* they were exiled or sent to the hulks by arbitrary sentence of the King, the rumours of the White Terror that leaked out of France despite the censorship, the German Princes' evasion of their wartime promises to grant constitutions, and the plight of Italy, in the grip of a disastrous and bloody reaction, hemmed in by the Habsburgs in the North and the Bourbons in the South, confirmed Hazlitt's gloomiest predictions. In France during the White Terror scores of Protestants had been butchered at Nîmes by the *ultra* Royalists and at least one burned alive. Hazlitt was not alone in his horror at this massacre, despite the grotesque efforts of the British government-journals to persuade the public that

[74] 19. 144.

these were unfounded rumours spread by mischief-making Dissenters.[75] Southey, who had opposed the Restoration (the ideal Head of State he envisaged for France appears to have been neither a Bourbon, nor a Bonapartist, nor a Republican, but a nominee of the Church of England) was deeply shocked. And things were even worse than was imagined. Hazlitt like most of his contemporaries, both English and French, underestimated the full extent of the Terror, just as, influenced by the *Morning Chronicle*'s championship of the victims, he correspondingly overestimated the direct political responsibility of Louis XVIII's government for what was in fact a brutal campaign of personal vengeance. Yet he was right in laying it at the door of legitimacy, as is proved by the vindictive role of the King's nephew, the Duc d' Angoulême. The mild appeal to fair play of Hazlitt's correspondent seemed perilously out of keeping with the times, and when Hazlitt wound up the debate in his last letter he wryly adapted the irony of Swift and Voltaire to warn his readers against endorsing the complacency of his opponent:

He takes leave of me, by wishing me better principles and a better temper. I despair of either. For my temper is so bad as to be ruffled almost as much by the roasting of a Protestant as by the spoiling of my dinner; nor have I better hopes of mending my principles, for they have never changed hitherto.

That was his response to the first point: had Napoleon the right to kill? On the second — did he kill the right man in the Duc d'Enghien? — his tacit misgivings explain his unwontedly offensive tone, and, as *Fair Play* provokingly pointed out, his long-windedness. Twelve years later, when he was writing his *Life of Napoleon*, there was no longer any doubt of his hero's crime, or at least of his tragic error, and he must have known he had fallen into the same trap of assuming guilt by inference in D'Enghien's case as the prosecution in the Treason Trials of 1794, which had elicited his friend Godwin's decisive *Cursory Strictures*: his discomfort is betrayed in his pretended unconcern and in the bravado of his dreadful persistence:

The only instance in which he struck a severe and stunning blow was one into which he was led in the outset by a mistake and by some studied management . . . I mean the arrest and condemnation of the Duke d'Enghien. I have no wish to qualify that affair, nor do I quail at its mention. If it were to do over again, and I were in Buonaparte's place, it should be done twice over. To those who think that persons of Royal blood have a right to shed your blood by the most violent and nefarious means, but that you have no right to transgress the smallest form to defend yourself, I have I nothing to say: to others, the question nearly decides itself.[76]

[75] See 'The Dissenters', *Courier*, 30 Jan. 1816.   [76] 14. 213–35, *passim*.

'The smallest form . . . '? It needed the sorry story of twelve years of
Bourbon misrule in France, and three of stubborn reaction under
Charles X, to exacerbate Hazlitt's original detestation of legitimacy to
the point of enabling him to pen such a word about a murder. That was
in 1827. To return to 1815.

In the wake of Waterloo he found himself cast in the role of devil's
advocate and he clove to it through all changes of scene and subject. His
conscientious habit of preparing himself for theatrical revivals by
rereading the plays must have taken him back to *Comus* in the very
month of the battle, and this no doubt in turn impelled him to look again
at *Paradise Lost* and especially at the passage where the proud banner of
satanic majesty unsubdued in defeat is unfurled:

> What though the field be lost?
> All is not lost; th'unconquerable Will,
> And study of revenge, immortal hate,
> And courage never to submit or yield:
> And what is else not to be overcome?

During that summer those words must surely have confirmed him in his
intransigence, and in the autumn sustained him in the writing of the Duc
d'Enghien letters, whilst his rereading of Milton (if the expression befits
a poet whom he had almost by heart), his slow pondering of Milton's
works during those months, produced, in addition to a study of the
versification, the essay on 'Lycidas', that elegy to a departed friend
('Whilst thee the shores and sounding seas / Wash far away'), a dirge of
loneliness in which the solemn evocation of loss and death on the one
hand is joined to the call of duty and the silent spur of fame on the other.
And his own sombre mood is reflected in the pages he wrote, in his regret
for the lost leader but not less in his resolve to persist in his opposition to
arbitrary power.

8

The Examiner

FOR all his rage and gloom Hazlitt was constantly being restored to an even keel by his work, and the despairing plunge into dissipation that Haydon deprecates was checked by the need to support his family. Nor should we suppose that he was entirely unable to command his moods, that he was entirely without moments of self-regarding irony that inwardly whispered the advantage of cultivating indifference, and the inherent absurdity of the role of the coffee-house politician; on occasion, no doubt, he was able to regard himself dispassionately enough, and as he walked through St James's Park, conjure up those endearing, absurd, bewigged ghosts from the pages of the *Tatler* — the upholsterer and his cronies sitting on their bench as they had sat long ago in Marlborough's day, when they had exerted themselves 'upon the [then] present negotiations of peace, . . . deposed princes, settled the bounds of kingdoms, and balanced the power of Europe, with great justice and impartiality', to such neglect of their proper trade and business that they landed themselves in the bankruptcy court.[1]

He managed not to go bankrupt himself. Hunt had finally entrusted him on 28 May with the regular dramatic criticisms of the *Examiner* in the place of Barnes, Hunt's locum tenens, and he attended the theatre with unflagging assiduity, after a break of some three weeks following Waterloo. And all this time he was writing weekly essays for the *Examiner*. From 6 August to the end of the year, and beyond, as well as writing his dramatic criticisms he stirred up Hunt's readers with what his delighted editor called his 'cannonade reasoning',[2] under the rubric 'The Round Table'.

The number of essays he published under the rubric 'The Round Table' in 1815–16 further proves his tenaciousness. His first miscellaneous essay in the *Examiner*, 22 May 1814, carried a note anticipating the kind of exchange later meant to fill out the 'Round Table' series ('An answer to this article in our next'). Hunt's answer, prevented by illness, never appeared. Hazlitt went on to contribute articles on art, literature, politics, but no general essays until October, when he embarked on a series of 'Common Places'. When three had appeared, Hunt, who had evidently in his last months in gaol been casting about for ways to

---

[1] *Tatler*, 6 Apr. 1710; 7. 96.

[2] Hunt, 'Harry Brown's Letters to his Friends. Letter III. To W. H. Esq.', *Examiner*, 14 July 1816, 411 (Hunt, *Poems*, 229, 'To William Hazlitt').

increase the *Examiner's* circulation, asked him to collaborate in a different arrangement, and a 'prospectus' on 25 December promised a new series of miscellaneous essays to be entitled 'The General Examiner'. But it was under yet another rubric that the first appeared on New Year's day, and it was then explained that 'a small party of friends' were to supply the papers. The inappropriateness of the name of the volume, *The Round Table*, has escaped comment, although Hazlitt himself pointed out that it 'fell short of its title': there is no need of a round table for only two diners, one of whom is hardly ever present. The explanation is that there were originally five, of whom only four are identifiable: Hunt, Hazlitt, Barnes, and Lamb.[3] The unknown fifth knight never appeared, and the project almost collapsed in its opening month. Barnes dropped out, from ill health or pressure of work at *The Times*; so did Lamb, probably because of his very burdensome office-duties combined with Mary's relapse into insanity in mid-December. Hunt in the end managed less than a quarter of the contributions. But Hazlitt did not fail.

His inability to write well on what did not interest him is rarely apparent in his works because of the great range of those interests, but it breaks through in his article on Sismondi in the June *Edinburgh Review*. He had told Jeffrey that he would attend to his suggestions in 'manufacturing' the article and the word, although perhaps ironically chosen, was chosen well: in the event (unless we assume unidentified cuts by Jeffrey) he parted company with Sismondi less than a third of the way through the book and turned to a different subject of his own choosing — in English literature, thereby evincing that combination of catholicity and dislike of restriction which later brought Rabelais into his *Lectures on the English Poets* and introduced Molière and Cervantes among his *English Comic Writers*. His desertion cannot have been because of difficulty with the French text. His frequent and lengthy translations are good: besides, he found things in Sismondi's chronicle that plucked at his imagination — a passage on the decline of the Muslim Empire (a curious anticipation of Macaulay's New Zealander, which may have recalled to his mind a very romantic passage on the same theme in his old friend Fawcett's poem 'Change'[4]), and another on the brutalities of the Middle Ages which predictably struck a chord of political parallels and led him to wonder whether there had ever been any improvement on those times. But his quotations are often clearly meant to fill out space; they are both perfunctorily introduced and cumbersome. In general, however, he only arrests attention when he is dealing with the early

[3] E. Blunden, *Leigh Hunt's Examiner Examined* (1928), identifies Barnes, but not Lamb, despite a clear hint in the *Examiner*, 16 Apr. 1815, 255.

[4] *Poems* (1798), 90–5.

Italian writers he admired, and more particularly Boccaccio, and, in him, especially, the two pathetic love-stories of Federigo Alberighi and his falcon, and of Isabella and the pot of basil. He often disagrees with Sismondi, who, he says, was more historian than critic, an anticipation of the verdict of posterity entirely characteristic of 'the unerring Hazlitt'.[5] He rejects the view of Dante taken by Sismondi, who judged by system, by the application of 'the common rules of French criticism, which always seeks for excellence in the external image, and never in the internal power and feeling'.[6] It was not true that Dante's excellence consisted in natural description or dramatic invention, but rather in his capacity to infuse into his subject his own inward life and sentiment. The implication (which, according to Hazlitt's scale of values, would place Dante with modern poets like Wordsworth and Byron, and in another camp from Shakespeare) is not made explicit, but remains present in succeeding pages where he reproaches Sismondi with not having done justice to Boccaccio, that 'great painter of the human heart'. Apart from this hinted preference, and a page on progress in the arts, there is not much else of interest here except for the lines on Petrarch and Laura.

We do not know whether reviewing the Sismondi was his idea or Jeffrey's, but we know that among the topics he proposed were the following: Castlereagh and the Congress of Vienna, Scott's edition of Swift, Rousseau, *The Sorrows of Young Werther*, *La Princesse de Clèves*, Mrs Inchbald, Mrs Radcliffe, Spurzheim, Bühle's *Geschichte des neueren Philosophie* (in a French translation).[7] The first was clearly impossible. We cannot see Jeffrey admitting at Holland House that he had entrusted it to an outsider like Hazlitt rather than to someone within their own circle, like Brougham. Spurzheim was more promising. In the previous October he had made the first attempt to interest England in, 'the craniological system of Gall, which [was] well-known on the Continent',[8] and Hazlitt's scepticism had been confirmed in conversation with the celebrated Doctor, who, as he gleefully notes, had found a want of ideality in Coleridge's phrenological development.[9] He nevertheless remained fascinated by the absurdity of the claims, and still more by the fanatical determination of these disciples of the 'modern philosophy' to weigh and measure the human mind as in a grocer's scales, and by their perversity to deny the freedom of the human spirit. In Edinburgh, in 1822, we shall find him remaining politely silent when introduced into a nest of believers at a friend's house, although this was at a time when he had just been impelled to take up in an essay the reasoned refutation he

---

[5] V. S. Pritchett's phrase, *The Living Novel* (1946), 21.

[6] 16. 41.   [7] *Letters*, 140–1, 144.

[8] *Courier*, 21 Oct., cf. *MC*, 20 Oct., 1814.

[9] 12. 137, 347; 20. 253.

had meditated in 1815 (and which he did not publish until four years after 1822), and finally in 1829, reading their ludicrous pronouncements on the phrenological characteristics of Burke the resurrectionist and on that monster's bumps of piety and benevolence, he loses all patience and abandons serious argument for ridicule.

The 'modern philosophy' itself was another matter, and he proceeded with a paper on Jourdan's translation of Bühle, *Histoire de la philosophie moderne*, which he sent to Jeffrey.[10] It came to nothing and is now lost. Jeffrey did not follow up any of his other suggestions, so he went on with his 'Round Tables' and his dramatic criticisms, attending the Haymarket and the Lyceum in July and August, and returning to Covent Garden and Drury Lane in the new season, exchanging the heat of the summer theatres for the noise of the winter houses. What with the rowdyism in the gods and the incessant chatter in the boxes, the play was often inaudible and his restless mind attended instead to the conversations around him. Overheard disapproval of the morality of a play by Fletcher prompted reflections on the refinement of manners, which he intended (so he said) to put into his article on the performance, but which did not survive the conflicting evidence of the lobbies crammed with leering prostitutes, pimps, and dissolute loungers, engaged in the blatant trade that survived in the theatres right through the nineteenth century and prompted a celebrated letter by Shaw in 1894. The cynicism of the obtrusive attendant 'madams' repelled Hazlitt, weak enough though he was to hearken to the more discreet midnight solicitations of his timid friends in Horse Guards Parade. The talk of the numerous cross-channel visitors among the audience chimed with his brooding preoccupations of that summer to confirm him in the view that the origin of Napoleon's downfall lay in the incurable levity of the French,[11] and a month later, in a 'Round Table' on 'Manner', the trait (but now softened into 'grace') is again linked to the reverses of fortune of 'the greatest man in modern history'.[12] This essay evidently grew out of his endless musings over his own failures with men and with women, in society and in love, musings which emerged not as self-pity, but objectified into moral statements of unswerving intellectual honesty. The implications were plain: if he was unloved it was because he was unfitted for love; he had been ruined by romantic books; and, over-simplifying his own case into one of pure alienation, he proceeds to sit to himself for a portrait of the Disappointed Sentimentalist:

a person who has spent his life in thinking will acquire a habit of reflection; but he will neither become a dancer nor a singer, rich nor beautiful. In like manner, if any one complains of not succeeding in affairs of gallantry, we will

[10] *Letters*, 322.   [11] 5. 239.   [12] 4. 45.

venture to say, it is because he is not gallant. He has mistaken his talent — that's all. If any person of exquisite sensibility makes love awkwardly, it is because he does not feel as he should. One of these disappointed sentimentalists may very probably feel it upon reflection, may brood over it till he has worked himself up to a pitch of frenzy, and write his mistress the finest love-letters in the world, in her absence; but, be assured, he does not feel an atom of this passion in her presence. If, in paying her a compliment, he frowns with more than usual severity, or, in presenting her with a bunch of flowers, seems as if he was going to turn his back upon her, he can only expect to be laughed at for his pains; nor can he plead an excess of feeling as an excuse for want of common sense. She may say, 'It is not with me you are in love, but with the ridiculous chimeras of your own brain. You are thinking of *Sophia Western*, or some other heroine, and not of me. Go and make love to your romances'.[13]

This is the corrosive self-distrust we saw in the 'Baron Grimm' essay of October 1813, and we shall find it again, implicit in the 'Advice to a Schoolboy' of 1822, and overt in the *Liber Amoris* of 1823. It is an anticipation, an exceptionally clear-sighted diagnosis of what was increasingly seen, as the nineteenth century advanced, to be the characteristic disease of Romanticism, hardly known (except to prophets like Cervantes) until the generation influenced by Rousseau, portrayed in one of the key novels of the century, *Madame Bovary*, and defined in its last years as 'le mal d'avoir connu l'image de la réalité avant la réalité'.[14] It is plain that his reflections on defeat and failure turned his mind once more towards withdrawal from the world, as is implied in his disapproval of Wordsworth's cant about the shiftless gypsies (those 'living monuments to the first ages of society', as Hazlitt called them). Wordsworth regarded gypsies with the eye of one who, at the close of the war, had £2,000 to invest in French Funds; and Hazlitt, as one capable at any moment of giving away his last penny to join those wanderers on the face of the earth. We hear in his sharp reproof to Wordsworth for omitting 'The Female Vagrant' from the 1815 poems, as well as in his wry amusement at a solemn friend's proposal to utilize the useless slabs of Stonehenge to build houses, and in his envy of the Hindus, who 'wander about in a luxurious dream . . . , [who] hardly seem to tread the earth, but are borne along in some more genial element and bask in the radiance of brighter suns', the note of one who, hurt by the archers, would have been willing to imitate the Brahmin's contentment in his unexpected simian avatar.[15]

The particular circumstances of this major crisis in his life, the defeat

---

[13] 4. 44.

[14] P. Bourget, 'Essais de psychologie contemporaine: Gustave Flaubert', in *Oeuvres Complètes* (Paris, 1899), i. 117.

[15] 5. 233 n.; 4. 45 n., 46.

of his public and private hopes, merged, as he pondered his case, into the more general enigma of inadequacy, failure, inadaptation. He reflected again that perhaps the fault lay not in the world but in himself. What accounted for this recurrent feeling of being different? And again (in the familiar, inevitable, self-revelatory movement away from the personal and practical to the universal and ideal, a movement that was itself an illustration of the very tendency it was intended to analyse) why was it that a certain type of mind veered naturally, like a stormy petrel, into the windy skies of idealism and abstraction, of lonely disinterested specula tion and of exacting fidelity to principle, instead of settling down to the unquestioning comfort of a social life, and keeping an accommodating and wary eye open for the main chance? This nagging problem, which had been with him for years, was started no doubt by his love and admiration for his father. Why had the world not recognized in his father what he saw in him? Why was his father a failure? The *Essay on the Principles of Human Action* is in an important sense an oblique justification of his father's life, and ultimately of his own. His was a mind which could not rest without establishing its bases, and the book is the rationalization of a tendency he felt strong in his own nature — the tendency towards abstract, disinterested speculation, the unselfish devotion to certain principles: justice, civil and religious liberty. Even setting aside the constant bias towards principle within himself, how could he fail to ponder over, and seek a justification for, his father's standpoint? If the sensationalists were right, then his father was wrong. Nay, if they were right, his father ought not to have been; he was a mirage and an illusion. But not only did he exist; his life seemed to Hazlitt the kind of life a man ought to lead, one anchored in principle — in his father's case, the principles of civil and religious liberty, and of justice (the American prisoners); and in his own, of liberty in the public sphere, and of beauty in the private (painting, literature, love). In both cases a certain inertia, an apparent indifference to well-being and even to the welfare of dependants (not in the sense of love and affection bestowed, but of the provision of worldly goods) required to be explained. Hazlitt had attempted such an explanation at the highest level of abstraction, in the *Essay*, which is, in one view, a philosophical, a logical, justification of a particular temperament.

The same problem, at a lower level of abstraction, lay behind the familiar essays he was now writing, in 1815. He does not here speak directly in his own person; that comes later (the wonder is that he could persuade his early editors to accept writings as personal as his own were, even at that time; and he became even more direct as his reputation grew and he moved up from daily to weekly papers, and from monthly and quarterly magazines to books), but he is still speaking for himself and for all men. He might at any time in his career have addressed his readers in the same way as Victor Hugo addressed his: 'Hélas! quand je vous parle de moi, je vous parle de vous. Comment ne le sentez-vous pas? Ah! insensé qui crois que je ne suis pas toi!' Or as Baudelaire: 'Hypocrite lecteur, mon semblable, mon frère.'[16] It is the principle that informs his adaptation of Hamlet's words, 'For a man to know another well, were to know himself'.[17] He continues to pursue the familiar path of the Romantic writer: the conversion of his individual experience into universal laws. All his persistent, minute, dispassionate reflection is directed towards this end. His rigour never abates. After his ruthless self-rejection in his essay 'On Manner', he now attempts in 'The Tendency of Sects' an explanation of his failures in terms of his origins. Not for the first time; nor yet for the last. Obviously a recurrent problem, the question of belief had formed the subject of a paper on 'Religious Hypocrisy' in the *Examiner* the previous autumn. His present condemnation, probably out of tenderness for his father, was not specifically directed to his own upbringing, but bore on Nonconformity in general, including Puritans and Quakers, and extended even further, to Freethinkers. It was not until after his father was dead that he explicitly blamed, in the 'letter' to his son that he entitled 'Advice to a Schoolboy', the sect in which he had been brought up:

lt was my misfortune (perhaps) to be bred up among Dissenters, who look with too jaundiced an eye at others, and set too high a value on their own peculiar pretensions. From being proscribed themselves, they learn to proscribe others; and come in the end to reduce all integrity of principle and soundness of opinion within the pale of their own little communion. Those who were out of it and did not belong to the class of *Rational Dissenters*, I was led erroneously to look upon as hardly deserving the name of rational beings. Being thus satisfied as to the select few who are 'the salt of the earth', it is easy to persuade ourselves that we are at the head of them, and to fancy ourselves of more importance in the scale of true desert than all the rest of the world put together, who do not interpret a certain text of Scripture in the manner that we have been taught to do. You will (from the difference of education) be free from this bigotry, and will, I hope, avoid everything akin to the same exclusive and narrow-minded spirit.[18]

These views, expressed in 1822, he has already arrived at now, in 1815. Unitarianism, considerably less narrow though it was, as to dogma, than other varieties of Dissent, had the same fatal flaw of isolation. Common sense persuaded him to admit that the world might be right, and modesty helped him conclude that the world must be right, in that any narrow view, however defensible, however logically impeccable, must,

---

[16] Preface to *Les Contemplations*; the poem 'Au lecteur' in *Les Fleurs du Mal*.
[17] 7. 203, 8. 316, 11. 2, etc.   [18] 17. 88.

of its very narrowness, be wrong. To hold a liberal view meant looking abroad into universality, and not inward into singularity; Shakespeare was right; Wordsworth (despite the beauty of his poetry) was wrong. Universal sympathy, the outward-looking imagination, was the basic vital principle. Hence his delight in that oddity, *The Life and Adventures of John Buncle*, which he made the subject of his next article. He envied the fictitious Buncle's fine constitution, robust health, and happy nature as much as he admired his easy transition from the cold, quaint subtleties of theology to brisk love-making, and he saw in his trick of applying philosophy to give a relish to life the enviable opposite of his own moroseness.[19]

And yet, however much he deprecated the cramped life at Wem and the mark it had left on him, he would not deny its positive values. And here we approach the obscure question of his attitude to religion; and it is strictly a question, a matter of conjecture rather than assertion. The probability is that he turned deist at Hackney, like most of his fellow-students, and then agnostic in his early years in his brother's house in London, and that he changed little thereafter. Of his attitude towards the beliefs of others we can speak with more confidence. There is no sign in all his writings of violence or disrespect, with the one exception of his distaste for the frenzy of the Methodists. It is not merely that with fewer prejudices than most men he was the last to shock those of others;[20] nor is it that he was simply tender of his father's feelings (where he had already grievously disappointed him). He was too intelligent to assume that simply because he had lived more in the world he was wiser than his father; he was intermittently too conscious of the mysterious hinterland of human existence to despise any man's apprehension of its shadowy reality, or indeed to despise the mode of recognition sanctified for him by his father whose whole life was a paradigm of the necessity of morality and religion, and of the inescapable claims of the spirit. If he had in fact merely been deferring to his father, the tone of his allusions to religion should have changed after 1820. If there is such a change it is in the sense of a deepening reserve, of a confirmation in the refusal to judge. 'My First Acquaintance with Poets', written in 1823, casts no shadow of scepticism, but rather shows comprehension of such spirits as his father's: 'My father's life was comparatively a dream; but it was a dream of infinity and eternity, of death, the resurrection, and a judgment to come!'[21] There seems at times to be a positive nostalgia of belief in him, a very modern hankering after the green paradise of childhood faith, ' "when I was in my father's house, and my path ran down with butter

[19] When a new edn. of *John Buncle* appeared in 1824, the advt. carried a long quotation from *The Round Table* essay of 1815 (*The Times*, 29 Oct.).

[20] Patmore, ii. 276.   [21] 17. 110.

and honey" '.[22] Certainly he was sensitive to the feelings of others, morbidly considerate even, and Haydon (who had all the heat of a revivalist) bears witness that of all their circle he was the only one who discussed the question with the respect it deserved, and this, from an emotional Christian and devout Sabbatarian, meant that he was open-minded indeed. Just as he was always ready to acknowledge that there were 'livers out of England' he was always ready to concede that in this matter he might be mistaken, and his volume of *Notes of a Journey through France and Italy* is proof that he was prepared to push this to what was then the damnably heretical extreme of conceiving that there might be something to be said even for the Roman Catholics. In the present essay he concludes his criticisms of 'The Tendency of Sects' with a word of admiration for those 'retired and inflexible descendants of the Two Thousand Ejected Ministers':[23]

There is one quality common to all sectaries, and that is, a principle of strong fidelity. They are the safest partisans, and the steadiest friends. Indeed, they are almost the only people who have any idea of an abstract attachment either to a cause or to individuals, from a sense of duty, independently of prosperous or adverse circumstances, and in spite of opposition.

He seems reluctant to abandon his theme, and promises in a footnote to correct what might be its one-sidedness, by giving 'at some time or other' the 'reverse of the picture: for there are vices inherent in establishments and their thorough-paced adherents, which well deserve to be distinctly pointed out'. He could not have foreseen that before long these vices would shed their theoretical guise and come home to him in the thundering attacks of his *Quarterly* and *Blackwood's* critics.

By a coincidence he went on 12 September, two days after 'The Tendency of Sects' was printed, to see Bickerstaffe's *The Hypocrite*, an adaptation at one remove of *Tartuffe*, in which that monster of guile is metamorphosed into Dr Cantwell, a Methodist who preys on the stupid pretentious enthusiast Mawworm, a tradesman with aspirations to the pulpit (two characters who brilliantly anticipate the Revd Mr Stiggins and Uriah Heep). This play gave a new twist to his thoughts on Christian sects, traces of which survive in his last notice of the year, where he deplores the 'methodistical moral' in *The London Merchant* and blames Lillo for dragging the theatre into the service of the conventicle. 'On the Causes of Methodism', which appeared the month after he saw *The Hypocrite*, draws an acute and persuasive analogy between Methodism and Roman Catholicism (which answers, we may observe, to historians' subsequent independent explanation of the success of

[22] 12. 222.   [23] 12. 322.

Methodism in such a country as Wales, where the Church of England had never taken firm root, had never effaced the lingering memory of Catholicism). Reflection on his own case, on his loss of faith at Hackney years ago, on the disappointment and distress of his parents,[24] and on the sadness of his father's old age, darkened by those sins and weaknesses he vainly strove to conceal from him, did nothing to relieve his dejection. Years later, someone who asked Hazlitt about his father got this reply: 'Say nothing about my father; he was a good man. His son is a devil; let him remain so.'[25] The image of a debased religion, set over against the honesty and austerity of his father's life and its few, unvarnished beliefs, led him to advert once more to the relation between the real and the ideal, between reason and imagination. His reflections, shuttling between the general theme and his personal situation, inevitably settled upon his own case and took on the familiar note of self-condemnation:

Poets, authors, and artists in general, have been ridiculed for a pining, puritanical, poverty-struck appearance, which has been attributed to their real poverty. But it would perhaps be nearer the truth to say, that their being poets, artists, etc. has been owing to their original poverty of spirit and weakness of constitution. As a general rule, those who are dissatisfied with themselves, will seek to go out of themselves into an ideal world. Persons in strong health and spirits, who take plenty of air and exercise, who are 'in favour with their stars', and have a thorough relish of the good things of this life, seldom devote themselves in despair to religion or the Muses. Sedentary, nervous, hypochondriacal people, on the contrary, are forced, for want of an appetite for the real and substantial, to look out for a more airy food and speculative comforts.[26]

Not less depressing was the ill success of another devotee of the Muses, his brother John, for whom 1815 was also a bad year. As an extreme radical he too no doubt felt the defeat of Napoleon as much as his brother, and in his case also there were personal setbacks. That autumn he put his name forward for an associateship of the Royal Academy. He did not get a single vote. At the same time he did what no candidate ever did: advertise in the newspapers for sitters.[27] He was in very low water. Hazlitt must have felt for him. Family piety was deeply, inextricably intertwined with his spiritual roots. We have seen some of his visits to his parents at Wem and at Chertsey; it would be strange if there had not already been others; and we shall later see more, to Bath and to Devon. And in 1822, in a desperate year, he found time to concern himself with his brother's difficulties. Margaret Hazlitt says that 'while they lived,

[24] Moyne, 107.
[25] Haydon, *Corr*. ii. 401–2.
[26] 4. 58.
[27] W. T. Whitley, *Art in England, 1800–1820* (1928), 251, quotes the advt.

their bond of brotherly love was never broken',[28] and John's present troubles could not have found him indifferent. We tend naturally to assume that the absence of surviving letters implies an absence of relations between the two families in York Street and Great Russell Street but the absurdity of the assumption appears when we reflect that they lived a mere mile and a half apart; and we then begin to get a glimmering of the shadowy figures and journeyings that lurk beyond the gaps in the biography of a man like Hazlitt whose extant correspondence is so slight.

His visits to the Montagus in Bedford Square helped to shore up his sagging confidence. His embarrassing abandonment in mid-stream of the informal lecture on philosophy in their house did not affect Montagu's admiration. He thought so highly of his skill in debate that he tried at one time to persuade him to read for the Bar.[29] These visits marked the highest point he was ever to touch in the social scale. The square was then the abode of lawyers at the head of their profession (Lord Chancellor Eldon lived at No. 6) and ever since his return to London Hazlitt had been made welcome at No. 25, an elegant house in the north-west corner. Although he was intimidated by the supercilious footmen and shy of the elaborately gracious presence of the 'Noble Lady', Montagu's third wife, the widow of a Yorkshire lawyer, he thought her conversation as fine-cut as her features, and he liked, as he put it, 'to sit in a room with that sort of coronet face'. What she said left 'a flavour, like fine green tea'.[30] We cannot doubt that he enjoyed sitting in the handsome drawing-room over which she presided, in front of the Adam fireplace, or in the summer looking out at the trees through the tall south-facing windows that reached almost from the floor to the medallioned ceiling. He had a great respect for the culture, amiability, and generosity of mind of Montagu himself, and probably also a good deal of sympathy for that turbulent helter-skelter household wherein dwelt the children of four marriages (at least in that early period before the turbulence got out of control and turned sour: it is doubtful whether he was aware of the bad blood that the wilfulness and arrogance of Basil Montagu jun. were fomenting between father and stepmother, compounded later by his dissipation and professional irregularities when he too became a lawyer).[31]

His multiplying activities as dramatic critic, and as contributor to the

[29] *Four Generations*, grangerized copy, Buffalo, letter of 10 Jan. 1844, Anne B. Procter to William Hazlitt the younger, 'I have a letter from your Father written to Mr Montagu in reply to one urging him to go to the Bar.'

[30] 12. 41–2.

[31] Robinson Diary, 26, 29 Apr., 2 May 1817, 28 Jan. 1821; Carlyle, 243.

Examiner, to the *Champion*, and finally to the *Edinburgh Review*, earned
him fresh consideration in the Montagus' eyes and more than compensated for his awkwardness and tenterhook diffidence. Not that they
patronized only the successful; Hazlitt would have seen through them
immediately and would not have called a second time; no, they also had
a happy knack of applying soothing treatment and benevolent flattery to
the unknown and discouraged. He met new faces there. Montagu's
professional contacts and private interests had brought him a variety of
friends. It is unlikely that Hazlitt was invited at the same time as either
Wordsworth or Coleridge, but in addition to people he already knew,
such as Lamb and Crabb Robinson, there were others like Dr Parr, 'the
Whig Dr Johnson', who made Montagu's house his *pied-à-terre* when he
came up from Warwickshire, or lesser figures like Anthony Carlisle and
John Fearn, whom until now he knew but slightly or not at all.

There is a hint in *The Spirit of the Age* that he may have seen Dr Parr at
Horne Tooke's in the years before Winterslow.[32] The old Latinist was
now in his mid-sixties, but he seems to have retained enough freshness
and curiosity to add some of Hazlitt's books to a library which at his
death in 1825 harboured by contrast none of the works of the Lake Poets,
nor yet of the contemporary giants, Byron and Scott; he not only
acquired the *Characters of Shakespear's Plays*, the *Lectures on the English Poets*,
and *The Round Table*, but was even indulgent enough to describe the last
as 'very ingenious'.[33] The pleasure was not reciprocal: Hazlitt did not
think much of his style: he found it 'dull and commonplace — a little
bordering on fustian'.

Carlisle, surgeon in Westminster Hospital for twenty years, had been
Coleridge's doctor, and may already have met Hazlitt.[34] Hazlitt must at
least have known his name, from Holcroft's diary, as well as from the
conversation of Captain Burney, Rickman, and especially Ayrton, to
whom Carlisle was both physician and life-long friend (when he died in
1840 Ayrton described him as 'a brilliant genius and an honest man' but
'eccentric and mistaken in his notions of self-conduct. Hence lost all his
practice and died insolvent'[35]). Carlisle was also acquainted with
Godwin and with Lamb, who thought him the best raconteur he ever
heard, and used one of his stories in the *Essays of Elia*.[36] He might for
these reasons have been congenial to Hazlitt, and also from his
connection with art, having been Professor of Anatomy since 1808 at the
Royal Academy, where in his very successful lectures he exhibited

[32] 11. 49
[33] *Bibliotheca Parriana* (1827), 521.
[34] *Courier*, 16 Oct. 1816; Marrs, iii. 61.
[35] BL Add. MS 60373 at 2 Nov. 1840.
[36] Lucas, ii. 373.

prize-fighters, naked guardsmen performing sword exercises, and Indian jugglers tossing up balls, to illustrate the movements of the muscles and joints.[37] Parr thought him a profound philosopher, a valuable friend, and an animated writer, notable for his taste and various erudition, but he inscribed this in a presentation copy he received of one of Carlisle's books so we ought perhaps to suspect exaggeration.[38] Robinson observed on meeting Carlisle that 'his significant face and the deep and sentimental tone of his voice' seemed to promise much, but the diary carries no hint that the promise was kept; young William Bewick disliked his grandiloquence, and he is said by another witness to have been hard-headed, utilitarian, and a contemner of poetry.[39] This confirms Hazlitt's experience, who heard him at a Bedford Square conversazione (possibly the same occasion as that referred to by the previous witness) proclaim the 'uselessness of poetry'. Nor would Hazlitt have been mollified to find him quoted in the press as gratified by the violent seizure of the art-treasures at the Louvre in September 1815, when, after Denon, the director, had made a show of resistance, Wellington ordered the 71st Regiment to march from their camp in the Champs-Élysées and deliver an ultimatum: he was to hand over the keys or have 'the whole building shattered about his ears'.[40]

But Hazlitt was curious to find out if he really had anything in him, and attended one of his lectures at Somerset House with Bewick. (He had taken a fancy to the lad, who was just such another guileless youngster — 'a raw country lad. . . . so sensitive that [he] blushed at [his] own name'[41] — as he himself had been when he first came up to town.) His going in with Bewick rather than getting himself formally invited by some Academician (say, Northcote, or Flaxman) is characteristic: by temperament he preferred the informal to the formal, the student to the professor, the back row to the reserved seats. He was horrified and nearly fainted that night when the artistic medico passed a cerebrum on a dinner-plate around the audience of apprentice painters, and he rightly saw in this intempestive presence of brains an absence of common sense. As he and his young friend left, descending the great staircase, he said he was now persuaded that his 'unpoetic acquaintance' was really 'an arrant puppy' who could never 'find anything good, or of *use*, or beauty in poetry'. Years later he was glad to see Flaxman quoting the dogmatic and now knighted professor as admitting the uselessness of his own anatomical teachings, and thus by implication the inadequacy of science

[37] *Examiner*, 5 Mar. 1809, 157–8. Much of what follows is taken from Bewick, i. 140–3.
[38] Maggs cat. 381 (1919), item 1922.
[39] Robinson Diary, 5 Sept. 1819; Carlyle, 252.
[40] *Courier*, 26 Sept., 14 Nov. 1815; Farington, 18 Nov.
[41] Bewick, ii. 248.

and the superiority of art.[42] This is the only time he mentions him in his writings. He liked his views no better than he did Parr's prose.

Fearn's case was different. Hazlitt found him, as a writer, almost as uncomfortable and unsatisfactory as Parr, but for a different and perfectly respectable reason: it was because Fearn was genuinely striving, clumsily and without much success, to articulate new ideas, his own ideas, an enterprise which was bound to command Hazlitt's respect. Eleven years older than his fellow-guest, Fearn had been from 1785 to 1809 an indigo-planter in Bengal,[43] where, with no special training, he had discovered a passion for metaphysical speculation. His spontaneous and natural self-initiation into philosophy was particularly likely to commend itself to Hazlitt's own empirical approach, although he does not say so. It was by accident that, in the abundant leisure of life on a plantation, his mind had fastened on certain phenomena of consciousness as curious and calling for explanation. When bad health sent him home to England he brought with him and got published an *Essay on Consciousness* (1810), which Hazlitt, no doubt justly, claimed to have been one of the few people in the country to have read, and certainly one of the few to have made anything of. The Russell Institution lectures of 1812 — a series which Fearn would be quite likely to have attended — were not Hazlitt's last word on philosophy; he contemplated a further work, and continued to read all he came across, including this *Essay on Consciousness*. It was something of an enigma in parts, but he was glad to find in it a good deal to admire; and there is no doubt also that Fearn proved a useful foil in conversation at the Montagus'.

The repetition at Bedford Square of that first lecture of the series was not an isolated performance. Mrs Montagu was fond of asking a guest to open the conversation on her 'evenings' with a half-hour paper, and Hazlitt gave more than one such on metaphysics. We may assume that he made some contribution to the discussion of others, including at least one on painting, and another on Italian literature. The letter of 1826 in which Mrs Montagu mentions this in describing to Jane Welsh Carlyle (newly married, and ambitious to entertain) her former mode of receiving is somewhat laconic but worth quoting:

At the beginning of the season I sent one card to say the day I should receive my friends — they came without ceremony. The men formed themselves into little groups for conversation and the evenings went off most pleasantly. When the assembly was very large some gentleman read a short paper which took about half an hour. Hazlitt gave a few papers upon metaphysics and others

---

[42] 10. 350 .

[43] *East India Kalendar* for 1799, and for 1808, where he is last shown, at Pabna. *GM*, NS. 9 (1838), 216, records his death aged 70 on 3 Dec. 1837.

contributed upon painting, upon the Italian novelists etc. This half hour of
constrained silence did wonders for the general attractiveness of the rest of the
evening. We had a little, and only a very little music. As my health got more
delicate by degrees our parties were given up and those whom we liked *much*
were told to drop in on any evening . . . These evenings were pleasant and
inexpensive. You may have the choicest society of any place if you will divest
your invitations of the formalities that include full dress — Men, literary men,
are always ready to entertain and be entertained if they are not expected to visit
in full dress — being satisfied that 'life is too short for buttoning and
unbuttoning.'[44]

The solitary Hazlitt had learned, if not at his brother's, then at the
Lambs', how conversation breeds thought, how the mind strikes out
truth by collision, and in all likelihood he deliberately introduced into
the talk in the Montagus' drawing-room the themes of his *Examiner*
essays and *Edinburgh* reviews, as he did later in Haydon's studio. Fearn
did not open out in the same way to Robinson, who met this
'metaphysician of some note' in 1815, but who considered that he 'had
not the air of a deep thinker'; and Carlyle thought him 'utterly dull and
dry'.[45] The fact is, he was shy, although he had confidence enough on
paper to challenge the opinions of the eminent Dugald Stewart,[46] but his
native diffidence emerges in a letter of 1814 to Montagu where he
ingenuously deprecates the involuntary harm done to his theme (immortality) in one of his recent books by 'the poverty and lameness with
which his scanty means obliged him to treat it'. Parr's comments on this
letter are not entirely convincing, but they curiously confirm Hazlitt's
judgement of Fearn, as well as showing the affection in which the man
was held:

Scanty might be the external means which books afforded him; but his own
vigorous and reflecting mind supplied him with materials copious and most
useful . . . a very original view of many circumstances in the moral world . . . I
grant that his style, in the scientific part of his book, is very uncouth, and yet I
prefer it to the rhetorical diction of the second part . . . I love Mr. Fearn.[47]

Hazlitt, who had pushed his own philosophical speculations further than
either the lawyer Robinson or the young schoolmaster Carlyle, and who
never consulted the opinions of others before hailing as good what struck
him as good, no matter how unlikely the setting, was able to discern the
vein of originality in this dull ore, and said of the *Essay on Consciousness*:

---

[44] NLS MS 1776, fo. 31, *c*.20 Oct. 1826   [45] Robinson Diary, 17 Dec. 1815; Carlyle, 252.

[46] *A Review of the First Principles of Bishop Berkeley, Dr Reid, and Prof. Stewart* (1813); *A Letter to
Professor Stewart* (1817). A 'Second Letter to Professor Stewart' was printed in the *Sunday Times*,
22 Apr. 1827.

[47] The letter, and the comments dated 13 Oct. 1814, were in a copy of Fearn's *Essay on
Immortality* (1814) in Dr Parr's library: see *Bibliotheca Parriana* (1827), 434–5.

There are notwithstanding ideas in this work, neglected and ill-treated as it has been, that lead to more curious and subtle speculations on some of the most disputed and difficult points of the philosophy of the human mind (such as *relation*, *abstraction*, etc.) than have been thrown out in any work for the last sixty years, I mean since Hume; for since his time, there has been no metaphysician in this country worth the name.

Also, and perhaps more important, Hazlitt too was fond of him, and made a point of saying so, neglected or underestimated though he might be, and made sure that his affection would be apparent by giving him a niche in that touching gallery of unworldly idealists portrayed in the essay 'On People with One Idea', where Fearn floats for ever down the Ganges clutching his metaphysics in 'bamboo manuscript', nursing that yearning aspiration to fame which, says Hazlitt, being real and immediate, is incomparably greater than fame itself.[48]

In this autumn of 1815, himself haunted by the idea of fame, and at the same time by the threat of failure, Hazlitt would have been glad to shake off the oppressive miasma of a year of defeat by a change of scenery. He was invited to the Montagus' country house at Bolton Abbey in Yorkshire, but lacked the coach-fare. The letter of excuse we now quote bespeaks not only a closer relationship with the Montagus than has hitherto been suspected, but hints also at an extension of his travels — an acquaintance with that beautiful stretch of Wharfedale below Barden Tower which is now a National Park. If, as is likely, the Montagus usually holidayed there before autumn term opened at Lincoln's Inn, he may well have journeyed there the previous October (we know nothing of him between 6 October and 4 November 1814), but this is conjecture:

Addressed to Mrs Montagu, Bolton, near Skipton, Yorks. (postmarked 30 October, redirected to Basil Montagu at Lincoln's Inn)

Dear Madam,

I am quite ashamed of having not answered your obliging letter before, but I have put it off to the last in hope of being able to fix a time for coming down. I am at last prevented by money matters — 'The Fates & Sisters Three.' I have been going to set off several times with Bardyn Tower right in the horizon before me, & not to stop until I got there. It will, however, stand where it does next year, and so I live in hope. It is now too late for I hear you return on Saturday. With best respects to Mr. Montagu.

I remain, Dear Madam, your obliged and faithful friend and servant

    W. Hazlitt

October 23rd, 1815.[49]

---

[48] 8. 65, 64.   [49] Unpub., in the possession of Mr Michael Foot, MP.

The birth of a son to the Hazlitts on 28 November implies a reconciliation at the beginning of the year, which in turn suggests that the love-affair which Hazlitt connects with the eclipse of his public hopes must have ended with Napoleon's first overthrow and departure for Elba and not with Waterloo. The word reconciliation is probably inapt. The intelligent, tolerant, cultivated, and easy-going woman we recognize in Sarah Hazlitt, both from the Lamb letters and from her 1822 journal, is not likely deliberately to have made their life together irreconcilably difficult. When they separated she did not want to be divorced, and only agreed for her husband's sake, after much pleading; they met amicably for years after, and she continued her connection with the family and her visits to his mother and sister. Hazlitt himself must have been loath to let his parents see that, after the failure of their first plans to make him a pastor, and then of his career as an artist, his marriage too was running into trouble. Nearly seventy years later, upon the publication of *Memoirs of William Hazlitt*, his brother, John's, son, writing from Mauritius to his cousin William Hazlitt the younger, was scathing about his correspondent's father but sympathetic towards his mother. He said he had kept all knowledge of the book from his Mauritian friends because he did not wish it to be known that he was the nephew of the 'poor, weak, inconsistent, half-mad subject of the work', and went on to give his memory of his aunt as he knew her in his teens and his twenties: 'I knew your Mother very well, having so often seen her in her own house in London, and I must say that I never heard anything but that she was a remarkable and learned woman, and one whose society was sought.'[50] We have seen the impression Hazlitt's love for young William (more like a mother's than a man's) made on Robinson, Haydon, Bewick, and Mrs Montagu. On Sheridan Knowles also, who said he 'never saw a father who was more wrapt up in a son'.[51] The birth of a second son, if it brought increased care, must also have brought an addition in happiness to Hazlitt and his wife. They named the child John in token of Hazlitt's regard for his brother (although Sarah no doubt made the name serve a double purpose).

It is likely that they had resumed their visits to the Humes at Notting Hill, and it may have been there that late in 1814 or in 1815 they met the Bells, new acquaintances with whom they got on very good terms, but of whom we learn nothing at the time, and very little even when they briefly figure in the Hazlitts' affairs in Edinburgh in 1822. Though they frequently appear in Sarah's journal of that period no reference is made to their London origin and background. The narrow basis of this

[50] BL Add. MS 38904, fos. 37–8, 31 Dec. 1883.

[51] R. B. Knowles, *The Life of James Sheridan Knowles* (privately printed, 1872), 11.

suggestion of an introduction at Notting Hill is that John Robertson Bell's business activities made him acquainted with Hume at Somerset House, and that he and his wife may have been guests at Montpelier House, but it may just as well have been that Hazlitt met him at the Montagus'. There is mention in Robinson's diary, on an occasion when Hazlitt was present, that 'Fearn a metaphysician and one Bell, manager of the Gas Light's etc. came in and disputed on philosophical subjects'.[52] Nothing else links Bell with the new gas industry, but he did have sufficient intellectual pretensions for the dispute, and we do have evidence that in a few years' time Montagu acted as Bell's lawyer.[53] However, there is not much purchase for conjecture here, and we have to take into account yet another possibility: Bell was a drinker, and it may even have been that Hazlitt met him, as his brother had formerly met the offensive Brown, in a tavern, although socially Bell was a cut above Brown. At all events, obscure though the manner of their meeting may be, it is certain that the two couples were on visiting terms at about this time, and we must assume, even in the total absence of details, a fairly close connection over the next five or six years. Mary Ann Bell deposed in 1822 that she was 'intimately acquainted' with William and Sarah Hazlitt, and that the Hazlitts were 'in the practice of visiting in [her] family'.[54] We are faced here, as so often, with an awkward blank in the accepted story of Hazlitt's life, but the importance of the Bells' role in the divorce makes it necessary to fill it in at this point with some acccount of this obscure pair.

Bell was the son of Adam Bell, Master Cooper at the Royal Victualling Yard, Deptford, who, born about 1754, had been bound apprentice at fourteen to a Mr Young, then Master Cooper to the Victualling Office, and apart from a twelve-month period in 1779–80 at Tower Hill, spent the rest of his life in the Deptford Yard. At the time of his son John Robertson's birth, about 1783, Adam Bell and his wife Elizabeth were living in Butt Lane in the parish of St Paul's, Deptford, and he was second foreman at the cooperage.[55] By the time the boy was apprenticed cooper at the Yard on 1 January 1799 the father had become Master Cooper, and at the turn of the century he no longer appears on

---

[52] Robinson Diary, 22 Dec. 1816.

[53] PRO, B.7.27, p. 272, and B.7.29, p. 18.

[54] RHE, 9 docs. recording 13 separate oaths, depositions, proofs, etc. headed 'Divorce Stoddart v. Hazlitt, 1822'. Six of the docs. are reproduced in J. A. Houck, 'Hazlitt's Divorce: The Court Records', *Wordsworth Circle*, 6 (1975), 115–21. The present doc. 9 is quoted in the writer's 'Hazlitt's Mysterious Friend Bell: A Businessman Amateur of Letters', *EA* 28 (1975), 152. Bell deposed that the Hazlitts 'occasionally' visited his family, but stated, more precisely than his wife, that he had been acquainted with them for seven or eight years.

[55] *Parliamentary Papers: Reports from Commissioners*, 1806, vii. 639. John Robertson Bell's name is wrongly given as Adam Bell in Bonner, 179, 184, etc., and in Wardle, 318, 320 ff.

the Pay Lists as plain Mister but as important Esquire.[56] He is soon, at £250 a year, the highest paid of the dozen officers in the Yard, with the exception of the Clerk of the Cheques, the bold flourish of his signature suggests an aggressive character, just as an incident of 1808, when he was compelled by the Victualling Board to reinstate several workmen he had dismissed for refusing to work on Sundays, suggests a harsh one.[57] He kept his eye on the main chance, for himself and for his son, and the order of entry in the list of apprentices, determined strictly by seniority in the books of 1795, '96, '97, and '98, is abruptly dislocated in the first quarter of 1799 by the placing of the new lad J. R. Bell first on the list, and in the autumn of 1802 young John again leaps over the heads of the Second and First Foremen's servants to become the first of the two Master Cooper's servants. On 31 December 1804, when he figures for the last time in the Deptford pay-lists, the Yard employed 175 block coopers, 85 day coopers, 50 sawyers, and 400 labourers, for all of whom his father was responsible.[58] This gives some notion of the importance of Adam Bell's position, which in turn has, as we shall see, a bearing on the predicament that later forced his son to leave London for Edinburgh.

John Robertson Bell seems to have remained in the service of the Victualling Board in some capacity for another four or five years. By the end of 1805 he was also a liveryman of the Coopers' Company, and got his first apprentice. He took another in 1806 and another in October 1808.[59] The registers of St Paul's, Deptford, tell us that on 8 December 1808 (the same year that saw the union of William Hazlitt and Sarah Stoddart on the other side of the river) he was married to Mary Ann Tebbut, of the parish of St Ann, Limehouse, in the presence of Dorothy Bell — 'presumably a sister — and Ann Rott.[60] His wife was probably one of the shipbuilding Tebbuts, whose yard was at Limekiln Dock, Limehouse,[61] and whom John Robertson is likely to have met through the transactions of the Victualling Department; in any case it was fitting for so ambitious a man to have married into a family engaged in so enormously profitable a business as shipbuilding was during the Napoleonic wars. Later on, the baptism is recorded of his son Adam, whom we shall meet in Edinburgh; the child was born on 28 October 1809 when Bell, living in Evelyn Row, was described as a 'gentleman'.

[56] This, and the facts that follow, will be found at the PRO under the relevant dates in Adm. 113/62, pay-lists for 1796–1802.

[57] PRO, IND 4888/102–3.

[58] PRO, Adm. 113/63.

[59] Guildhall Lib., London. MS 5629, vol. 2, Cooper's Company, Register of Apprentices, 1 Oct. 1805, 4 Feb. 1806, 1 Nov. 1808.

[60] Parish Register.

[61] The firm of Tebbut, Hitchcock & Batson (*Kent's Directory of London*, 1808).

Later, on 3 February 1811, there was born a daughter, Mary Eleanor, who also came to Edinburgh, although not mentioned by name in Sarah Hazlitt's diary, where, however, Mary Ann's 'children' are said to be with her. About the time of the birth of his son, Bell resigned from the Victualling Department to set up as a merchant and agent in the City, making good use no doubt of contacts established while in Government service, and moving away from his birthplace where his parents had lived for fifty years. In 1811 we find him in partnership with one William Wilkinson as ship and insurance broker at 18 Change Alley, Cornhill.[62] His main business was the victualling on a big scale of naval and merchant vessels, but by turning to his own account the crafty attempt of an MP, also director of a timber-importing firm, to use him to capture the timber contract at Deptford, he had also been shrewd enough to get control of the entire import of Canadian staves for use in the Yard, at a time when Napoleon's blockade of the Baltic had cut off that traditional source of supply. The volume of business he handled between 1811 and 1815 must have been very great, and if, as his entrepreneurial enterprise makes not unlikely, he was also connected with the Gaslight Company, more considerable still.

At all events, by 1816, when he moved his office to 53 Old Broad Street, another step up in the commercial world,[63] he was a rich man, with a house in fashionable Blackheath (where lived Angerstein the banker, whose collection of pictures later formed the nucleus of the National Gallery). Bell had pretensions to culture, and his extensive library of some five thousand volumes, which Hazlitt may well on occasion have found a convenience, included the classics, philosophy, many works in French, all the standard authors, and, among contemporary writers, first editions of almost all of Byron, Crabbe, Godwin, Moore, Southey, and Mme de Staël (but no Wordsworth and little Coleridge), as well as history, travel, atlases, sheet music, prints, and engraved portraits of Shakespeare, Garrick, and Godwin.[64] His books seem to have been bought if not by the yard, then on the prescriptions of accepted contemporary taste, and not from personal preference. It was a Gentleman's Library, complete with indispensable plaster busts. The only apparent and surprising bias was towards education, and it was perhaps this that later, after the divorce, prompted Bell to advise Sarah Hazlitt to open a school for young ladies.[65]

Bell may well have stood in the same helpful relation to Hazlitt as

[62] *London Post Office Directory*, 1812.
[63] *Johnstone's London Commercial Guide*, 1817.
[64] BL SC, 1240, cat. of Bell's bankruptcy sale, 21–6 July 1819.
[65] Le Gallienne, 275; Bonner, 210.

Joseph Hume to Godwin. If it was through Hume that he met Hazlitt (Victualling Office records show Hume as Chief Clerk for Examining Agents' and Storekeepers' Accounts, accounts which certainly involved Bell in visits to Hume's office in Somerset House[66]), he may have imitated that humble Maecenas by extending loans to his hard-pressed friend. He would have been flattered by the interest of a writer, and perhaps particularly by an *Edinburgh* reviewer. On his library shelves stood a complete run of that journal from its inception in 1802, and also first editions of *The Round Table*, *Characters of Shakespear's Plays* and the *Lectures on the English Poets* (the last of Hazlitt's books published before Bell's bankruptcy and sale), of which more than one may have been a presentation copy. On the flyleaf of the *English Poets* Bell was able to point to the inscription, 'J. R. Bell, Esq. with the Author's respects'.[67]

---

[66] PRO, IND 4939, 13 Jan. 1817. See also *Parliamentary Papers*, 1806, vii. 622–3.
[67] CUL Keynes Coll.

9
The Round Table

MONEY had again become a problem, with the birth of this second son. It was useless to resort to portrait-painting, as in the last crisis: he must bring out a book. His spare time, given the necessary tranquillity, was in fact now being devoted to writing his work on 'metaphysics', to be entitled *The Philosophy of the Human Mind*, an adaptation, no doubt, of his Russell Institution lectures, and ultimately deriving, like them, from his prospectus of 1809. It seems to have been of the same order as his *Essay on the Principles of Human Action*, and an equally serious bid for fame, but, on the face of it, no more likely than that 'chokepear', or his father's volumes of sermons, published by subscription, or Fearn's *Essay*, to bring in money. If the *Principles of Human Action* of which he was so proud had not succeeded, what guarantee was there of the success of a further attempt in the same vein? His style in the new treatise was no doubt unchanged, but journalism had persuaded him to adopt elsewhere a brusquer, more direct, and more highly coloured manner. It had also made him better acquainted with the market. He was looking for a younger and more adventurous publisher than Joseph Johnson had been, who was in his late sixties when he published Hazlitt, and who, Hazlitt felt, had not pushed his work, or his father's, as much as he ought. An admirable character, widely respected, a Dissenter and a Whig, Johnson, who had died in 1809, had made so much money (over £10,000) by Cowper's works that he had little incentive to fight for his other authors. Hazlitt's letters, and Lamb's, at the time when the *Principles of Human Action* (1805) and the abridgement of Tucker (1807) came out, show that he was slow to move.[1]

For the *Reply to Malthus* (also 1807) Hazlitt had moved to Longman's, and in 1808 he had failed with Murray, but he was now hoping for someone more responsive and more dynamic still. He probably did not know that the not entirely implacable Robinson had recommended him, apparently unprompted, on 1 November 1815, to Cosmo Orme, one of Longman's partners, to write a preface for an edition of Burke's speeches, perhaps even to compile the work, but Orme did not take up the recommendation and the commission went to the editor of a collection of Fox's speeches published the year before.[2] But he was crestfallen at the silence that greeted the *Reply to Malthus*, and he was on

[1] *Letters*, 88; Marrs, ii. 199, 208.   [2] Morley, 176.

the look-out for someone quite new to undertake his latest project. There was apparently no end to the delays, objections, alterations, injunctions, dogging the *Memoirs of the Late Thomas Holcroft*. Hazlitt's introduction, or 'advertisement' is dated January 1810. The book was announced as being on sale on 20 April 1816[3] (the imprint is dated 8 April 1816); but on the very day that that advertisement appeared in the *Morning Chronicle* Kenney, presumably as one of the committee, wrote to William Shield promising a proof of a letter from Shield to Holcroft printed as part of the *Memoirs*, and inviting him to strike out anything he did not wish published.[4] It seems Shield did so. We do not know what Hazlitt thought of these setbacks to the 'Life everlasting' as Lamb called this book, but what he now envisaged was a collection of essays. The best of the eighty-odd pieces contributed to the *Morning Chronicle* and the *Champion*, added to the 'Round Tables' already published in the *Examiner*, plus those projected, would fill out a tidy book.

Jeffrey had accepted his proposal to review Schlegel's *Lectures on Dramatic Literature*, and although the book, announced as an imminent publication on 23 September, did not appear until 9 November,[5] Hazlitt, armed with an advance copy obtained from the translator himself, his friend John Black, hoped to be in time for the next *Edinburgh Review*, the belated 50th number. He was not, however, and on 20 November wrote to excuse himself, promising to forward the notice in another three weeks. It seems not to have occurred to Jeffrey, who had passed through London at the end of October after a month in Paris (he dined at Holland House on the 29th),[6] to look up his humble contributor, but there was compensation for Hazlitt in being approached by another important figure at this time, the publisher of the *Edinburgh Review*, Archibald Constable, who was also then in London. In the previous year Constable had appointed Macvey Napier, an Edinburgh lawyer and a contributor to the *Review*, to be editor of the Supplement he was planning to issue so as to bring up to date his recent purchase, the *Encyclopaedia Britannica*. Napier visited London in May 1814 in search of collaborators, and met with little success. One of his immediate concerns was to find someone to write on the fine arts. Twelve months later he still had no one, but in the mean time Hazlitt had so effectively made his name in that field that it had procured him entry into the *Review*, and when Constable left for London on businesss towards the end of October, Napier, who was just about to write the Advertisement to the

---

[3] *MC* advt.

[4] ALS, Kenney to Shield, Sotheby cat., 30 Nov. 1892, lot 44; 3. 274.

[5] *MC* advts.

[6] Cockburn, i. 238; BL Add. MS 51952, at that date; BL Add MS 52181, fo. 75.

Supplement, must have asked him to offer the commission to Hazlitt.[7] He did so, and Hazlitt accepted for February 1816, but the task evidently held little appeal for him. The pretension to authority and to finality implied in contributing to encyclopaedias was alien to his temperament, but any commission was better than none at a time when money was short, and in any case, by putting him in touch with the firm of Constable, it enabled him to make the first move towards getting the *Round Table* published.

He made the proposal to Constable, who proved agreeable. And a week before Christmas 1815 Hazlitt drafted the following proposed list of contents (including essays by Hunt) and delivered it, together with the contract dated 18 December, at Walker's Hotel, Blackfriars, on the eve of Constable's journey north after his long stay in London. His covering letter corrects the impression he had given that he would send the essays in manuscript: what he meant was, he said, *in print* (i.e. cut out of the *Morning Chronicle*, *Champion*, and *Examiner*), and he would forward half the collection in that form 'at the end of next week'.[8]

I agree to give to Mr. Constable & Messrs. Longman & Co. the right to print one edition of a thousand copies of a work to be entitled The Round Table, or a collection of miscellaneous Essays, in consideration of Fifty Pounds to be paid in half a year after the publication, the price of any other edition remaining to be arranged afterwards, & the copyright remaining with the authors.

Wm. Hazlitt.

The subjects to be as follows or nearly so

1.	Introduction	L. Hunt
2.	On the love of life	
3.	On classical education	
4.	On character of women	L. Hunt
5.	On egotism	L.H.
6.	On the Tatler by Steele	
7.	On the imagination	
8.	On the passions	
9.	On the character of Iago	
10.	On decline of modern comedy	
11.	On the literary character	
12.	On a country life	

[7] NLS MS 789, fo. 441. The 'Advertisement', or preface, to Vol. 1, Part 1, by Macvey Napier, is dated Nov. 1815, and Part 1 itself appeared 4 Dec. 1815 (*Courier* advt., 20 Nov.), but, as the advt. said, 'the list of contributors had long been before the public', and the *prospectus* had appeared at the end of 1814, announcing Vol. 1, Part 1, for the spring of 1815 (BL 1879b. 1, Dawson Turner Coll.). The list of contributors, mostly Edinburgh men, included Jeffrey, Leslie, and Brewster, but also 'Mr Stoddart (London)'. Vol. 1, Part 2, containing 'The Fine Arts, by William Hazlitt, Esq.', was advertised in the *Courier*, 25 June 1816.

[8] First pub. NNHL, 265–6. The version in *Letters* is faulty: Hazlitt did not write 'When I said I could not let you have the manuscript . . .'.

13.   On patriotism
14.   On Milton's Lycidas
15.   On Milton's general character as a poet
16.   On importance of manner
17.   On Chaucer Squire's Tale   L.H.
18.   On Dissenters
19.   Character of John Buncle, a theological romance
20.   On causes of Methodism
21.   On poetical character   L.H.
22.   On means of making a sick-bed comfortable   L.H.
23.   On Midsummer Night's Dream
24.   On the Beggar's Opera
25–8. On Sir Joshua Reynolds's Discourses & pictures
29.   On posthumous fame
30.   On religious hypocrisy
31–2. On Mr. Wordsworth's Excursion
33.   On Hogarth's marriage à la mode
34–6. A day by the fireside   L.H.
37.   Whether public institutions promote the fine arts
38.   On taste
39.   On beauty
40.   Comparison between Henry VI & Richard II
41.   On pleasure derived from tragedy
42.   On dramatic illusion
43.   On merit
44–6. On modern philosophy
47.   On good-nature
48.   On commonplace people   L.H.
49.   On commonsense   L.H.
50.   On acting, & actors

A glance is enough to show that this does not correspond to the contents of the *Round Table* as published. Instead of ten essays Hunt ultimately provided five, one of which was the excellent but disastrous 'On Washerwomen'. It is not unlikely that Hazlitt's reputation would have grown faster if his essay 'On modern philosophy' (Nos. 44–6) had been retained, and more securely if it had not been replaced by Hunt's piece, which was to be the laughing-stock of the *Quarterly* and *Blackwood's* for years to come, and which cast ridicule on Hazlitt by association. Two explanations are possible: 'On modern philosophy' may have been a new, generalizing title for his *Morning Chronicle* paper on 'Mme. de Staël's Account of German Philosophy and Literature', and may have been sacrificed to Hazlitt's doubts about the commercial value of his 'aforesaid logical way of writing'. On the other hand it may have been an entirely different essay which Hazlitt eventually decided to incorporate in (or not to detach from) his *Philosophy of the Human Mind*.

The first twenty-four papers had already appeared in print. The titles

in general are the same as their eventual titles in the *Round Table*. Others
are recognizably similar: No. 9 is unlikely not to have been 'On Mr.
Kean's Iago', and No. 12 'On Love of Nature', No. 15 must be 'On
Milton's Versification' (called 'Milton's Works' in the *Examiner* index),
and No. 16, 'On Manner'. No. 17 is Leigh Hunt's 'On Chaucer'. No.
18 is generalized into 'On the Tendency of Sects', for reasons we have
glanced at, and No. 19 contracted into 'On John Buncle'.

Two of the first dozen titles failed to appear in the book: they are Nos.
7 and 8. The first is 'Round Table No. 9,' published in the *Examiner*,
26 February 1815, where it is indexed as 'On the Predominant
Principles and Excitements in the Human Mind'; and the second,
'Round Table No. 13', of April 1815, indexed as 'Love of Power or
Action as Main a Principle in the Human Mind as Sensibility to
Pleasure or Pain'. These abstruse themes suggest the motives for
omission we have already brought forward: namely, diffidence or the
divarication of plans (or perhaps a combination of both); they are in any
case permanent themes in Hazlitt's work. The two essays, conflated,
were eventually published by William Hazlitt the younger in *Winterslow*.
As to diffidence, Hazlitt's financial situation was precarious, and the
success of the book vital. A contribution that could be sent at a venture to
a newspaper, and might there pass, required to be carefully weighed as
part of such a book.

Hazlitt speaks of the 'collection' as though it were all newspaper
cuttings, but this could not have been true of the second batch. Hunt's
four essays had already appeared, but of the remaining twenty-two to be
contributed by Hazlitt only fifteen were at this time in print; fifteen, that
is, if we take the modern philosophy essay to correspond to the Mme de
Staël articles; but if they were part of the projected *Philosophy of the Human
Mind* then only twelve of those intended for the second volume were in
print.

Leaving aside 'On modern philosophy', then, as uncertain, there
remain seven not yet printed at the time of the letter. Of these, Nos. 40,
47, and 50 did later appear in the *Examiner*, one of them, 'On acting, &
actors' (as 'Round Table, No. 48'), as tardily as the day on which the
'Advertisement' to the *Round Table* was signed, 5 January 1817.

If they had not yet appeared it was perhaps because they were not yet
written, and existed only as projects, a word or two, or a sentence,
scribbled on that fantastic wall around the chimney-piece in that bare
living-room. This would account for 'On common sense', 'On pleasure
derived from tragedy', and 'On dramatic illusion', which, although
entirely characteristic, as themes, of Hazlitt's preoccupations, seem
never to have achieved existence as essays. The second is reminiscent of
the opening of 'Othello' in *Characters of Shakespear's Plays*, and the third, of

'Mr. Kean's Richard II' or of 'The Tempest', in *A View of the English Stage*, or even of the last page of 'Whether Actors ought to sit in the Boxes?' in *Table-Talk*. However, the two disembodied titles turned up years later in an intriguing coincidence: in Hazlitt's volume of *Characteristics* (1823) two contiguous aphorisms, Nos. cclxxxviii and cclxxxix, begin respectively as follows: 'The pleasure derived from tragedy is to be accounted for in this way . . . ', and 'The question respecting dramatic illusion has not been fairly stated . . . '. The essays not yet printed nor perhaps even written are evidence of permanence of interests and steadiness of purpose, but we ought also to look at those which were abandoned. We saw that he either feared some of them might prove, in words he later used of the Scottish economist J. R. M'Culloch's writings, 'a load to sink a navy', or else he decided to put them into *The Philosophy of the Human Mind*: at all events he dropped all the philosophical essays, including one he had first thought to disguise behind the title 'On merit'. This is plainly no other than 'Round Table No. 27', 10 December 1815, indexed as 'On the Doctrine of Philosophical Necessity'. Recovered and printed in 1904 by Waller and Glover under this index designation, its proper and significant title is now restored.

Others, as we shall see, he was obliged to sacrifice to Macvey Napier and the *Encyclopaedia Britannica*. Just after Christmas he received a letter from the editor, who was unable to wait for Constable's return: the publication of Part 2 of the Supplement was pressing. Hazlitt delayed answering until 10 January, either from lack of enthusiasm for the commission or on the assumption that Constable, who had in fact arrived home on Christmas Eve,[9] would have conveyed to Napier his acceptance, and it was not until the next day that he sent off his promise to supply by mid-February 'what was wanting on the subject of the Fine Arts' in the original *Encyclopaedia Britannica*, taking care as he did so to forestall any suggestion that he was expected to incorporate in his article a digest of available information and a summary of received ideas.[10] He was the last person to make an efficient contributor to an encyclopaedia.

At the end of January or the beginning of February he sent the first batch of half the *Round Table* essays to Edinburgh. Constable promptly set up the type and on 10 February got ready to send the first proofs to London, but some mysterious difficulty arose, and the proofs, together with Constable's covering letter, were set aside.[11] Hazlitt's anxiety at their non-appearance and at Jeffrey's silence over the Schlegel review which was now in his hands, is apparent from the letter he wrote to Constable on the 19th: 'I sent off the first half of the Round Table some

---

[9] NLS MS 789, fo. 466.
[10] *Letters*, 151 (the *p—k*, not there noted, is 11 Jan.).
[11] NNHL, 270-1.

weeks ago, and I begin to fear some accident has happened to it or that you do not approve of the contents. Neither have I heard of the Review from Mr Jeffrey, so that I am getting *blue-devilish* on that score also.'

The same day, with noticeably greater promptitude than in accepting the original commission, he wrote to Napier to thank him for the extension.[12] It is apparent from his relief that the commission was proving even more exacting than he anticipated, and this accounts for his being obliged in the end to sacrifice some of the articles he earmarked for the *Round Table* to fill it out. Apart from the articles on philosophy it is those on painting that were finally omitted from the December list.[13] No. 37 (the piece that had attracted Lady Mackintosh) and Nos. 25–8, the study of Reynolds and his discourses, were cut up and fused with the articles on Gainsborough and Wilson, and with parts of others on various exhibitions, published in the *Champion* and the *Morning Chronicle*. He later stopped the gap in the *Round Table* with two other pieces on art, 'On the *Catalogue raisonné* of the British Institution', and 'Why the Arts are not Progressive'.

His anxiety about his book was understandable: Leigh Hunt's *Story of Rimini*, although accepted by Murray as recently as New Year's Day, had already appeared, on the previous Tuesday, four days before this letter, and Hazlitt had written warmly to congratulate his younger and more favoured friend, but he must have envied the ease of his success and would have been more anxious still if he had known that he himself had such a long wait ahead of him. He later turned to account his reflections on *Rimini* in the *Edinburgh Review* for June, in an article so drastically revised by Jeffrey that it is usually excluded from the Hazlitt canon.

Nothing is known of his life during the winter of 1815–16, beyond what may be inferred from his articles and his few letters, but we may take it that he continued his visits to Bedford Square, and with less formality to the Lambs in Inner Temple Lane, and the Humes out at Notting Hill, that he dropped in from time to time at Haydon's studio in Great Marlborough Street, and in the spring walked out of a Sunday to call on Hunt, who had moved to the Vale of Health. On 9 December we glimpse him at Alsager's, where Robinson found him 'sober, argument-ative, acute, and interesting'. Walter Coulson, in his early twenties at this time, was a protégé of Bentham, who was a friend of the Coulson family and had brought him up from Devonport to spend three or four years in his house as one of his secretaries.[14] A clever man of wide

[12] Wrongly dated 9 Feb. in *Letters*, and with no indication that 'obliging' was an interlinear addition.

[13] My former opinion (NNHL, 268), mistaken, as I now believe, was that these *Champion* articles were omitted on account of Hazlitt's quarrel with Scott.

[14] As late as 16 Feb. 1822 Robinson Diary has, 'he lives much with Jeremy Bentham'.

information (Hunt called him The Admirable Coulson), he was now a reporter on the *Morning Chronicle*. He later became editor of the *Traveller* newspaper, and then a successful lawyer, and at the same time a contributor to the *Edinburgh Review* on the recommendation of M'Culloch.[15] He knew most of Hazlitt's friends, and was constantly in and out of his house, but although godfather to William Hazlitt junior, in 1814 he was, as we shall see, closer to Sarah Hazlitt, whom it is pleasant to find, like Mary Lamb, mothering the youngsters in their set. Hazlitt mentions him only once, and then only to complain of his never having said a word at a critical moment that might have served him with his landlord.[16] He was a sober, long-headed young man; Bentham's final, and surprising, word was that he was 'forward, but cold'; Robinson was puzzled to reconcile his reputation with his dry manner;[17] and his preferred companion was the dour, obsessively practical, and tetchy James Mill.

With Mary once more at home, discharged from her asylum, Lamb was in high spirits, 'good-humoured and droll, with great originality', whereas two months earlier, following her breakdown in September (brought on, he feared, by his own worries over his work at East India House which threatened to submerge him), he had been desperately lonely and depressed.

Towards the end of winter, on Saturday, 10 February, Robinson again met Hazlitt at Alsager's: 'After calling at the Colliers' I went to Alsager's, with whom I spent an agreeable [time] without cards or music: the party, Burrell, Godwin, and Hazlitt. The conversation was very shrewd on Hazlitt's part, captious and minutely critical and even rude by Godwin.'[18] We are not told what the occasion of the rudeness was, nor the target, but it was probably politics. The other topics were Wordsworth, and *The Fable of the Bees*. Rumours of the Keswick gossip put about by Wordsworth may by now have reached Hazlitt's ears:

Hazlitt was bitter, as he always is, against Wordsworth, who, he says, is satisfied with nothing short of indiscriminate eulogy, and who cannot forgive Hazlitt for having passed him off with a slight reserve of blame . . . Mandeville was praised. Hazlitt asserted, and on a reference to the book it appeared, that the leading ideas of Mandeville are contained in an essay by Montaigne entitled 'One Man's gain another's loss.' Both Godwin and Hazlitt expressed themselves very strongly in admiration of Montaigne. Godwin, Burrell and I left Alsager's at ½ past 12. As usual, Hazlitt stayed behind. Alsager had liquor behind which Hazlitt wished to drink, and Hazlitt's sense Alsager enjoys and appreciates.

[15]   BL Add. MS 34614, fos. 146, 158, 469.
[16]   *Memoirs*, i. 213; BL Add. MS 38899, fos. 390–2; Howe, 291.
[17]   Robinson Diary, 9 Nov. 1817.
[18]   Morley, 179, where 'cards' is misread (very improbably) as 'conversation'.

But he did not enjoy and appreciate it for much longer. Hazlitt cannot have failed to bring up Wordsworth's politics, which were those of his brother-in-law, and (as he must have known) of his brother-in-law's friend Burrell, there present. We may find here another instance of his want of discretion, leading to the alienation of a friend. Burrell seems to have become as frequent a visitor at Alsager's as Hazlitt had previously been, and at the end of the year Robinson met him at Alsager's when they were the only three present. They talked about Jacobinism, and then 'We talked about Hazlitt and were agreed in opinion. Alsager is a very tolerant man but Burrell says he would not meet him in company. This is going beyond me somewhat.'[19] Alsager's tolerance was not proof against the influence of Burrell and Robinson (and of others, Barnes in particular) and we never see Hazlitt in his house again, the house of 'the kindest of neighbours, a man of business, who contrived to be a scholar and a musician'.[20]

In the interval Hazlitt had had, personally if not professionally speaking, a good season at the theatre, at first with the *Merchant of Venice*, *The School for Scandal*, *As You Like It*, *The Beggar's Opera*, *A Midsummer Night's Dream*, *Venice Preserved*, and *A Trip to Scarborough*. The snowy Christmas season brought the pantomimes; indifferent this year, except at Covent Garden, where he was glad to see Grimaldi, recently rumoured dead, come on and exuberantly let off a culverin at his enemies; his enjoyment of this turn is typical of the simplicity and naturalness of his taste.

In February the theatre-goers saw a dramatic episode prepared by no author's pen. Not, so far as we know, witnessed by Hazlitt, but certainly by the Lambs, it was one of those shocks that so often shatter a certain image of the Regency as an urbane and colourful pageant, one of those brief explosions of violence, madness, and despair that burst on to the stage from the wings in that ceremonious and glittering civilization, an incident that must have brought to mind, for Lamb and his sister, when they recovered themselves, the personal story of their friend Basil Montagu as well as that moment of terrible violence in their own past. They were in the pit at Drury Lane watching Miss Kelly in a farce, when an unbalanced young man in the audience who had sent her several love-letters and who imagined himself a rejected lover rose, a few seats behind them, and fired a pistol at that 'divine plain face', beloved also of the solitary Lamb. Some of the shot fell in Mary's lap. Hazlitt, who must have heard the story from them and whose mind was no stranger to the emotional frustrations behind that kind of violence, saw in it a confirmation of a very personal theory of what he took to be a singularity

[19] Robinson Diary, 28 Dec. 1816.   [20] Hunt, 246.

of modern love, that since it was now more often cerebral than directed at mere possession of the person, lovers who had formerly avenged themselves on their rivals now did so on their mistresses.[21]

The play which is a type of sexual jealousy, *Othello*, was always very close to his imagination. He had seen it again, a few weeks before, with Kean as the Moor, and pondered again 'the tug of war between love and hatred, rage, tenderness, jealousy, remorse'.[22] Kean, he thought, had too much passion, energy, relentless will; he lacked imagination, that grandeur of pathos which he occasionally expressed so well, as in 'the tone of voice in which he delivered the beautiful apostrophe, "Then oh, farewell!" ', which 'struck on the heart like the swelling notes of some divine music, like the sound of years of departed happiness'. Yet it was, without exception or reserve, his best character; it was the highest effort of genius on the stage. So he thought on 7 January. Within a week he was to find Kean surpassing himself in the role of Sir Giles Overreach which raised the town's enthusiasm to a pitch it had not reached since his début, and electrified even Kean's fellow-actors into presenting him with a commemorative cup. Immense crowds swamped the repeat performances, and, most startling of all, the great Sheridan came back one night to sit in his box at Drury Lane, for the first time, it was said, since the old theatre was burned down in 1809.[23]

This new access of celebrity stood Kean in good stead at the end of March when he played in a further Massinger revival, *The Duke of Milan*, and got so drunk that he had to invent a spectacular accident with an upset gig to account for his failure to appear on the night of the 26th. Hazlitt seems to have known him at least well enough to invite him to dinner.[24] Not that this, any more than his acquaintance with other players, invalidates his assertion that he had no wish to penetrate the secrets of the green-room, and preferred not to see actors except across the footlights: Kean was in any case and in all ways an exception. He was as great a drunkard as he was a genius, and Hazlitt is more likely to have met him at the Harp, or at the Coal Hole in the Strand, than backstage.

[21] This incident was reported in *Courier* and *MC*, 19 Feb. 1816. See Hazlitt, 'Characteristics', Nos. ccxlvii and ccxliii–ccxlvi (9. 203).

[22] 5. 271.   [23] *Courier*, 3 Feb. 1816.

[24] In 1819 Hazlitt said that the only actor with whom he ever had any personal intercourse was Liston (5. 177). Perhaps he did not mention Kean because he was conscious of the rumours that he had been paid to write him up. On the other hand, the dinner invitation to which Kean replied on 30 Mar. 1816 (Sotheby cat., 26 Feb. 1906) may have been from John Hazlitt. Hazlitt was invited to meet Mrs Siddons at Haydon's, probably after the exhibition of *Christ's Entry* (*TLS*, 22 Jan. 1971, 101), but refused, from, as he said, reluctance to imperil a cherished illusion ('I shall not come, for I have been accustomed to see Mrs Siddons only on the stage, and to regard her as something almost above humanity, and I do not choose to have the charm broken'). He certainly in 1819 knew John Howard Payne, whom he probably met through Godwin: a letter of 8 Feb. is quoted in the writer's 'Some New Hazlitt Letters', *NQ*, July–Aug. 1977, 339.

These were houses that Hazlitt doubtless knew well, and often visited
when he had seen the play and written out his notice in the *Examiner*
office.[25] Kean's accident had the happy result of impelling Hazlitt to
write a subtle, sympathetic, and convincing vindication of the improvidence, shiftlessness, and drunkenness of players against the self-righteous
flings and sneers of the solid citizenry, a vindication also, by implication,
of his own life, which was based on feeling and on the hazardous exaction
of complete freedom in the pursuit of excellence: 'A man of genius is not
a machine.'[26] He was on familiar ground, and his position in this
endlessly recurrent collision between prejudice and sympathy, restraint
and liberty, narrowness and generosity, spontaneity and respectability,
self-regard and disinterestedness was one he had elaborated over many
years of self-examination, self-mistrust, remorse, and defiance. Patmore's perceptive and penetrating account of his method of dramatic
criticism is in point here:

To the play itself . . . he paid scarcely any attention, even when he went there in
his critical capacity as a writer for the public journals; for, notwithstanding the
masterly truth and force of most of his decisions on plays and actors, I will
venture to say, that in almost every case, except those of his two favourites,
Kean and Liston, they might be described as the result of a few hasty glances
and a few half-heard phrases. From these he drew instant deductions that it
took others hours of observation to reach, and as many more of labour to work
out. In this respect his facility was, I imagine, never before equalled or even
approached; and his consciousness of, and confidence in it, led him into a few
ridiculous blunders. Still, upon the whole, he was doubtless right in trusting to
these brief oracles and broken revelations, rather than pursuing them to their
ultimate sources — as most others must do if they would hope to expound them
truly and intelligibly: for his was a mind that would either take its own course
or none; it was not to be 'constrained by mastery' of rule or discipline. It was a
knowledge of this truth, and his habit of acting on it, which constituted the
secret of his success as a writer.[27]

Patmore might have added some of his favourite plays to the two
favourite players he named, but for the rest he is right, and the proof is in
Hazlitt's frequent errors of fact, inconceivable in a minutely conscientious critic.

   Among the papers he wrote at this time were 'Mr. Locke a Great
Plagiarist', and 'On Pedantry'. The first, in which he uncovers Locke's
debt to Hobbes, was no doubt a by-product of his reading for his book on
metaphysics, a labour of which we get a wry glimpse when he jokingly
remarks, in noticing *The Duke of Milan*, upon 'the systems of ethics

[25] See *The Times*, 5 Sept. 1816, on public houses and gin-shops near the two theatres.
[26] 5. 293.   [27] Patmore, ii. 317.

continually pouring in upon us from the Universities of Glasgow, Edinburgh and Aberdeen'.[28] His point of departure was Dugald Stewart's 'Dissertation' on the history of post-Renaissance philosophy, in Vol. 1 of the *Supplement to the Encyclopaedia Britannica*, but it is clear that he was again immersed, not only in Hobbes and Locke, but also in Berkeley and Hume, Priestley and Horne Tooke, Malebranche, Condillac, and Helvétius.[29] The second paper is a defence of exclusive interests and self-absorption against his own uneasy sense that they were a betrayal of the principle of sympathy, of the liberal imperative to look abroad into universality; it is a hint of that long, slow, reluctant withdrawal into a private world of memory, increasingly perceptible in his writings after Waterloo, a decline he was not fully aware of at the time, but which was plain when he looked back on that crisis twelve years later: 'Since then, I confess, I have no longer felt myself young, for with that my hopes fell. I have since turned my thoughts to gathering up some of the fragments of my earlier recollections, and putting them into a form to which I might occasionally revert. The future was barred to my progress, and I turned for consolation and encouragement to the past.'[30]

He now asserted that the power of attaching an interest to any pursuit in which our whole faculties are engaged is one of the greatest happinesses of our nature, and is so far to be desired, although liable to charges of prejudice and submission to habit. Among the examples he adduces of this broad class of pedants are (need it be said?) essayists, artists, authors in general, and readers of the *Fratres Poloni*. The memory of these seventeenth-century Socinians in his father's library at Wem brings a pang of that nostalgia which began to well up after Waterloo and became more and more characteristic as the years pressed in on him. Speaking of the veneration he had in his childhood for this 'learned lumber' (and unwittingly anticipating Proust's romantic paradox of the superiority of soap-advertisements over Pascal's *Pensées*), he says, 'We would rather have this feeling again for one half-hour than be possessed of all the acuteness of Bayle or the wit of Voltaire.' There is the same sense of loss in a dramatic criticism of 5 October 1815, the sense of the irrecoverable uniqueness of past experience that Proust expressed so

---

[28] 5. 290.

[29] In the first part of 'Mr. Locke a Great Plagiarist' he promised to return to the subject in three separate papers, 'On Imagination, Wit, and Judgment' (20. 72). They did not appear. The first is evidently No. 7 of his Dec. list, and the second may be the 'Definition of Wit' pub. in *Literary Remains*.

[30] These are the words of a manuscript of 'On the Feeling of Immortality in Youth' pub. by his son in *Winterslow* (1850). In Hazlitt's final version of the essay he adopts a bolder, a Miltonic expression of the aspiration to fame, saying instead 'and putting them in a form that might live' (17. 197).

poignantly in looking back on the form and colour of his childhood, when he says of the roads and paths he trod at Combray:

C'est parce que je croyais aux choses, aux êtres, tandis que je les parcourais, que les choses, les êtres qu'ils m'ont fait connaître sont les seuls que je prenne encore au sérieux et qui me donnent encore de la joie. Soit que la foi qui crée soit tarie en moi, soit que la réalité ne se forme que dans la mémoire, les fleurs qu'on me montre aujourd'hui pour la première fois ne me semblent pas de vraies fleurs.'[31]

Hazlitt's closer anticipation of this passage had to wait for another five years, but the seed is visible in what he now wrote: 'Why can we not always be young and seeing The School for Scandal? This play used to be one of our great theatrical treats in our early play-going days. What would we not give to see it once more, as it was then acted, and with the same feelings with which we saw it then?'[32]

On 18 March the last part of his article on the Fine Arts was ready and he sent it off with some misgiving: was it too long? was it offensive in its political allusions? The weeks had slipped by without a word from Constable. The following day he wrote to Constable again, saying he had begun to be apprehensive, from not having heard anything of the *Round Table*, that some objection or difficulty had occurred, and offering to make any alteration Constable might suggest 'as to particular passages that might require to be softened'.[33] The notion of blue-pencilling the essays will seem odd unless we remember that a subsequent reviewer in the *Literary Gazette* raised horrified hands at their 'gross indecency, libertine principles, and . . . spirit of irreligion and scepticism'.[34] In that era the right-thinking adherents of Church and Throne were not slow to establish a connection between political subversion and religious scepticism, or rather to re-emphasize a long-established identity, by assiduously pointing out that *The Age of Reason* came from the same pen as *The Rights of Man*. And in any case, not even the most scrupulous expurgation of these essays of Hazlitt's could sufficiently atone for the crime of their having first appeared in that sink of republicanism and free thought, the *Examiner*.

He had to wait another six weeks for the February *Edinburgh Review*, but, as we have seen, Constable had already on 10 February made *The Round Table* the subject of a letter which for some reason was put aside

---

[31] *A la recherche du temps perdu* (Pléiade edn., Paris, 1954), i. 184.

[32] 5. 250.

[33] The date 21 Mar. has been proposed for this letter (Wardle, 165), evidently in the belief that 20 Mar. is the date of the preceding letter (*Letters*, 157 and 156, respectively) whereas '20 March' is an Edinburgh receipt-docket. There is no doubt that this letter was written on the 19th, whatever delay may account for the Edinburgh arrival-postmark of 23rd.

[34] 3 May 1817, 228.

until well after the above letter had been received. It was finally sent off on 27 March together with proofs of the first four sheets and a list of the papers received, which Constable said differed a little from Hazlitt's original list. The printing, he hoped, would be finished in April, and he asked Hazlitt to provide a title-page, which should not, he felt, mention that the essays were '*From the Examiner*'. In conclusion he said, 'I have not overlooked your work on the Philosophy of the Human Mind. I hope to be able to get you a few subscribers in this quarter.'[35]

Here is one reference to a work on metaphysics which was presumably based on the prospectus of 1809 and the Russell Institution lectures of 1812. Altogether he delivered four courses of lectures on different subjects in seven years. We may wonder why he published the last three as soon as they were delivered, but not the first. Part of the answer may be that in their case he had learned from an initial fiasco that the only thing to do was to publish the lectures just as they stood as soon as the courses were over, since it now appears that he tried for years to turn the philosophy lectures into a book, and that here, in 1816, he was still canvassing subscriptions. A year later it seems he had still not completed the work to his satisfaction, if, as is probable, he refers to it in a letter of 20 April 1817 where he invokes Jeffrey's help over the launching of his latest book, *Characters of Shakespear's Plays*: 'If I could dispose of the copyright of the *Round Table* and of this last work, I could find means to finish my work on Metaphysics, instead of writing for three newspapers at a time to the ruin of my health and without any progress in my finances.'[36] Four years later he was offering 'articles on Modern Philosophy, 8 in number, at 5 guineas a piece' to the publisher Baldwin.[37]

Nothing came of all this, but it is apparent that the birth of his son and the signing of the *Round Table* contract started or revived as many 'plots and projects' in his head as had churned there in 1810. Macvey Napier had not found 'The Fine Arts' entirely satisfactory, and with reason. Hazlitt used a freedom of tone, took liberties of expression, that accorded ill with the abstract impersonality proper to an encyclopaedia and to the exiguousness of the space available. He hardly altered a word of the passages transferred from his *Champion* articles, and in no way modified the familiar tone. He retains, for example, in one paragraph, the following discordant gambit: 'Shall we speak the truth at once? In our opinion Sir Joshua did not possess either that high imagination, or . . . '.[38] He countered Napier's criticisms in a letter of 2 April, which we have already quoted, explaining his tendency to paradox (a letter written, unexpectedly, we may note, from Southampton Buildings, at a moment

[35] NNHL, 271.   [36] *Letters*, 171.   [37] Ibid. 203.   [38] 18. 58.

when he was perhaps on difficult terms with his wife in York Street):

I daresay that your objections to several of the observations are well-founded. I confess I am apt to be paradoxical in stating an extreme opinion when I think the prevailing one not quite correct. I believe however this way of writing answers with most readers better than the logical. I tried for some years to express the truth and nothing but the truth, till I found it would not do. The opinions themselves I believe to be true, but like all abstract principles, they require deductions, which it is often best to leave the public to find out.[39]

Clearly this turn of style was in no sense an affectation, but an irreversible organic development, and henceforth, whether he was writing for the *Morning Chronicle*, *The Times*, the *Examiner*, the *Yellow Dwarf*, the *London Magazine*, the *Edinburgh Review*, or the *Supplement to the Encyclopaedia Britannica*, it was all one, he had 'found out a secret' (as he later said of Sir Walter Scott) and wrote as he felt.[40] He had harkened to the Romantic imperative, and shaken off the trammels of rules. Just as the authors of the *Lyrical Ballads* had rejected the empty dessicated nobility of the post-Augustan convention in verse, so he, in prose, had turned his back on the lifeless fixed cadences of elegant composition and moved towards the natural rhythms and idioms of common speech. Not for him the padded and polished periods of the study, elaborated at the ornate old-fashioned writing desk, behind curtains drawn against the world, but the plain speech of a man talking to men, across a cluttered dinner-table.

An observant Scots lawyer said very shrewdly of the article on the Fine Arts that it was 'full of erroneous principles, bad taste and obstinate prejudice; but was evidently written by a man of talent, and interspersed with some fine passages'.[41] Robinson also was dazzled by Hazlitt's plea for the imitation of nature as against the cultivation of the Ideal, although it had been something he was 'not inclined to receive'.[42] Nor is it what we, imbued as we are with a divergent anti-representational prejudice, are any more inclined to receive. To us, Hazlitt's 'imitation of nature' looks like an old-fashioned and exploded heresy, whereas it was merely a call to freedom, to a liberation from the tyranny of a mid-eighteenth-century conception of the ideal, which claimed not only superiority to nature but also the finality of perfection, and closed the door on experiment and originality. As always with Hazlitt, it was a turning away from the lifeless and artificial, and a looking abroad into universality. This is proved by two additional papers for the *Round Table*

---

[39] *Letters*, 158.   [40] 19. 97.

[41] J. W. Burgon, *A Memoir of Patrick Frazer Tytler* (1859), 134. Tytler's journal entry is of Mar.–Apr. 1816. He probably saw Hazlitt's article in MS or in proof; the *Supplement* was not ready to be shipped to London until 11 June (NLS MS 789, fo. 595).

[42] Morley, 186.

based on his reflections whilst making out his article for Napier, 'On Imitation', and 'On Gusto', and particularly by a footnote to the first, where, after having asserted in his article that 'true genius . . . combines truth of imitation with effect, the parts with the whole, the means with the end', he laments in Turner the greater importance accorded to treatment as opposed to the subject, but, adverting to what he had said two years earlier, he yet recognizes in him

the ablest landscape painter now living, whose pictures are, however, too much abstractions of aerial perspective, and representations not so properly of the objects of nature as of the medium through which they are seen. They are the triumph of the knowledge of the artist, and of the power of the pencil over the barrenness of the subject. They are pictures of the elements of air, earth, and water. The artist delights to go back to the first chaos of the world . . . All is 'without form and void'.[43]

Napier had objected to the style, which Hazlitt defended, but he did not, it seems, object to the political allusions, hardly appropriate in an encyclopaedia, which Hazlitt was ready to sacrifice. He was a Whig, and among his many Whig contributors were John Playfair, John Leslie, James Pillans, and J. G. Dalyell, and to these were added, after the founding of the Whig *Scotsman* in 1817, the names of two of its editors, J. R. M'Culloch and Charles Maclaren.[44] This political weighting ensured that the *Supplement* and its editor, like the *Edinburgh Review*, would in a few years' time come under ceaseless attack from *Blackwood's*.

Hazlitt received an advance of £15 on the Fine Arts article in April, and it is likely that this was put to use when, at the end of the month (sometime after seeing Kemble as Sir Giles Overreach on the 26th), with the fine days returning, he went down to see how his parents had weathered the winter in Bath.[45] They had moved there the previous spring, probably in April. We cannot doubt that he had helped with the expense of the removal, nor that he lent a hand with packing up their goods and chattels.[46] He may even have accompanied them on their

---

[43] 4. 76 n.

[44] Revised list of contributors, Vol. 1 of 1824 edn. Stoddart seems not to have contributed after all.

[45] Hazlitt's biographers, following Howe, deduce a visit to Bath in Sept. 1816 from an article dated 'Gloucester, October 1, 1816' (19. 355, n. to 176) and a ref. in Sept. 1817 (18. 245) to having seen Stanley 'in some characters at Bath, about a year ago (Young Mirabel was one of them)'. The Theatre Royal, Bath, did not re-open until 5 Oct., and Hazlitt was at Covent Garden at the end of September to see Tobin's *Yours or Mine*, and on 2 Oct. to see *As You Like It*. At Bath, Stanley only played Mirabel on 7 May. See the writer's 'Hazlitt and the "Theatrical Examiner": Two Additions to the Canon', *EA* 38 (1985), 429–31.

[46] They left Addlestone between 2 Feb. and 3 May 1815 (Chertsey Parish Records P2/3/6, Surrey Record Office). Hazlitt saw *The Unknown Guest* at Drury Lane *c*.1 Apr., was at Alsager's on the 15th, and wrote to Jeffrey from York St. on the 20th, and again on 1 May, so that there would have been time for him to take his parents down either in the first half of April or in the last ten days.

journey to Combe Down. His mother was seventy, his father seventy-nine. Such an exhausting upheaval can have been forced upon them only by bad health and the hope of an improvement by taking the cure. The move was apparently not a success. They did not remain long in the neighbourhood of the fashionable and expensive spa (where there was nevertheless an active Unitarian congregation in Trim Street), but moved south to a warmer and cheaper home among another little community of Dissenters at Exeter.

When Hazlitt got to Bath on this 1816 visit he found that Kemble's sister-in-law, his admired Mrs Charles Kemble, the best Lucy Lockitt he ever saw, was at the Theatre Royal for her 'season', a week starting 4 May, and so, killing two birds with one stone, he indulged his passion for the stage while fulfilling his duty as a son.[47] Since his appointment as dramatic critic of the *Examiner* he had regularly attended at Covent Garden and Drury Lane, and at the minor theatres, for twelve months, week after week with hardly a break, and now he was able to relax and enjoy his evenings at the Theatre Royal knowing he need not write about them. On Saturday 4th he saw Mrs Kemble as Julio in Holcroft's *Deaf and Dumb*, a part he had long commended her in. Also in the cast was a Mr Stanley, with whom he was much taken, either because he was in an indulgent holiday mood, warmed by memories of the provincial theatres of his youth, or from a literary association, seeing in him the likeness of young Wilson, who had trod the very same boards in *Humphry Clinker*.[48] He went back on the Tuesday to see him, again with Mrs Kemble, as young Mirabel in Farquhar's *The Inconstant*. A year later, when Stanley came up to town, he modified his opinion and leaned to the less enthusiastic view the *Bath Journal* had expressed at the time.

He went to the theatre, at Bath, but not to Trim Street. If the Revd Joseph Hunter, pastor there from 1809 to 1823, had met him he would have given his own personal impressions in the notes he left of what were evidently conversations with Margaret Hazlitt about the family's life at Wem.

Returning to town (he was back by 9 May, when he saw Maturin's *Bertram*), he found that the long-awaited 51st number of the *Edinburgh Review* had arrived in his absence.[49] But not a word from Constable about his *Round Table*, the first volume of which was to have been ready by the end of April. His frustration had no doubt been increased by the appearance of *The Story of Rimini* (however loyal he was to Hunt, fond of him even, despite his absurdities). It was further exacerbated by a work published on 10 May. Of the two new books that burst into notoriety on that day — Lady Caroline Lamb's *Glenarvon*, known to be a *roman-à-clef*

[47] *Bath Chronicle*, 2 May 1816.   [48] 18. 246.   [49] *Courier*, 3 May 1816.

with Byron as the master-key, and Coleridge's *Christabel and Other Poems* — the second held much the greater interest for him. It was Coleridge's first volume of verse since 1798 and his well-wishers expected it to vindicate his reputation as a poet, sustained for years past solely by his conversation. As best it was a disappointment: the two major poems were unfinished, hardly touched since first conceived and abandoned years earlier; at worst a disaster: they readily lent themselves to parody and derision. Even an otherwise favourable anonymous review in *The Times* (which so rarely noticed verse that it was probably written and personally inserted by his old friend Stoddart, who had known him at Malta) carried wounding reservations: 'It is well known to many of Coleridge's friends that *Christabel* as it now stands has remained with scarcely the variation of a line ever since the year 1800, a singular monument of genius, — shall we add, of indolence or of those wayward negligences by which genius is often characterized?'[50]

Hazlitt, reviewing the volume in the *Examiner* on 2 June, made no attempt to conceal his vexation at encountering nothing but 'two unfinished poems and a fragment'. In another seven years, when he himself had been badly buffeted by fate, his exasperation at the non-fulfilment of promise would be sufficiently mellowed by the golden light of memory to enable him to speak more mildly, in 'My First Acquaintance with Poets', of Coleridge's indolence. That exasperation is sharply expressed here, but so also is that admiration for Coleridge's genius, of which it was indeed a function. He was still close enough to the Augustan prejudices of, on the one hand, substantial meaning, to underestimate 'Kubla Khan' (although he said, 'We would repeat these lines to ourselves not the less often for not knowing the meaning of them'), and of, on the other, poetic diction, to accuse Coleridge of deliberately shocking his readers at the outset with the lines

> Sir Leoline the Baron rich
> Hath a toothless mastiff bitch.

At about the same time Moore, writing to Jeffrey to propose a derisive review of *Glenarvon*, told him that the gossip of the town was currently applying to the unfortunate Caroline Lamb a later line, 'Oh, what can ail the mastiff bitch?' Jeffrey declined the submitted review at the urging of Lord Holland who was concerned for William Lamb, Caroline's husband, a member (like Moore, of course) of his circle, but it seems certain that in compensation he accepted from Moore the jeering anonymous review of *Christabel*, which Coleridge, like many others until

[50] 20 May 1816.

very recently, laid at Hazlitt's door.[51] Hazlitt's view of Coleridge's Gothic fantasy in this *Examiner* notice is mild and sober:

> The fault of Mr. Coleridge is, that he comes to no conclusion. He is a man of that universality of genius, that his mind hangs suspended between poetry and prose, truth and falsehood and an infinity of other things, and from an excess of capacity, he does little or nothing . . . In parts of *Christabel* there is a great deal of beauty, both of thought, imagery, and versification; but the effect of the general story is dim, obscure, and visionary . . . The sorceress seems to act without power — Christabel to yield without resistance.[52]

Between visits to the theatre, including the oratorios at Covent Garden (Haydn's *Creation*, that year) he was keeping a close watch on the quarrel now coming to a head over the acquisition of the Elgin Marbles by the nation. The absence of his name from Haydon's fitful if copious diary is no reason to suppose that they did not meet and discuss the question. Haydon attacked the expert Payne Knight in the *Examiner* of 19 March for casting doubt on the value of the sculptures. Named by Lord Elgin as one of his witnesses, he had not been called to give evidence before the Select Committee, and the violence of his attack was no doubt prompted as much by pique as by concern for Art. The raising of the question in Parliament on 7 June brought two articles from Hazlitt. In the first, of 16 June, resuming the theme of his articles on Sir Joshua Reynolds and on the Fine Arts of the *Supplement*, he praised the Marbles as perfect examples of the highest principle of art, which was fidelity to nature, without the interposition of a preconceived Ideal: 'It is to be hoped . . . that these Marbles . . . may lift the Fine Arts out of the Limbo of vanity and affectation into which they were conjured in this country about fifty years ago, and in which they have lain sprawling and fluttering, gasping for breath, wasting away, vapid and abortive ever since.'[53] He followed this with a diatribe against the hypocrisy of the English, who were preparing to keep what was not theirs, after condemning the French for wanting to do the same with the spoils of conquest in the Louvre; and in the course of it he listed once again the works of Correggio, Domenichino, Raphael, Titian, whose memory was still fresh in his mind after fourteen years. More than an academy, more than rules, more than

---

[51] BL Add. MS 52181, fo. 84, Jeffrey to John Allen, 25 June 1816: 'It is rather provoking that I should have received the other day a very clever and severe review of it [*Glenarvon*] from Tommy Moore (pray do not mention this however as he may wish it not to be known) which I should have been very much tempted to insert. It is full of contempt and ridicule rather than of serious refutation . . . You adhere however I suppose to the objection to any notice being taken of it . . .?' For a definitive discussion of the authorship of the *Christabel* review see Elisabeth W. Schneider's masterly 'Tom Moore and the *Edinburgh* Review of *Christabel*', *PMLA* 77 (1962), 71–6; Kathleen Coburn's letter, 'Who Killed Christabel', *TLS*, 20 May 1965; and J. Beer, 'Coleridge, Hazlitt, and "Christabel" ', *RES* NS 37 (1986), 40–54.

[52] 19. 33.

[53] 18. 101.

annual exhibitions by academicians, more than prating of the ideal, young artists required the permanent example, kept before them in a national gallery, of the greatest works of the past. Remembering no doubt also his dogged journeyings to private collections in his youth, the long miles he tramped, and his brazen persistence in gaining admittance (the only time, as he said, he was ever able to assume the character of impudence), he asserted that 'The only possible way to improve the taste for art in a country, is by a collection of standing works of established reputation, and which are capable by the sanctity of their name of overawing the petulance of public opinion'. And here he let slip something which was sure to offend Haydon.

This result can never be produced by the encouragement given to the works of contemporary artists. The public ignorance will much sooner debauch them than they will reform the want of taste in the public . . . It was in this point of view that the Gallery of the Louvre was of the greatest importance not only to France, but to Europe. It was a means to civilise the world.[54]

Eleven weeks had passed since Constable's letter, and, so far from the printing being finished, in April he had received no further proofs. Despite ill-health he had worked steadily on, at his 'Round Tables' and his theatrical notices. On 8 June his depression was not lightened by the spectacle of Mrs Siddons emerging from retirement to play Lady Macbeth, and revealing a disastrous decline, as startling as it was saddening, from that glorious figure, 'one of nature's greatest works', that he had first seen — a first sight that was 'an epoch in everyone's life' — when he arrived in London twenty years earlier. Coming as it did, just twelve months after 'the worst, the second fall of man', this eclipse of another of his early idols marked a further stage in the apparent dissolution of his world, in the darkening twilight of his youthful gods. Not that in his *Examiner* notice his near idolatry of this actress stayed his hand: as usual he faced, and spoke, the disappointing truth. And in the next month, as though to prolong the gloomy note of loss and decay, there faded away another ghost from his early days in town, the once brilliant and glittering figure of Sheridan, wasting down into a solitary and helpless death on his befouled truckle-bed in the garret of his once elegant Mayfair house, with his sick wife, and the bailiffs, in the bare echoing rooms below.

The weeks went by. On 15 July he wrote to thank Napier for £15 he had received. He asked diffidently for the balance, another £8, and suggested further articles, including one on beauty.[55] Napier's wary

[54] Ibid.
[55] *Letters*, 160, undated, where neither the London postmark, 15 July, nor the Edinburgh arrival-postmark, 18 July, is noted. Napier's endorsement, 17 July, is evidently a slip on his part.

reply on the 19th suggests that a certain mistrust of his contributor's docility, if not of his objectivity, had crept in.

I do not know if you have any new theory upon the subject [of beauty] but if you write the article you must treat the Scotch philosophers with respect. Stewart and Alison are men whom I hold in high esteem and I should wish to see a clear and fair account given of what they say, whatever adverse opinions you may yourself entertain. Meantime there are two lives which I must have from you as soon as possible, viz. of Barry the painter and Basedow the German philosopher. There is an account of the latter in the *Biographie universelle*, and of his system, writings, etc. in the 6th volume of Bühle's *Histoire de la philosophie moderne*, both of which books you will easily get access to in town.[56]

Hazlitt's review of Sismondi had evidently earned him a reputation as having a command of French. Just as evidently Napier's tone is that of an editor writing to a journeyman hack who supplies articles to order as a joiner supplies tables and chairs. It is plain from Hazlitt's letter of 2 April that he hoped to follow up his article on the Fine Arts with others in a comparable vein of theory: 'I was thinking just now that the words *Colouring*, *Drawing*, *Ideal* and *Picturesque* would make proper articles under the head of the Fine Arts, the metaphysics of which is in a very confused state at this day.' Napier, equally plainly, had decided (to our loss) that he should not, and condemned him to the menial compilation of biographical summaries, which he was forced to continue because he needed the money. It seems from the same letter that he had been in debt for a long time.

Napier's insistence that the authority of Stewart and Alison (the first of whom was one of his contributors) must never be challenged was a hard directive for a man like Hazlitt. Ironically, it was by thus allowing himself to be treated like a journeyman joiner that he gave the *Blackwood's* gang the chance later to bring two additional charges against him, of plagiarism and of fraud (they accused him of imposing on his editor by concealing his borrowings from the *Biographie universelle* and from Bühle). Either he conceived that there could be no infringement of copyright in a bare narrative of the facts of a man's life and a bald summary of his philosophical system, or that, if there might be, the responsibility was his editor's (a lawyer) to look into. He endorsed the letter with the possibly naïve but certainly exculpating words (the verb leaves no doubt) 'Mr Napier's letter ordering the translations'.[57]

[56] NLS MS 674, fos. 67–8, Napier to Hazlitt, 19 July 1816 (London arrival-postmark 22 July), enclosing a remittance of £9.

[57] In addition to the endorsement, we find in Hazlitt's hand the following: 'There do I find a never failing store / Of personal themes & such as I love best', a recollection of Sonnet III of Wordsworth's 'Personal Talk' (1807).

Napier must have told him that Constable was in London (he had in fact been in England since early June, but was apparently in no hurry to get in touch with Hazlitt). Hazlitt wrote an aggrieved note to the publisher at Walker's Hotel on 24 July.

Dear Sir,
   May I hope to see some more *Round Tables* before I die? I have been and am exceedingly ill, but I think it would do me good, in several respects, to have these volumes out. I have heard from Mr Napier, with a remittance. I remain, Dear Sir, very respectfully,

your obliged humble servant,
W. Hazlitt.

P.S. I can send the whole of the second volume.[58]

His appeal had its effect. The publisher was evidently embarrassed, or at least prompted to action. Proofs were sent. If it was an embarrassment it was compounded when he found the proofs returned corrected within the week. However, some obscure, stubborn, quite unexplained difficulty remained, and the prospects of publication, so close in the February, were not realized for another twelve months. It may have been the economic climate. Constable found the publishing trade in the South affected by the general stagnation, and may have decided to draw in his horns.[59] In August he declined to undertake a second edition of *The Story of Rimini* (originally published by Murray).[60] Hazlitt met him in London again before the end of the year, but there is no hint of any explanation that was offered him of this extraordinary delay. At the end of September, in a letter to a Dublin bookseller, Constable names half a dozen titles he is about to publish, but not *The Round Table*.[61]

   Hazlitt's bad health was far from being the only additional trial in that cold wet summer. Nor was it the worst. He was visited by a common calamity of those days, one which had befallen Southey the previous April, and had also at one time been the lot of poor Wordsworth twice in one year, the death of a child. On 19 June his little son John, barely seven months old, succumbed to the measles. It must have been agony for so affectionate a father as Hazlitt. The memory of it is faintly echoed in the essay 'On the Fear of Death' he wrote in 1822 for *Table-Talk*, when he was probably thinking of the six-month-old child they had lost in 1809, and of his grief then, and of how in 1816, wanting a lock of the child's hair who had just died, he had asked his wife to cut it off, so as to avoid the pain of looking on the little lifeless face.

[58] NNHL, 272. The *Letters* version is inaccurate.
[59] NLS MS 789, fo. 615, Constable, reported to Walter Scott, 9 Aug. 1816.
[60] Ibid., fo. 622, Constable to Leigh Hunt, 27 Aug. 1816.
[61] Ibid., fos. 643–4, Constable to J. Cumming, 28 Sept. 1816.

I have never seen death but once, and that was in an infant. It is years ago. The look was calm and placid, and the face was fair and firm. It was as if a waxen image had been laid out in the coffin, and strewed with innocent flowers. It was not like death, but more like an image of life! No breath moved the lips, no pulse stirred, no sight or sound would enter those eyes or ears more. While I looked at it, I saw no pain was there; it seemed to smile at the short pang of life which was over; but I could not bear the coffin-lid to be closed — it almost stifled me; and still as the nettles wave in a corner of the church-yard over his little grave, the welcome breeze helps to refresh me and ease the tightness at my breast!

Hazlitt always held that Southey wanted the patience to think that evil is inseparable from the nature of things. It is illuminating to compare what they each have to say in this connection. Southey says of his dead ten-year-old son, 'I could not help feeling that when a creature of this kind came into the world, it was not likely that he should be suffered to remain in it; he lived in it long enough to know all that was good, — and nothing but what was good; and he is removed before a thought of evil has ever risen in his heart, or a breath of impurity ever tainted his ears.'[62] This pious, pathetic illusion, a projection, transposition, or rationalization of grief anticipating the Victorian mind, little Nell, and the *Basket of Flowers*, contrasts strongly with Hazlitt's words, which, although just as pathetic, are clear-sighted and concrete, encompassing both good and evil, as is implied in the apposition between the encroaching nettles and the returning breeze, proclaiming at the same time the inexplicable bitterness of loss and the compelling, and welcome, but equally inexplicable resurgence of life.

[62] Southey, *Selections*, iii. 29.

The Claims of Barefaced Power

Liberty is short and fleeting, a transient grace
that lights upon the earth by stealth and at
long intervals . . . But power is eternal.

Hazlitt

HIS main preoccupations in the next few years, literary criticism and
politics, are typified by two articles he published in the same month, July
1816, 'Shakespear's Female Characters', and 'The Lay of the Laureate'. Not that these interests were exclusive. Dramatic criticisms and
papers on art continued to flow from his pen. Any circumscription would
have been irksome to such an ardent far-reaching mind, animated by a
half-deliberate, half-unconscious aspiration to universality, the mind of
a moralist who took the whole world for his province, moving with no
sense of strangeness from the philosopher's study to the pit of the theatre,
turning with equal vigour to economics or civil and criminal law, and (in
a few years' time) reporting a prize-fight and analysing the state of the
periodical press with equal absorption, or embarking with enthusiasm
upon an extensive four-volume biography of Napoleon. All this might be
mistaken for the professional versatility of the journalist were it not for a
powerful unifying principle that inspired every part of it and gave it
identity and cohesion, and determined also the very nature of the style in
which it was all written.

These two essays were subsequently incorporated in *Characters of
Shakespear's Plays* (1817) and *Political Essays* (1819) respectively. The
success of the first book enabled him to give up writing for the daily
papers, and by the time the last appeared he was able to do without the
weeklies also. However, his modest triumph was due to talent alone,
without benefit of worldly calculation and in neglect even of the
safeguards of discretion. The reputation he gained by his criticism was
constantly jeopardized by the enmity stirred up by his political writings,
and it was owing to these that he was pursued with abuse for the rest of
his life, and abandoned to disparagement and neglect after his death. To
see why this was we must look at the context in which these political
writings appeared.

On 23 September 1816 Lamb, in one of his gossipy letters, informed
Wordsworth that Coleridge had written a sermon for 'the middling

ranks of the people' to persuade them that they were not so distressed as they believed or was commonly supposed, and continued wryly, 'Methinks he should recite it to a congregation of Bilston Colliers — the fate of Cinna the Poet would instantaneously be his.' If Lamb, who hitherto had seldom had much to say about public affairs, was aware of the colliers' plight and felt the untimeliness of Coleridge's pamphlet, the one must have been notoriously desperate and the other grotesquely unapt. In his own case, if he discouraged political topics at his Wednesday nights in 1806, ten years later he read Cobbett's *Political Register* every week, and not solely on account of Cobbett's prose, as we may guess if we turn to his seldom-mentioned poem, 'The Three Graves', with its hovering spectres of Oliver, Castles, and Edwards. We soon find an exasperated Hazlitt elaborating the same contrast between worker and pamphleteer. The colliers of Bilston Moor in Staffordshire, oppressed by unemployment since 1814 (one of their feeble attempts at public protest had been broken up the previous November by the cavalry charges of the county Yeomanry[1]), had that summer been driven desperate by the high price of bread. Their plight was not unique; the whole nation was in such dire straits that the 1812 slump, according to Brougham in the House of Commons, might almost be called prosperity in comparison. There were over two million paupers in a population of nineteen million.[2] But the Bilston men became an emblem of the times through an ingenious device they adopted: rather than resort on the one hand to violence or lie down and starve on the other, they advertised their misery by a demonstration that went home to the public imagination, and pricked the conscience of the country. Groups, sometimes as many as eighty, harnessed themselves to ponderous wagons loaded with coals and petitions, and dragged them along the dusty country roads into the streets and squares of cities as far apart as Birmingham, Liverpool, and Leicester[3], implying by this a destitution so extreme and inhuman that an animal yoke was its only appropriate symbol. But not everyone was sympathetic; nervous magistrates met these human beasts of burden at the city boundaries and turned them away. Generally, however, they were regarded with pity and fear as the visible proofs of bad times present and worse to come.

The bad times had arrived abruptly with the end of the wars, in 1814, and had rapidly worsened. Their chief manifestation just before men's eyes were turned in another direction by the landing at Fréjus, was in the Corn Bill riots of March 1815, when the houses of those Members held responsible, Lords Grenville, Darnley, and Hardwicke, and the detested

---

[1] *Courier*, 20 Nov. 1815.

[2] *Examiner*, 26 Feb. 1815, 'Increase of the Poor'.

[3] *MC*, 8 July 1816; *Courier*, 15 July.

Croker, as well as Lord Eldon, at No. 6, Bedford Square, were stoned and their windows shattered. The military were called out — even Stoddart, we remember, was shocked by this — and they shot dead an innocent bystander and sabred others, acts symbolic of the ineptitude of the rulers and their crude efforts to put down by force disturbances that should have been deflated and prevented through reforms. The 1815 Corn Bill was a protectionist measure prohibiting the importation of corn at under 80s. a quarter. Possible hardship was at first cushioned by the good harvest of 1815, which lowered the price of grain, but this in its turn ruined the farmers, who could pay neither rent to the landowners nor interest to their bankers; a steady market for their crops throughout the long war had accustomed them to ploughing their profits back into their farms and borrowing capital from the banks. They were suddenly unable to pay the interest. The banks were forced to call in the capital or close their doors. Between 1814 and 1816 nearly a third of the seven hundred-odd country banks failed. Similarly the prosperous manufac- turers, who were confident of being able to make good the post-war shrinkage of the war-stimulated home market by exporting to a Europe which was again accessible, found that the Europeans were too poor to buy from them. They had so little understood the situation as to calculate in 1814 that peace would increase demand.[4] The American market, reopened at the end of the war of 1812, helped temporarily, but by the end of 1815 they were forced, if not into bankruptcy, then at least to cut production, and lay off their workers. The ranks of these numerous but law-abiding unemployed were further swelled by the swarms of rough unruly disbanded soldiers and sailors, numbering some 300,000. It was in these conditions that the enormous increase in taxation, hitherto masked by wartime prosperity and the rise in public revenue, became for the first time monstrously plain. There followed upon this the bad harvest of 1816 which first raised the price of corn and then created a scarcity exacerbated by the trickle of imports allowed by the Corn Laws. By December the price rose to 103s., but there was little to import, the harvest having been just as bad in Europe. The very real threat of starvation drove even the most docile labourers to desperate action. Luddism became widespread and sustained, particularly between the Thames and the Wash; night after night, that autumn, the sky reddened to the burning of ricks and even of farmhouses; night after night the country roads echoed to the marching and countermarching of the military and the clattering hooves of the yeomanry. The land's decay was repeated in the factories; the decline in the demand for iron threw large numbers on the parish, and the running down of the foundries brought the collieries in turn to a standstill.

4 'Public Distress', *MC*, 29 July 1816.

The ramshackle economic and political machinery of the State, most of it hopelessly old-fashioned, although in the industrial sector revolutionary and far ahead of its time, had been kept going for twenty years by the artificial stimulus of a great war. It had rattled on, prodded in turn by defeats, victories, coalitions, continental blockades, and orders in council. After Waterloo the levers no longer operated, and in the sudden lull it ground to a shuddering halt. The experience was unique in history, the results totally unforeseen. For the next five years its movements, uncertain, ponderous, erratic, and understood by none, gave rise to terrifying rumours. Threatening to collapse and crush everybody, wheezing, whining, coughing, it would from time to time break down completely, and the ensuing silence was soon filled with the furious or pitiful clamour of the bewildered machine-tenders, and the angry, self-righteous threats of the equally mystified owners. Thereafter, when trade had gradually recovered and prosperity returned, the increasing momentum of the economic parts of the machine distracted attention from the inertia of the political, until a slowing down at the end of the twenties forced the introduction of the Great Reform Bill, and the machine picked up for a time until another lull brought on the Chartist Agitation; and so it went on in a perpetual see-saw throughout the nineteenth century.

Such was the social and economic state of the country in 1816. If we ask what remedies were being proposed by the political parties, the answer is, by the Tories, who were in power, none. They were persuaded, like the Prince Regent, that 'these severe trials were chiefly to be attributed to obscure unavoidable causes', and piously hoped that 'the people would continue to sustain them with exemplary patience and fortitude'.[5] They were determined at all costs to hold on to what they had; the only things they could be induced to part with were specious sympathy and judicious charity.[6] What the Whigs offered was what they had been offering for over thirty years — themselves as an alternative government. Apart from minor differences in taxation their domestic policy was as supine as the Tories'; they parted from them only on foreign policy, in their criticism of repressive foreign governments. At home their attitude was almost purely negative, a disapproval of the repressive measures adopted by the Tories, although in the eternal and inevitable dilemma of dim political foresight and clear hindsight there is no saying that they would not have adopted the same measures had they been in power. Almost purely negative, but not quite: their sympathy,

---

[5] The Prince Regent's reply to the 'Humble Address and Petition of the Corporation of London on the Subject of the National Distresses' (*Courier*, 10 Dec. 1816).

[6] The Prince Regent subscribed £500, Angerstein £300, and Robert Owen £100 to the Association for the Relief of the Manufacturing and Labouring Poor (*MC*, 30 July 1816).

backed as it was by concrete proposals for moderate fiscal, social, and judicial reforms, was somewhat more convincing than that of the Tories. But it was plain that both parties stood for privilege.

The third force in the country, the Reformers, saw only this difference between them, that whilst the Tories were for King against the Whig aristocracy and the people, the Whigs were for the Great Families of the Glorious Revolution against King and the people, so that the common enemy of both was the common people, who therefore had nothing to expect from either of the two aristocratic factions. In February 1819, when John Cam Hobhouse stood as Reform candidate for Westminster, Sir Francis Burdett declared from the hustings that the Whig and Tory factions were two thieves between whom the Constitution had been crucified; and on the night of Hobhouse's defeat the angry mob smashed the windows of the *Courier* office and the *Morning Chronicle* office with equal and impartial enthusiasm.[7] Hazlitt was of like mind, but put it differently, distinguishing between the 'modern Whigs' and the 'old Whigs' who had been, as he was, 'opposers of the Divine Right'. His analysis of the political stance of the Whigs was extraordinarily acute, and not less witty:

. . . the Opposition have pressed so long against the Ministry without effect, that being the softer substance, and made of more yielding materials, they have been moulded into their image and superscription, spelt backwards, or they differ as concave and convex, or they go together like substantive and adjective, or like man and wife, they two have become one flesh. A Tory is the indispensable prop to the doubtful sense of self-importance, and peevish irritability of negative success, which mark the life of a Whig leader or underling. They 'are subdued even to the very quality' of the Lords of the Treasury Bench, and have quarrelled so long that they would be quite at a loss without the ordinary food of political contention. To interfere between them is as dangerous as to interfere in a matrimonial squabble . . . They will not allow Ministers to be severely handled by anyone but themselves, nor even that: but they say civil things of them in the House of Commons, and whisper scandal against them at Holland House . . . they rather wish to screen the Ministry as their *locum tenens* in the receipt of the perquisites of office and the abuse of power, of which they themselves expect the reversion.

Strange that such difference should be
Twixt Tweedledum and Tweedledee.

The distinction between a great Whig and Tory Lord is laughable. For Whigs to Tories 'nearly are allied, and thin partitions do their bounds divide.'[8]

No improvement was possible while privilege remained. And here was the dilemma. The symbol of privilege was the rotten borough, carrying seats that could be, and were, auctioned at Christie's; the instrument of

---

[7] *Courier*, 17 Feb., 4 Mar. 1819.   [8] 7. 20.

privilege, Parliament. Parliament was unlikely to vote for its own
extinction, yet nothing could be done until Parliament was reformed,
and the people more effectively represented. 'The people', here, was of
course a crude, undefined term which misled the real lower classes. After
the successful passage of the Reform Bill of 1832, it turned out, to their
dismay, that the term had meant simply the rich merchants and
manufacturers who had been excluded from the traditionally based
suffrage because they lived in the industrial towns of the Midlands and
the North, which politically had no existence. The franchise qualification, in the event, was still property. Householder suffrage had to wait
another fifty years, manhood suffrage another ninety.

There were five main centres of activity among the amorphous body
of Reformers. The first and oldest was an individual rather than a
centre. This was Major Cartwright, active in the cause since the distant
year of the Declaration of Independence, and now, in his seventies,
undertaking endless coach-journeys between the towns of the industrial
North to found the Hampden Clubs, which, following on the Westminster election victory of 1807, may be regarded as having initiated the
new movement of reform which culminated after a hollow period in the
1820s, in the triumph of 1832. Hazlitt respected the gallant and worthy
Major as an honest man, and admired his industry and dedication, but
saw in him a fatal flaw. His activities, he thought, were vitiated by the
fallacy of retrogression; his declared aim was to recover an imaginary
epoch of law and liberty that never was, but which he supposed to have
existed in our remote past.[9] It was an obsession, and Hazlitt put him in
his gallery of bores, of 'people with one idea'.

The second was a man who had worked with the Major in founding
the Hampden Clubs in 1812, but who afterwards went his own way, Sir
Francis Burdett. An energetic, fox-hunting, classics-quoting man of
wealth and of parts, the type of the English *honnête homme* of the
eighteenth century, imprisoned for his principles in 1810, he was
enormously popular. Plain, honest, straightforward, without vanity,
without affectation, he was one of the few of those contemporary figures
praised in *The Spirit of the Age* to be praised almost without qualification.
Hazlitt saw him as the traditional Englishman, endowed with the
traditional English virtues. It is certain that he did immense service to
the cause of Reform.

After these two came the demagogues Henry Hunt and William
Cobbett, the first of whom did, at vast open-air meetings, with his
stentorian voice, his white top-hat, and his tricolour, what the second did
for as wide a public with his farmhouse-table pen in his *Political Register*.

[9] 11. 141.

Hazlitt did not take 'Orator' Hunt, 'that stout-lunged gentleman', very seriously — in all his writings there is only one incidental reference to him[10] — but Cobbett was a horse of another colour. He represented the English yeoman in the same way as Burdett represented the English squire, and for that reason stood high in Hazlitt's estimation, firmly grounded as it was in English history and the real life of the past. But with reservations. Hazlitt was closer to the background of Cobbett's erratic changes of direction than we are, and saw more clearly how violent they were. If Cobbett was vehemently Tory at one moment, how dependable was his equally vehement advocacy of Reform at the next? It does credit to Hazlitt's perceptiveness that he did not doubt Cobbett's fundamental worth, despite its various perverse and contradictory disguises, nor his genuine concern for the interests of the country-folk.

Below these again was a small group of die-hard revolutionaries, the only really dangerous group in the country. They called themselves Spenceans, after Thomas Spence, a mild north-country schoolmaster, an agrarian socialist who had come to London in the 1790s to spread, from a street-hawker's barrow, the gospel of the nationalization of the land, and who had died in 1814. Their activities, far removed from the ineffectual half-crazed theorizings of poor Spence, were sinister, and not at all like the constitutionalist corresponding-society propaganda of the Major or the direct and open addresses of Cobbett or even the crude platform violence of Hunt. Their world was the congeries of dark courts and warrens and working-class taverns of London; it was in smoky tap-rooms that they recruited their followers in the fight for social justice, arguing with them in low urgent tones across scarred and beer-ringed table-tops under the covert side-glances of the government spy reading his paper in the corner. But their success was small: in the end they fought and died alone, a mere handful, among whom were Arthur Thistlewood, an ex-Army officer, and the shadowy Dr Watson and his son.

Ultimately more effective than any of these, although at this time working only in the background, were the Benthamites. Their impatient disdain of the Tories and the Whigs was exceeded only by their contempt for the rhetoric, or violence, or amateurism, of the other Reformers. Bentham was now sixty-eight, but he regularly continued his labours at his desk in Queen Square Place (and his mid-day 'circumgyrations' of the garden under Hazlitt's window),[11] whilst James Mill went doggedly

[10] In an article on Stoddart's reaction to the Spa Fields meeting, which will not be found in the *Complete Works*. It is in the writer's 'Three Additions to the Canon of Hazlitt's Political Writings', *RES* NS 38 (1987), 355–63.

[11] To the Tories, Bentham was at best 'that luminous expounder of unintelligible ideas', at worst 'a corrupter of morals and the inventor of an absurd Panopticon to deal with his own corruptions' (*Courier*, 23 July, 12 Mar. 1819).

on with his studies in political economy and his monumental *History of British India*; but neither the authority and fame of the older man nor the industry of the younger could alone have achieved the next important step towards the triumph of Reform that the cause was soon to achieve, and which was owing to the painstaking, limelight-shunning organizing capacity of the remarkable Francis Place, the radical master-tailor of Charing Cross. After twenty-five years' experience of committee-rooms (he was a member of the Corresponding Society at the time of the Treason Trials), and ten years' association with Bentham and Mill, he succeeded, after he retired from business in 1817, in engineering the election of a second Reform Member, Byron's friend John Cam Hobhouse, to represent Westminster with Burdett. The twin bastion of Reform thus established in March 1820 endured without a breach until the victory of 1832. Place never grew tired, nor discouraged, nor ever in his preparations neglected any aspect of the campaign permanently mounted by himself and his friends. His only weakness, often a strength, was that he was humourless and had the self-assured rigidity of the autodidact.

However, this refers to a later period. In the mean time, with the growing troubles of 1816, another and opposed campaign was about to begin which goes far to explain the drift and tone of Hazlitt's political writings in the next few years. It involved the Government's deliberate manipulation of the Tory press to excite public fear of civil disturbances and revolutionary movements in order to cloak their own selfishness, ineptitude, and impotence, and also to justify the coercive and repressive measures they saw they would have to take as the inevitable alternative to showing generosity and initiative. This campaign involved fabricating rumours of seditious activity, exaggerating reports of tumult and violence, as well as employing *agents provocateurs*; it resulted in the suspension of Habeas Corpus, in the passing of the 'Gagging Acts' of 1817, and rose to a climax with Peterloo in the August of 1819, which led to the Six Acts of the 'Savage Parliament' at the end of that year, restricting among other things public meetings and the freedom of the press.

Hazlitt's attitude is singular and complex: he appears less concerned with the distresses of the governed than with the iniquities of the governing body. His compassion for the poor did not take the form of violent indignation at their penury, and this is no doubt to be explained by his background. He was neither of the working class nor of the middle class (in the limited sense in which these terms can be used of that period) but he was closer to the second than the first, and was less fitted than Cobbett or Place to enter into the feelings and attitudes of the working-people. Himself brought up poor, aspiring only to the satisfactions of the mind, looking not at all to the world's glittering prizes, a non-consumer in fact, an intellectual who had no material interests beyond the common necessities, and who found himself on neutral ground, neither in the ranks of the oppressors nor of the oppressed, his compassion sprang rather from his allegiance to the principle of liberty (the creed of the nonconformist rebel) than from the kind of humanitarian zeal for the abolition of want and suffering that characterized the social conscience of the later nineteenth century. It sprang from a hatred of the indignity of their enslavement rather than from distress at their deprivation. His *Reply to Malthus*, like the publication it criticized, is in the eighteenth-century tradition of abstract political theory and is not immediately concerned with the pitiful concrete realities of the life of the poor. Certainly it shows him, and shows him clearly, to be on the side of benevolence, charity, and humanity, but his childhood in an unenfranchised self-supporting household of scant means was too close to actual penury for him to be sentimental about the evidence of it that came under his gaze. Nor, despite the hard times, had poverty in the capital yet taken its terrifying plunge into the vast anonymous abyss of Victorian destitution. Destitution there no doubt was, but largely of an eighteenth-century sort, dreadful enough indeed, but localized and half-concealed in the surrounding areas of general shabbiness and undernourishment, in a London still open to the fields and commons north of Oxford Street, west of Hyde Park Corner, and south of Pimlico. Its most evident sore was the night-cellar district of St Giles's, but the few rookeries, more notorious for their crime than for their squalor, had not yet proliferated, and the hopeless depersonalized poverty and neglect, the downright starvation that in mid- and late-Victorian times crept like the Black Death over whole districts of south and east London, had not yet begun to spread their faceless infection.

The detachment that marks off his interests and his concerns as essentially different from those of the poor was, as we have seen, symbolized, in 1814, by the evening star setting over a poor man's cottage, which in its turn symbolized the connection between hope and the heart of man. The occasion that supplied the image was linked with the battle of Austerlitz and with Napoleon as liberator, and it is plain that he was moved to resentment more by the kings who enslaved mankind than by the specific miseries of the Salopian cottager, and that the Test and Corporation Acts were generally more often in his mind than Speenhamland.

He had been brought up poor, and few such ever learn the value of money, and none the value attached to it by those brought up rich: their parents have not learnt this lesson, and since they have not the experience to transmit, the cycle is not easily broken. Hazlitt did not break it. Although twice married, he acquired neither house nor chattels; he never, except for one brief period, had a regular salary and, although he must have earned a good deal, he never saved, never kept accounts, never checked bills, and it all slipped away. He died poor, and in furnished lodgings. But it was not merely lack of training; it was also a question of principle. He abhorred self-interest.

How much, a small example will show: in order to discourage it and inculcate liberality he would often give his small son a pound note with instructions to get rid of it as quickly as possible, spending it or giving it away, in whatever manner he pleased. Inevitably, and in view also of his upbringing among the excluded, disabled intelligentsia of Dissent, politics for him would be less a practical matter of rates and taxes, wages, hours of work, and so on, than a matter of human dignity, human rights and liberties, and the constitutional means of their assurance and preservation.

Freedom was an instinct in him before it was a principle, and when that instinct first asserted itself early in life it was at such a cost in remorse to himself and in distress to those he loved, that it burned itself forever into his mind, and thereafter remained, a dearly bought prize, at the forefront of his scale of values. We can only guess at the occasion and circumstances of his rejection of a career in the Unitarian ministry, but his father's earnest piety, his own sensitiveness and the love that bound the two of them together, make it certain that it was a painful time. There can have been no quarrel, no rift, but there must have resulted a deep bruise of disappointment and misgiving on the one side and an inward gnawing of frustration and regret on the other. What he had bought so dear he would henceforward always place first in all the affairs of life; any public threat to freedom he would attack wherever it arose; any desertion of the cause by friends or allies would stir him to bitterness and even savagery. And it was at once a happy and an unhappy irony, a bitter and consoling paradox, that his refusal to fall in with his father's plans was justified by his father's own principles and it was equally no doubt a comfort to him to be able to cleave to his father in one regard, civil freedom, even while he was moving away from him in another, religious belief. The principle of civil and religious liberty was the rock on which they both took their stand, and the inevitable corollary, in their actions, of their commitment to freedom was, as we have noted, that all their lives long they remained, and chose to remain, on the fringes of society. They spent their whole lives in serving the human spirit, the elder through his pastorate, the younger with his pen, but, with regard to the Establishment and the conventions of organized society, the lifelong motto of the son, whether explicit or not, was a Miltonic *Non serviam!* It is plain in all his actions and attitudes: in his choice of a career as artist; in his marrying a small independence and retiring to the country; in his escape from the gallery of the House of Commons as soon as he could turn dramatic critic, and then shaking off that nightly bondage to become free-lance writer; in his divorce from his first and his separation from his second wife; in his indifference to money, his open-handedness, and his improvidence; in his avoidance of the encumbrance of property; in his impatience with the formalities of social life; even in such a slight but significant detail as his refusal to carry a watch. And explicitly it is seen from first to last in all his writings on politics.

Above all it was apparent in his steering clear of all political parties. For the greater part of his life, like the majority of his fellow-countrymen, he was disfranchised, but during six years as ratepayer in York Street he was on the electoral roll of the City and Liberty of Westminster. His political writings nearly all fall within this period. The volume into which he finally collected them was published a few days before the militia attacked the crowd at Peterloo. That wild climactic spasm of fear and hatred in the ruling classes was followed by several events that discouraged overt political activity and speculation, beginning with the legislation of the last months of 1819, the repressive Six Acts of the 'Savage Parliament'. In the following year the Cato Street conspiracy hardened public opinion against violence. The King's attempts at divorce and the Queen's sordid trial disgusted those who still harboured illusions about the Establishment, and replaced them with a cynical though uneasy indifference. Above all, a rapid increase in trade and prosperity led to a suspension of hostilities between Parliament and the people, and dissipated for a decade the threat of revolution that had constantly loured, although often half-neutralized by internal divisions among the Radicals, during the five years after Waterloo.

We should now label Hazlitt a radical, but he criticized the radicals as much and as often as he castigated the Whigs and the Tories. It is difficult to determine where he stood on a precise question like Reform. He states his general position thus, at the opening of the *Political Essays*:

I am no politician, and still less can I be said to be a party-man: but I have a hatred of tyranny, and a contempt for its tools; and this feeling I have expressed as often and as strongly as I could. I cannot sit quietly down under the claims of barefaced power, and I have tried to expose the little arts of sophistry by which they are defended. I have no mind to have my person made a property of, nor my understanding made a dupe of. I deny that liberty and slavery are convertible terms, that right and wrong, truth and falsehood, plenty and famine, the comforts or wretchedness of a people, are matters of perfect indifference. That is all I know of the matter . . .

His political writings are indeed almost purely the expression of this feeling. Not that they are divorced from actuality. Nothing, for example, could be more concrete than his analysis of the effect of the war on the economy ('On the Speeches in Parliament on the Distresses of the Country'). But it is in vain to search them for dogma, system, policy. There are two reasons for this, in addition to the explanation we have suggested above. In the first place, since his theme was liberty, any formal exposition of dogma, system, or policy would be inconsistent; it would introduce a limitation, a circumscription, and thus far be a denial or a weakening of his theme. In the second, it was not his function. He was not concerned with the equitable financial and economic organization of the State, nor even specifically with a just parliamentary representation. Indeed, he was not, in fact, concerned with organization at all, but with freedom, the freedom of all men of good will to follow their bent, whatever it might be, cheated by none, bullied by none, and above all with freedom from the degrading oppression of barefaced power, symbolized by the throne. The equation was simple: he hated self-interest and oppression; he loved disinterestedness and freedom. He was entirely alive to the difficulties and complexities of turning the principle of freedom into practice, but he was equally aware of the dangers of keeping silent, during the endless discussion of these niceties, in an urgent and shameful situation. When you saw a man being beaten and robbed, finicky deliberations regarding the legally permissible length and weight of the cudgel seemed a desperate insanity. Destructive criticism (he held) has its merits no less than constructive, and if the basis of the first is plainly right it should be applauded as useful, without the second, in all its detail of propositions and corollaries, being also demanded at the same time of the same man. His sole function he conceived to be to protest as often and as loudly as he could against selfishness, injustice, and inhumanity. And this he did.

Perhaps it was because he was not as closely involved in the mêlée as his fellow radicals of the intellectual middle class — Leigh Hunt, John Scott, and Thelwall, who, after all, owned or edited newspapers — that his articles provoked so little response from the politicians of the day. But even the voices of these editors were little heard beyond the newspaper offices of their rivals. The quality, now perfectly evident, of such a newspaper as the *Examiner* is no index of its circulation and influence at that time. A letter of John Stuart Mill to Leigh Hunt twenty years later, attempting to palliate a resented reference of Mill's to the Radical journalism of the Regency, perceives, in looking back, a deliberate policy of neglect by those in power, a contemptuous passing over in silence:

You will not find it said [in my article], even by the most distant implication, that the *journalism* of the party, as such, was an object of contempt to anybody: what is said is, that radical *opinions* were an object of contempt to almost all persons of station and consideration . . . It was not yourself only, and Hazlitt, and Cobbett; Godwin, and Bentham, and my father, and various others, had laboured for radicalism with more or less of acceptance, and had gained or were gaining reputation to themselves individually, but the cause had not yet profited much by them: it has since, and we are now benefiting by what was then done.[12]

The contempt is casually conveyed in an incidental phrase in a letter of the time to the editor of the *Courier*: 'The seditious, or, as they are technically termed, "the Reformers" . . . '.[13]

There were, of course, violent attacks by the opposing *press*, but there were also attacks by the Reformers' presumptive allies, the other factions among the Radicals. In Hazlitt's case the assault did not come from the *Quarterly* and *Blackwood's* alone. He was misjudged and misunderstood also in his own camp, or at least in one closely allied. In a letter to Hobhouse, correcting a falsely favourable impression left in the candidate's mind by Hazlitt's cheering for him at the Westminster hustings in 1819, Place dismissed him as 'a crazy kind of a fellow wholly impelled by his feelings'.[14] And again, in all Hobhouse's voluminous journal of his long life there are only one or two references to Hazlitt, and scarcely more to Leigh Hunt, the companion and collaborator of Hobhouse's friend Byron. It is very striking. This evident lack of solidarity is confirmed by the fate of the able T. J. Wooler's newspaper *The Black Dwarf*. Founded in 1816, it succeeded Cobbett's *Political Register* as the most widely read Reformist journal, but was closed down in discouragement in 1824 with the valedictory lamentation that there was at that time 'no public devotedly attached to the cause of parliamentary reform'.[15] In the last year of its life the author of an anonymous article on 'Modern Literature' entirely ignores Hazlitt's political writings proper and attacks him for what he says about Cartwright (who supported and probably subsidized Wooler) in the 1821 *Table-Talk*:

Mr. Hazlitt in the very frenzy of impertinent affectation of superior discrimination gave Major Cartwright as an instance of a man of one idea. This he did without any knowledge of the individual, from whom, had he been acquainted with him, Mr. H. might have gathered more ideas than he has ever yet favoured the world with. Another of the same school in a periodical publication, has been attempting, with a mind hardly capacious enough to estimate the worth of a jingle of a rhyme, to measure the mind, and estimate the effects of the talent of Mr. Jeremy Bentham . . . Mr. Bentham happening to be opposed to the corrupt system under which he lives, is not attended to by the

[12] *The Collected Works of J. S. Mill*, ed. F. E. L. Priestley *et al.* (Toronto, 1963–   ), xii. 359.

[13] 5 Dec. 1817.

[14] BL Add. MS 36457, fo. 340, Place to Hobhouse, 20 Aug. 1819.

[15] E. P. Thompson, *The Making of the English Working Class* (Harmondsworth, 1968), 891.

agents of that system. Therefore this critical minim says his works have no
influence at home — that he is better known to the savage of the torrid zone
than to the sage of Europe; and that he is *more* popular in Siberia than in
Britain. The real object of this species of writing has no reference at all to its
apparent subject. The writer has no purpose upon earth but to write a smart
essay — to say something that has not been said before — and the further he
gets from the truth, the more likely he is to accomplish what he aims at.[16]

As to Bentham, while we may be sure that Hazlitt never aspired to set
foot in Holland House, it is odd that there was no contact with his
next-door neighbour and landlord of six years; unless it was owing to his
dislike of putting himself forward. Bentham, in his 'hermitage', was
hospitable but somewhat dry and prickly: in the intervals of his endless
labours, apart from Mill, Ricardo, and Place, and a few young disciples
who acted as secretaries, he saw only visitors who came to do homage,
and, more frequently, those of the same exclusive and absorbing
interests as himself — the studious James Mill, the political organizer
Francis Place. But he had a good deal of simplicity, even of playfulness,
in his nature, as is seen in his treatment of his young protégés. It is plain
from the warm tone Hazlitt uses in the *Spirit of the Age* that he knew a lot
about Bentham's personal life, and this he must have got from one, or
perhaps two, of these young men. We cannot be sure he had met Joseph
Parkes at this time, but he knew Walter Coulson, and if he never stepped
next door (did his landlord send a servant for the rent? or employ a
collector?) it must have been because Bentham never asked him, and not
because he refused. In 1819 Bentham professed not to know who his
next-door neighbour and tenant was.[17] It was probably a question of
temperament: Hazlitt was both shy and proud, and the old man was
totally indifferent to the literature of his time and, even in political
writings, preferred reports and Blue Books to Burke or Paine.
    Perhaps Hazlitt on public affairs would have created more stir if, like
Bentham or Robert Owen of New Lanark, he had had a system. But
system he had none. As the avowed enemy of fixed, immovable,
arbitrary power, he conceived it to be his sole task to rouse his
fellow-countrymen to a like opposition, and nourish the vacillating,
transient, and threatened flame of liberty. His struggle in the next few
years was precisely the opposite of Southey's, who hoped to avert the
'bellum servile' he saw approaching by educating the masses into
acquiescence in their poor lot. Hazlitt hoped to alert them to the
Government's more or less covert assault upon their liberty, and at the
same time to awaken men of good will in the slumbering middle classes

[16] *Black Dwarf*, 14 Jan. 1824, 55 (the failure to identify the author of the *New Monthly* article on
Bentham betrays unfamiliarity with Hazlitt's writings).
[17] *Letters*, 205.

to the chronic convulsive rapacity of the rich. His aim was to deprive the power-wielders, the main-chancers, of the ally they found ready to hand in the inertia natural to the human mind, and which was, he believed, more especially characteristic of the educated minds of his time, fog-bound as they were by the mists of the deterministic 'modern philosophy' billowing up from the old century. 'It is the people at this time', said Southey (anticipating the mischief of the human engineering of the twentieth century), 'who stand in need of reformation, not the government.'[18] Hazlitt thought they stood in need of liberty, and the constant and consistent aim of his political essays, from first to last, was to enable them to perceive the danger, to break the ossified attitudes and forms of subservience. 'The service we have proposed ourselves to do', he said half-way through his campaign, 'is, to neutralize the servile intellect of the country. This we have already done in part, and hope to make clear work of it, before we have done.'[19] Southey's son, years later, describing his father's mood in 1816, said, ' . . . he had contracted a gloomy misanthropical way of speaking, because circumstances had forced upon his unwilling mind the fact that human nature was not as good as he had fancied it.'[20] This corroborates Hazlitt's independent contemporary judgement: 'Mr. Southey has not fortitude of mind, has not patience to think that evil is inseparable from the nature of things.'[21] It also explains why the malignant disreputable Hazlitt is a better writer than the loyal respectable Southey. Although continually buffeted by life Hazlitt threw himself again and again into the mill-race; although perpetually solicited by the temptation to withdraw he never permanently succumbed; at bottom he had faith in humanity and trusted to freedom ultimately to encourage right action; if he was angry with his former friends it was because he conceived they had gone over to the enemies of life; life ought to be and could be better, if men of good will would stand firm. But Southey lacked faith; he turned his back on life; he was without hope; he set up his tent in the narrow enemy camp; and his writings remained nerveless and stultified, yellowed like the withered leaves that betray a severed tap-root.

Certainly at this period Southey typified the reactionary establishment's ominous compound of pessimism, nervousness, and angry obstinacy. The needle of what constitutes at such times a delicate barometer, the degree of freedom enjoyed by the press, gave a significant leap in a small paragraph of the *Morning Chronicle* on 4 July, reporting the seizure at Dieppe of English newspapers carried by travellers from England, a species of inquisition to which Hazlitt himself

[18] Quoted by Hazlitt, 7. 204.   [19] 7. 189.   [20] Southey, *Corr.* iv. 199.
[21] 11. 79.

would have to submit in a few years' time, on entering the Kingdom of Sardinia at Pont de Beauvoisin. Such fearfulness did not agitate the bosoms of continental authorities alone. In September the level-headed Robinson called on Southey in Keswick and found his views on the situation of the country almost gloomy: 'He thinks that there will be a convulsion in three years! I was more scandalized by his opinions concerning the Press than by any other doctrine. He would have transportation the punishment for a seditious libel!' (Nor was Southey any the less timorous at the end of those three years when Haydon, passing through Keswick, found 'he feared in case of a revolution he should be the very *first man hung by the Radicals*'.[22])

In one respect, however, Hazlitt, although unwittingly and for different reasons, agreed with Southey, and that was in his scepticism over the remedy to the evils of the time offered by Robert Owen in 1813–14 in his *New View of Society*. In August 1816 Southey recounted to Rickman a visit from the proselytizing Owen who was, he said, a visionary 'neither more nor less than such a Pantisocrat as I was in the days of my youth'.[23] Owen's scheme would not work because it presupposed a goodness and good will that men did not have. Southey spoke of the people as in need of reform. If he did not really believe in the possibility of moral reformation, if he was sceptical of Owen's Institution for the Formation of Character, then his intentions in positing the need for such reform were somewhat sinister. At best the proposals of both were tainted with a paternalism that Hazlitt rejected out of hand: to suppose that men were children, whether potentially good or wholly bad, seemed to him an overweening insult. He was as incredulous of the pedagogic calculus of Owen as he was impatient of the felicific calculus of Bentham (which was to order the forces of pain and pleasure as Newton's laws had ordered gravitation), as suspicious of the Utopia of New Lanark as he had been of the Chrestomathic School that Bentham had contemplated erecting for the 'cloven-hoofed rabble of Westminster' in the garden behind No. 19 York Street in 1814, after demolishing Milton's house and the fine trees that had stood since his day, to provide access.[24] In the *Examiner* in this summer of 1816 he too dismissed Owen's scheme, but from a quite other standpoint than Southey's; he did so because it implied the efficiency of a deterministic system of control over human behaviour, because it assumed that the 'new philosophy' was older than the creative human spirit, because it aimed at substituting benevolent direction for the dignity, even if also the dangers, of freedom.

Not that he ever supposed it likely that Owen's schemes would

[22] Morley, 189; BL Add. MS 57976, letter of 12 Feb. 1824 to Miss Mitford.

[23] Southey, *Corr.* iv. 195–6.

[24] 11. 6; 12. 249.

become effective. They continued for some years intermittently to attract those who were sufficiently concerned with the distresses of the country to give any solution a hearing, and sufficiently impressed by the specious logic of the sensationalist philosophers to believe mankind readily malleable. In 1819 when Owen returned to the charge (and Basil Montagu served on a committee formed to consider his plan, with Peel and Ricardo),[25] Sir William de Crespigny, MP, voiced at a public meeting one of Hazlitt's objections when he said that Owen's plan was opposed on the ground that by it 'man was made too much a machine'. He went on to say, to applause, that 'the more we were machines the better, provided those machines were properly directed'.[26] Hazlitt was unlikely to have been present (he was probably at Winterslow), but if he read the report in *The Times* he must have smiled wryly at this obduracy, and smiled also to find the attitude towards the lower classes, which he had condemned in his *Reply to Malthus*, crudely and ingenuously exposed in one of the objections Owen mentioned as having been raised to his plan, viz., 'that it would make the poor and working classes too comfortable and happy, and that the population in consequence would increase too rapidly'.

It is clear from the newspaper reports that Owen's schemes were equally mistrusted by the Tories, the Whigs, and the Radicals. Cobbett dismissed his project of Villages of Co-operation, based on the experiments in New Lanark, in which the Government were to employ the poor, with a phrase at once blunt and acute, as 'parallelograms of poverty'. Hazlitt's criticism of Owen is more explicit, and even more discerning. He might easily have accepted him at his humanitarian face value, recognizing as he did that here was no New View of Society but the old vision of the French Revolution resuscitated, 'the doctrine of Universal Benevolence, the belief in the Omnipotence of Truth, and in the Perfectibility of Human Nature'.[27] But he also saw the emerging fallacy that had turned the early hopes of the Revolution into the Terror of 1793, the hair's-breadth gradations that divide persuasion from direction, direction from manipulation, and manipulation from oppression. The disinterested ideals of that revolution he still clove to, but its illusions, nurtured by the fallacious Modern Philosophy, the inadequate theories of the Enlightenment, he rejected. Now as always he bore in mind the Shakespearian quotation which constantly reappears in his work, from *The Round Table* to the posthumous 'Aphorisms on Man': 'The web of our life is of a mingled yarn, good and ill together.' Owen's

[25] *Courier*, 28 June 1819.
[26] *The Times*, 27 July 1819. See also the *Courier*, and in both papers Thelwall's remarkable reaction to the MP's speech.
[27] 7. 98.

philosophy, like Bentham's, ultimately derived from the sensationalist *tabula rasa* of Locke, through all its developments on both sides of the Channel in the eighteenth century. Universals, innate ideas, 'the fates and sisters three', good and evil, all these, according to the 'modern philosophy', were absurdities. There were no mysteries. Anything could be written on the 'blank sheet of white paper, originally void of all characters whatever', that the mind of man was at his birth.[28]

Another reason for the silence surrounding his political essays, discernible in the strictures of the *Black Dwarf*, may have been the novelty of his style. Worn smooth to familiarity now, it must have seemed affected, rebarbative, or downright incomprehensible to those who were lulled to reassurance by the conventional prose of his day. Political journalism, in any event, like political speech, is occasional and essentially evanescent, but his writings in this field are recognized by one of the foremost authorities of our day on the history of popular reform as having a distinctive note, an acuteness, a discrimination, and a breadth of vision that sets them apart from those of his contemporaries.[29]

Nevertheless his acuteness and breadth of vision are different in kind from Coleridge's. He was more strongly marked, and continued to be tightly held, by the close grip of the eighteenth century against which he struggled. In this revolt Coleridge and he were on the same side, although not in the same camp. He lacks the spiritual dimension that was the region of Coleridge's wanderings. It is, however, because he has a strong apprehension of the essential direction proper to be pursued in this dimension — freedom — that his frustration, when he regards Coleridge's activities, breaks out into anger. At the most immediate level it seems as though he is angry with Coleridge because he has allied himself with reaction and allowed himself to be enlisted by the *Courier*, but more profoundly it was because he regretted him as a lost champion, on the metaphysical plane.

There were, of course, a number of occasional reasons. His irritation with Coleridge, at this moment in 1816 in reviewing the *Lay Sermon*, may have been partly an effect of the anger generated during their commerce in 1812, when it is plain from Coleridge's letters, as well as from Hazlitt's comments on Coleridge's lectures, that they annoyed each other. Hazlitt then had a more leisurely opportunity than in 1808 to compare Coleridge with what he had been in 1798 and in the first years of the century, and to observe the ravages of opium upon his old mentor's will-power. He was no doubt as upset as Robinson by Coleridge's abject gambit as lecturer, of playing upon his bad health and

[28] 2. 147.
[29] Thompson (n. 15 above), 687–8, 820–2, 829, 862–4.

whining for the sympathy of his audience. It was too painful a reminder of the archangel on the road from Wem to Shrewsbury.

There is a strange note of intimacy in all he says of Coleridge, half-concealed by the kind of irritation, resentment, even, that a younger brother would feel towards an elder who had not fulfilled the early promise he saw and admired in him. His tone in speaking of Wordsworth is quite other. Wordsworth is a wooden figure in 'My First Acquaintance with Poets'. Hazlitt's feeling for Coleridge was deeper. And by Wordsworth he was at that time merely tolerated — a boy, a tiresome hanger-on of Coleridge's who had the assurance to pretend to an interest in Wordsworth's poetry. Iron will and unshakeable self-centredness enabled Wordsworth to accomplish more than the dilatory and amiable Coleridge. But Coleridge saw in Hazlitt what Wordsworth did not — 'a thinking, observant, original man . . . [who] sends well-headed and well-feathered Thoughts straight forwards to the mark with a Twang of the Bow-string'. And he here implies also that he attended closely to all of Hazlitt's conversation, to his clumsy as well as to his effective remarks: 'tho' from habitual Shyness & the Outside & bearskin at least of misanthropy, he is strangely confused & dark in his conversation & delivers himself of almost all his conceptions with a Forceps, yet he says more than any man I ever knew . . . that is his own in a way of his own.'[30]

But Coleridge also discerned in his companion (more original-minded, we may note, in his view, even than Wordsworth) an obscure self-confidence and energy of will that rather startled him and which he interpreted — in terms of his own irresolution and pliancy — as 'irritable Pride'. He seems not to have perceived that it was this pride that roused itself later on *his* behalf, and that, just as Hazlitt resented the damage he had inflicted on himself by drugs, he resented also his demeaning himself before his intellectual inferiors. If he appeared to be accusing Coleridge of 'servility to all men alike in conversation', he was in fact angry at a failure in self-respect all the greater for Coleridge's undoubted genius (these things do not matter now, but they were very real at the time). And he was angry not only at the lack of self-respect but also at the embarrassing obsession with reputation that accounted not only for this deference to others but for such things as the volte-face that Coleridge executed after their argument at Sir George Beaumont's, as well as his multifarious complaints of plagiarism and the theft of his ideas.

All this is borne out by Coleridge's own words. He not only says, in January 1817, that he had been a brother to Hazlitt, but makes the

---

[30] Griggs, ii. 990–1.

surprising assertion (ignored by his own biographers and by Hazlitt's alike) that Hazlitt was *the only one who knew him*, and this at a time when he was bitterly complaining of Hazlitt's criticisms. This astonishing statement, so astonishing that no one has been able to believe what it says, has passed unremarked, perhaps because it is overshadowed by a preceding theatrical exaggeration. It comes where Coleridge is speaking of the Keswick episode: '[among those who attack me is] one man whom I had saved from the Gallows, besides having previously behaved like a Brother to him for three years together, & *the only one who knows me . . .* '[31] This throws a quite different light on their relationship, places Hazlitt in an unexpected position in the circle of the Lake poets, and goes far (though this was not the writer's intention) to justify Hazlitt's present resentment.

Hazlitt distinguished between Coleridge and Wordsworth, and also between Coleridge and Southey. Wordsworth he saw as a man of fine sensibility, but with little awareness of contemporary life, a great poet, but a poet whose world, politically, was the isolated, out-of-date world of the Westmorland statesman, and who had reverted to his obscure forefathers' subservience to the feudal lord, underlined in his case by an occasional public assertion of loyalty to his lord's lord, the King. Southey he saw as a perfectly honest man,[32] but a dangerous, with all the rigid well-meaning righteousness, all the conscious disastrous good intentions of a sea-green incorruptible, and who also was made doubly dangerous by having an effective mouthpiece, in the Government-sponsored *Quarterly*.[33] Coleridge he knew to be different from either, and his politics to be not only free from corruption, as theirs were, but identified with and dedicated to the free-ranging creative imagination, a specifically, indeed an explicitly, religious imagination, it was true, but nevertheless a living force implying a wholeness, a total expression of the spirit of man that perceived the ineffectuality of partial and exclusive political systems. It was this recognition of a potential ally, or rather of a lost leader, that accounts for the extraordinary bitterness, quite disproportionate to the degree of Coleridge's dereliction or of his political prominence, the anger and ridicule which make us so uncomfortable in his article on the *Lay Sermon*, and which were owing not to the harm the volume might do but to the good it could have done. The desperate case of the country and the unregenerate implacability of its rulers called for a more urgent voice than the bumbling, obscure, apologetic ruminations of this volume. It demanded, rather, the sublimely confident voice he

---

[31] Ibid. iv. 699–700.

[32] Morley, 201.

[33] As Southey wrote to his brother when the *Quarterly* was founded in 1808, 'In plain English, the Ministers set it up. But they wish it not to wear a party appearance' (Southey, *Selections*, ii. 107).

had heard declaim from the pulpit of a gloomy chapel on a cold Sunday morning in the dead of winter in Shrewsbury nearly twenty years before, when in sounds that echoed from the bottom of the human heart poetry and philosophy had met together and truth and genius had embraced. His anger did not spring from contempt, but from disappointment. He felt betrayed.

Coleridge on his side was so deeply wounded by Hazlitt's attack that he failed to perceive, or refused to admit, its hidden spring. He set it down to an inexplicable animosity. The panic and bewilderment he displayed in letter after letter to his friends is extraordinary, and hints at an unavowed recognition of some justice in Hazlitt's reproaches. It hints also at his consciousness of a closer bond than is apparent in what is known of their relationship immediately after 1798. Hazlitt is in no doubt of the uniqueness of that relationship. Coleridge was the only man he ever knew who answered to the idea of genius. He was the only person from whom he ever learnt anything. There was only one thing Coleridge could have learnt from him in return, and that was steadfastness; and although he complained savagely at times, as now, that 'all this mighty heap of hope, of thought, of learning, and humanity . . . has ended in swallowing doses of oblivion and in writing paragraphs in the *Courier*',[34] he still hoped, as others, that he who had taught him speech when he was dumb would 'shake off the heavy honey-dew of [his] soul . . . start up in [his] promised likeness, and shake the pillared rottenness of the world'.[35]

Historically, Hazlitt was proved right (was this what Coleridge meant when he said he alone knew him?). Coleridge's ideas and idealism, of immense importance, were worthy of a stronger vessel. The spiritual element in the Romantic movement, or, from another point of view, the realistic, which proclaimed the inalienable values and freedom of the human spirit, failed to stem, or even divert, the powerful current of deterministic hedonism deriving from the sensationalist philosophy of the eighteenth century. Utilitarianism prevailed, and the ideal of material progress flourished virtually without a check until 1914, and beyond.

[34] 11. 34.
[35] 8. 251.

II

Apostates from Liberty

Jamais on ne fait le mal si pleinement et si
gaîment que quand on le fait par conscience.

Pascal

IN the summer of 1816 Lamb, who had been close to a nervous
breakdown from overwork, was glad to get away for a month's holiday,
which he was to spend in Wiltshire with Coleridge's former Hammersmith friends the Morgans. But others of Hazlitt's companions continued to call at York Street; Walter Coulson, Martin Burney, and no
doubt the Humes and the Bells, and he also saw much of Leigh Hunt.
On free days he would walk out to the Vale of Health of a morning, share
dinner and a bottle of wine with the Hunt family, discuss his next
Examiner article with his editor, look over the books on his shelves, listen
to him play Mozart on the piano, take a turn with him on the Heath,
back to a supper of bread and butter, lettuce, and hard-boiled eggs, and
then walk the six miles home by moonlight, accompanied for the first
few, through the fields, by his host, who knew his dread of footpads.[1]
Perhaps as he approached York Street the fireworks which at twelve
o'clock began to shoot up into the night sky above Vauxhall on the other
side of the river tempted him to walk on and cross the newly opened
gas-lit Vauxhall bridge to pursue a different and untranquil mode of
pleasure in the dark avenues of the Gardens, where the parading women
of the town were augmented by fresh contingents pouring in after the
theatres closed, transporting across the river 'the gaudy display and gay
fascination of [the] saloons and lobbies'.[2] Perhaps, on the other hand, he
responded to the solicitations of the humbler, less blatant, street-walkers
he knew in Horse Guards Parade. And perhaps again, he simply, if as
always reluctantly, went home to bed.

Thus, when he went out to Hampstead, or Notting Hill, or to the John
Dyer Colliers's at Hammersmith. In the 1790s when he lived with his
brother in Rathbone Place he had spent whole mornings at Old Lord's
cricket ground watching Robinson and other giants bat. Lord's had
moved north since then and he himself had moved south, so that that
pleasure had no doubt become rarer, but he was not any the less a

[1] Hunt, *Poems*, 228–30.   [2] *MC*, 12 Aug. 1816; 20. 207.

sportsman. The fives-court in St Martin's Street, Leicester Fields, saw him engage young John Payne Collier and other sporting gentlemen ten years or more his junior, playing with a passionate concentration, narrowing his whole universe to the three walls of the court, and in his violent enthusiasm too impatient even to wait for the surer stroke he could make when the ball had bounced, but taking it at full volley, shouting with glee at a hit, swearing furiously when the ball fell below the line, desperately hitching up his sagging trousers as he darted from wall to wall, and at the end of the game throwing his racket up to the roof if he had won, or, if he had lost, clutching his head in mingled exasperation and chagrin; and then, after taking off his sweat-soaked shirt, walking rapidly home through the park with his coat buttoned up to his chin, like a sporting and taciturn Mr Jingle. It is not until 1821 that we find the first dated reference to his playing fives, but it is no game to take up in middle life, and we can be sure he played from boyhood or youth up and resumed regular visits to St Martin's Street on his return from Winterslow.[3]

Such pleasant excursions as those to Hampstead cannot however have been too frequent in that cold wet summer. In town he visited Godwin, and it may have been at this time that he renewed contact with John Philpot Curran, the Dublin Master of the Rolls who had ably defended Finnerty in 1797. He had met him in the old days at Horne Tooke's house in Wimbledon,[4] but he seems to have formed a closer acquaintance with the celebrated wit in the twilight of his career and reputation, sometime before his death at the end of 1817. At the time of publishing the *Eloquence of the British Senate*, which includes a specimen of Curran's eloquence, he had not heard him make a speech in the House: he had since had professional occasion to do so, but he evidently appreciated him more in private conversation, where his *bons mots* and sallies (his reply to the fussily vain barrister friend who asked, 'Do you see anything ridiculous in this wig?' — 'Only the head inside it') were a recurrent delight, although he was puzzled by his dullness in argument and his indifference to reasoning and serious discussion. To explain this he perhaps discerned a clue in a curious and cautionary saying of Curran's: 'When I can't talk sense I talk metaphor.'[5] At any rate he had the worst taste he ever knew and he was irritated by his perverse and obsessive railing at *Paradise Lost* and *Romeo and Juliet*.[6] The Irishman was often at Godwin's in Skinner Street, and it was probably here that Hazlitt met

[3] Bewick, i.136–40. In 1817 we find him proposing to meet Procter at the fives-court (*Letters*, 177).

[4] 12. 41.

[5] T. Moore, *Life of R. B. Sheridan* (1825), i. 521 n.

[6] *Memoirs*, ii. 248.

him again. A letter of late 1816 from Godwin to Archibald Constable invites the publisher (who had returned to London) to dine on 'a beef-stake, with Curran and Hazlitt for sauce'.[7] But Hazlitt seems also to have been invited independently to Curran's house in Upper Grosvenor Street, and on one occasion, at least, he stayed overnight.[8] A few years later, after Curran had died, his daughter Amelia dined at Godwin's with Hazlitt at a time when she was attending his lectures on poetry.[9] There is an unexpected vignette of him resplendent in gilt-buttoned blue coat, white cravat, black breeches, and silk stockings, calling at the house of his friends the Reynells, who lived above their printing office at 21 Piccadilly, on his way to Mayfair.[10] This unwonted glimpse of a Hazlitt who 'looked well' in the garb at that time made fashionable by Fox, Burdett, and the Whig Club, only surprises because his biographers, taking their cue from his friends and certain essays, have over-emphasized his slovenliness. But there is one essay, 'The Fight', where he concedes that on occasion he was, like any other man, pleased to cut a dash.[11]

Godwin lacked Hunt's breadth of interest, and the conversation in Skinner Street was probably more exclusively political than in the Vale of Health. Wordsworth had long been resentfully antagonistic towards Godwin, and when the author of that *Political Justice* by which he had once been so strongly influenced called to see him at Rydal Mount on his way back from Edinburgh in April he rammed his present political views so insistently down his guest's throat that Godwin walked out of the house in bitterness and hostility. Even so faithful a champion of Wordsworth as Robinson considered him injudiciously obtrusive and uncharitably intolerant towards his guest.[12] Godwin called on Hazlitt the day after telling Robinson of the incident, and we may assume that he was still full of it and that the talk turned to the *ci-devant* republican triumvirate of the Lakes. It seems likely that it was Godwin who then told Hazlitt of Wordsworth's extravagant indignation at the escape of La Valette from his Bourbon jailers in Paris the previous December, a romantic episode, with Sir Robert Wilson as a Scarlet Pimpernel of the Restoration, which caught the public imagination.[13]

Hazlitt was even then writing for the *Examiner* his first paper on Coleridge, not indeed on his politics but on his poetry in *Christabel*.

[7] NLS MS 327, fo. 224, 22 Nov. The dinner was on 27 Nov. (Godwin Diary).
[8] Godwin Diary, 23 May 1817.   [9] Ibid., 29, 30 Mar. 1818.
[10] *Memoirs*, ii. 306.   [11] 17. 85, 86.
[12] Moorman, ii. 292–3, 335; Morley, 183.
[13] 7. 133. The incident as described by Hazlitt is clearly authentic (as the ironical inventions that succeed it are equally clearly not). No mention of it, to my knowledge, is made in Wordsworthian biography.

Wordsworth he had hitherto reproached only in asides and footnotes (it is striking that he never did write an article exclusively aimed at Wordsworth's politics), and the next of the three to engage his pen was Southey, whom he had let alone since 1814, but who had just recently published *The Lay of the Laureate*. He reviewed it in July, making as ruthless and effective fun of its inconsistencies and bad logic (somewhat disingenuously, it must be admitted, since it was a poem) as in his earlier squib 'Prince Maurice's Parrot', but what Lamb was acute enough to detect and identify in his later notice of Coleridge's *Lay Sermon* as 'a kind of respect [that] shines thro' the disrespect' is perceptible here also, despite his dislike of Southey's unshakeable self-righteousness. He had never admired Southey's talent as heartily as he had admired Coleridge's brilliance and as he still admired Wordsworth's poetry, but he thought highly of his prose, and he respected (with justice) his uprightness of character. What he now deplored, in addition to Southey's having moved over to the enemy camp, was his rigid and repellent self-complacency. He had satirized Southey mildly enough at the time of the laureateship, but that was before Waterloo and before Southey had veered round from a preoccupation with foreign policy to his own vehement brand of authoritarian domestic politics, with its high-pitched extremist and alarmist clamancy. Ten years were to elapse before Southey wrote a book to promulgate his views, but his influential anonymous articles in the *Quarterly* had immediately been recognized as his: the editor William Gifford said his prose was so good that everybody detected him;[14] and in any case the authorship of the *Quarterly* articles could have been no secret to Hazlitt because one of Southey's regular correspondents was Lamb (himself also, in that blandly impartial way of his, a contributor to the *Quarterly*). The most significant passage in Hazlitt's review is his definition of the bad Jacobin, alias the doctrinaire fanatic in one of his many disguises. Consistent with his own dislike of system and repugnance to the rigidity of dogma, it is to be placed side by side with his metaphor of the good Jacobin, in his image of the evening star setting over a poor man's cottage. Opening with a sighting shot — an allusion to *Joan of Arc*, the epic poem of Southey's revolutionary years — he straightway hits the bull's eye:

Mr. Croker was wrong in introducing his old friend, the author of *Joan of Arc* at Carlton House. He might have known how it would be. If we had doubted the good old adage before, 'Once a Jacobin and always a Jacobin,' since reading *The Lay of the Laureate* we are sure of it. A Jacobin is one who would have his single opinion govern the world, and overturn everything in it. Such a one is Mr. Southey. Whether he is a Republican or a Royalist . . . he is still the same

---

[14] Smiles, i. 260.

pragmatical person — every sentiment or feeling that he has is nothing but the
effervescence of incorrigible, overweening self-opinion. He not only thinks
whatever opinion he may hold for the time infallible, but that no other is even
to be tolerated, and that none but knaves and fools can differ with him. . . his
sentiments everywhere betray the old Jacobinical leaven, the same unim-
paired, desperate, unprincipled spirit of partisanship, regardless of time,
place, and circumstance . . .[15]

The old Jacobinical leaven fermenting in the young Southey had
produced more than one revolutionary work, and Hazlitt's reference to
the youthful indiscretion, *Joan of Arc*, was a harbinger of the uncomfortable ghost of *Wat Tyler* which in 1817 returned to haunt the Poet
Laureate, and it may even have drawn attention to the cupboard where
that skeleton lurked.

As to that other field of Hazlitt's interests, the sluggish world of
English art, it was not by the Elgin Marbles alone that it was unwontedly
stirred that summer. In August appeared the second part of the
Catalogue raisonné of the British Institution. The Institution had been
founded in 1804 for the encouragement of British art, and after holding
exhibitions of aspiring British painters for ten years it had now gone on
to organize an additional exhibition of works of the past, initially, in
1814, and tentatively, keeping close to the terms of its foundation, by
limiting it to British artists of the eighteenth century, but in 1815, for the
first time, extending it to Old Masters. This immediately provoked a
now hardly comprehensible outburst of nervous anger and jealous
spleen in the shape of a bogus, anonymous *catalogue raisonné* jeering at
the canvases. This crude pamphlet, now ascribed to Robert Smirke,
RA, was sent through the post to selected, influential addressees. It was
intended to protect the professional interests of living painters by
deriding the exhibits, and so by implication the taste of their owners, the
patrons of the Institution.

The Academicians had never dared officially to oppose the Institution, but their jaundiced, pretended indifference concealed resentment
of a presumptuous rival. The Directors of the Institution and their
friends, who perversely spent their guineas at Christie's auction-rooms
instead of at the Royal Academy exhibition at Somerset House or in
commissions, they regarded as impertinent busybodies who were now
further undermining the authority of the Academy by setting up an
exhibition of the great masters which was bound to unmask the painful
inferiority of its members.

The humour of the production was lumbering and gross, the satire
grotesque, but the ill-informed public, most of whom had never seen

[15] 7. 86-7.

such paintings, may have accorded it some credit and talked it into a measure of success. Its 'Part 1' (1815), on the exhibition of Dutch and Flemish works, made little stir, but twelve months later when the persistent Directors went on to exhibit Italian and Spanish paintings and the anonymous iconoclast as persistently returned to the attack, its 'Part 2' made a great deal. This second and more important exhibition at the British Gallery lasted from 22 May to 12 August 1816. The *Catalogue*, perhaps significantly, since it was thus not possible to compare it with the pictures, came out a fortnight after it closed.

Joseph Farington, RA, the *éminence grise* of the Academy, notes the speculation surrounding the authorship of the pamphlet.[16] The secret was well concealed; the ubiquitous Farington seems never to have ferreted it out; and beyond the question of the author's identity the pamphlet itself had no more interest for him than the Old Masters themselves. Painting for him had no ancestry, no history. It was an affair of trade practice and professional career, rather than of the grand, the sublime, or even the merely beautiful. The journal of this man of weight and authority unconsciously justifies the attacks Haydon and others made on the Academy. He has little to say about art, nothing about literature or music; never describes an exhibition, except to list by rank the notabilities who patronize it; never reads. His journal, in so far as it regards the Academy, is all investments, incomes, rents, decorations, honours, precedence, protocol. It is a careerist's diary. The astute Northcote put it succinctly: 'His great passion was the love of power.'[17] He might have sat for Hazlitt's portrait of the archetypal Academician in 'On Corporate Bodies', who was not 'open to the genial impulses of nature and truth' and neither '[saw] visions of ideal beauty, nor [dreamt] of antique grace and grandeur', but had a mind haunted by 'rules of the academy, charters, inaugural speeches, resolutions passed or rescinded, cards of invitation to a council-meeting, or the annual dinner, prize-medals, and the king's diploma'.[18]

The materialism of the Academy, implied in this man's ascendancy, explains equally the tone of the *Catalogue* itself and the violence of Hazlitt's attack on it. Haydon's later claim that Hazlitt attacked it to oblige him probably means simply that it was he who brought it to the attention of Hazlitt, who must have seen both parts together and for the first time in Haydon's studio in September or October 1816.

It is typical of the mystery surrounding this publication that Haydon not only forgot that it came out in two parts but could not say when it

---

[16] Farington, 9, 17, 21 June 1815.
[17] *Conversations of James Northcote R.A. with James Ward on Art and Artists*, ed. E. Fletcher (1901), 165.
[18] 8. 270.

appeared.[19] Hazlitt begins his first article, 3 November 1816, by quoting, he says, 'the *Catalogue Raisonné* of the pictures lately exhibited at the British Institution', but the quotations refer in fact to the first foreign exhibition of 1815, and not to the second. It is pretty certain that he had seen both and, starved as he was of great paintings during the war, they must have brought back all the high hopes and tremulous admiration that coursed through his veins in 1802 in his heyday at the Louvre, so that when he now came upon this brochure he was outraged. He must have seen in its parochialism, self-interest, and vulgarity the same assumption of privilege, of self-complacent superiority, the same expression of brazen power in the world of art as he saw in the political world, in Legitimacy and the Tory party.

It is ironically possible that Smirke may have modelled his tone on the tone Hazlitt himself once adopted in a rash moment in the past. Smirke, as an Academician and professional gossip, could not have missed Hazlitt's account in June 1814 of the President's picture of *Christ Rejected* and its informed but somewhat heavy irony.[20] Hazlitt's estimate of this picture has been confirmed by posterity, but this does not justify his derogation in the first part of his article, where his pleasantries may easily have encouraged the author of the *Catalogue raisonné* to launch into crudities of which the following is a typical instance: 'Portrait of Himself — Rembrandt — [Owner, the] Earl of Ilchester — What, is GOG come again? What mountain bulk of blubber and black jaundice is this? faugh, what a mass of filth and brutality is here! The very mimicry of which threatens us with disease, makes us shudder with alarm, and forewarns us of approach, lest by accidental collision the plaits of the drapery should unfold and disgorge upon our heads their horrible contents.'[21]

Smirke's sincerity is a question. We can hardly believe that these scurrilous jibes were serious and that he did not have his tongue in his cheek, but a doubt persists. Hazlitt preferred to take the clumsy sarcasms literally, and replied in a resounding article, a frothily inventive piece of vituperation which may have proved a second hostage to fortune: its vehement, scornful, rough-and-tumble tone is in much the same style as that adopted towards him and his friends in succeeding years by the *Blackwood's* gang, who were always close students of the London journals. Much, but not exactly: they adopted the vehemence but could not reproduce the leaven of creative fancy bubbling up in a string of droll variants on every satirical stroke. This verbally inventive vilification in the style of Falstaff was no doubt inspired by his rereading of Shakespeare at this time in preparation for *Characters of Shakespear's*

[19] Haydon, 352. (The editor failed to locate a copy of the *Catalogue raisonné* — both parts are in BL Print Room).

[20] 18.32.   [21] *Catalogue raisonné* (1815), 32.

Plays. (We find the same stringing of satirical beads in his contemporaneous review of the *Lay Sermon* where he calls the procrastinating Coleridge 'the Unborn Doctor; the very Barmecide of knowledge; the Prince of preparatory authors, . . . the Dog in the Manger of literature, an intellectual Mar-Plot', etc., and *The Friend* 'an enormous title-page; the longest and most tiresome prospectus that ever was written; an endless preface to an imaginary work', etc.[22])

A minor mystery of the *Catalogue raisonné* affair is Northcote's reaction: reaction: he thought it admirably done, the wit and humour exquisite, and so forth.[23] We do not recognize the wise moralist of Hazlitt's *Conversations*. Hazlitt was so upset by this strange perversity in his old friend that it was long before he could bring himself to cross his threshold, and five years later he still had not forgotten. And yet Northcote venerated Titian on the one hand, and on the other detested the insufferable collective insolence of the Academy, which he called 'a nest of vermin'.[24]

If the episode estranged Hazlitt from Northcote it put him on more intimate terms with Haydon, who saw a chance, during a long talk about art reported in the diary on 3 November, of harnessing him in the same team as Robert Hunt and James Elmes to whirl the Haydon chariot to victory in the great Historical Painting contest. Haydon, who quarrelled indiscriminately with fellow-artists as well as patrons, was a lonely man and needed not merely allies but friends. Not that any kind of social life was possible in his squalid surroundings, which were hardly fit to receive guests. The air in his cramped ill-ventilated lodgings (one room and a box-room off, which were his quarters for nine years) was so foul with the smells of paint and cooking and bedclothes that he nearly ruined his health before finally moving in autumn 1817. Hazlitt was one of his few visitors. Haydon found a guilelessness in him (who was as vulnerable as ever and had learned nothing from the treachery of John Scott) that reconciled him to what he deemed to be his wrong-headedness about art. He talked for three hours, while sitting to Haydon for one of the spectators in *Christ's Entry*, developing, as he went, an essay he was projecting ('On Commonplace Critics'), lamenting the present doldrums of British art, remembering his old painting days.

He said some fine things, things which when he writes them will be remembered for ever. I gave him a bottle of wine, & he drank and talked, told me all the early part of his life, acknowledged his own weaknesses and follies . . . When a man of genius is in a humour of pouring out, he should never be opposed. . . . Hazlitt is a man who can do great good to the Art. . . . My object is to manage such an intellect for the great purposes of art; and if he was to write

---

[22] 7. 115.   [23] Farington, 23 Aug. 1816.   [24] Fletcher (n. 17 above), pp. 164–5.

against me for six months, still would I be patient. He is a sincere good fellow at Bottom, with fierce passions and appetites. Appeal to him & he is always conquered and yields . . .[25]

He was indeed a 'sincere good fellow' but his sincerity did him injury. What followed is an instance of Haydon's unreliability, if not duplicity. He was blown about by every gust of vanity and ambition, even if equally by impulses of generosity and aspirations to good. When he later reported this conversation to Wordsworth, who was attempting to detach him from Hazlitt (attempting indeed to close all doors against Hazlitt), he gave (what Wordsworth wanted to hear) a warped and damaging account. Claiming that he saw Hazlitt scarcely ever, and then not in his own studio (he forgot that Wordsworth was also in *Christ's Entry*), he said it took place 'one night when I saw him half-tipsy and so more genial than usual', thus slily suggesting an accidental encounter.[26]

Hazlitt's relations with the two painters that winter may have been chequered but did not lead to a break. He could make concessions (could 'yield') for the sake of friendship in disputes on painting, which lay within the sphere of opinion, but not in politics, which involved principle, and there was no chance of reconciliation with another object of his attacks, and that was his brother-in-law, Stoddart. The rift between them, so fundamental that it seemed ordained in nature, and which had grown to a yawning gulf during the last months of Napoleon's Empire and during the Hundred Days, again widened with the increasing evidence of the real nature of the peace.

With a large family and an inadequate income, Stoddart had yet acquired no discretion — professionally, that is, in his journalism; socially, it seems, he was as astute as ever and knew when to hold his tongue. The radical Hobhouse, meeting this vehement reactionary at this time, thought him an agreeable man.[27] But when he took up the editorial pen the Bourbon zeal of this stubborn, opinionated, and aggressive man became, as we have seen, outrageous. It had long irritated his employer John Walter II, who had been alerted to the abuse by Robinson, and whose repeated requests to Stoddart to lower his tone and abstain in particular from attacking the *Morning Chronicle* had been flouted. Walter had for twelve months been obliged to employ Barnes to blue-pencil Stoddart's leaders.[28] For this fierce Tory editor the Restoration in France was not enough; its regime, reactionary in the eyes of most, was culpably democratic in his. He had not merely turned royalist after abandoning Jacobinism; as Hazlitt had predicted, he had swung round from ultra-Jacobinism to ultra-royalism. He now supported the

---

[25] Pope, ii. 64–5.   [26] Haydon, *Corr.* ii. 34 (15 Apr. 1817).
[27] Broughton, i. 342.   [28] Robinson Diary, 1 Nov. 1815.

French *ultras* against the French Government, and the ominous Comte d'Artois (later Charles X) against Louis XVIII.

If anything was needed to confirm Hazlitt's dislike of Stoddart it was certain cross-Channel manoeuvres that he embarked on during this summer. In the previous September Hazlitt had linked Coleridge with the reactionary Chateaubriand (of the French Chamber of Peers, and Council of State) in the article 'Chateaubriand the Quack'. He did so more on account of Chateaubriand's antagonism to Napoleon than of any apostasy in this *ancien régime* aristocrat, former *émigré*, and unrelenting foe of the regicides of 1793, although there is a hint even of this in his reference to Chateaubriand's youthful admiration of Rousseau. It was the *Champion* and the *Examiner* who first dubbed Chateaubriand quack, meaning hypocritical opportunist, foreseeing as they did that he and his intransigent associates aimed to sabotage the first Restoration.[29] Hazlitt, opposed equally to the restored Louis XVIII, the *ultra* Chateaubriand, the apostate Coleridge, and the reactionary Stoddart, returned to the attack twelve months later in a paragraph on his brother-in-law. 'Now that Louis XVIII has offended [Stoddart] too we think that he and Chateaubriand (we ask pardon of our friend Mr Coleridge for associating anybody but himself with Chateaubriand) had better go over together to Ferdinand [of Spain].'[30] At first sight the linking of Stoddart and Chateaubriand seems to derive from their political pamphleteering and kindred political views, with no suggestion that they knew each other. Chateaubriand had recently brought out *La Monarchie selon la charte*, a conjuring trick which made out the reactionary French Government to be actively Jacobin, Bonapartist even. Banned in France, it immediately reappeared in London. John Murray, proprietor of the *Quarterly*, issued a translation early in October; and soon after, the original was published by the newly established and enterprising Henry Colburn.[31] And before long the *Courier* was praising the 'resistless truths' of its account of France.[32]

But in fact this shaft of Hazlitt's, which has aroused no remark, evidently means that he had got wind (through his wife?) of a six-week visit Stoddart paid to Paris in August–September 1816, and of his transactions with the French *ultras*. Stoddart was planning a new periodical, the *Correspondent*, to consist of 'letters, moral, political, and literary, between eminent writers in France and England'. Two unpublished letters of Stoddart's to Walter Scott (survivors of a longer

[29] *Champion*, 5 Mar. 1815, 79; *Examiner*, 16 July, 460; 6 Aug. 1815, 504.

[30] 19. 165.

[31] *MC*, 5 Oct., *Courier*, 8 Oct. 1816. Extracts had appeared in *MC* on 27, 28, 30 Sept., and 2 Oct. 1816.

[32] 2 Dec. 1816. See also its article on Chateaubriand, 28 Oct. 1816.

exchange) describing these manoeuvres and inviting contributions, not only show that Stoddart was on friendly terms with Scott (whom he had met on his northern tour in 1800) but confirm the coincidence of their political views. Scott even had a hand in naming the new journal, which, ostensibly aimed at reconciliation, at building a bridge between the two countries, had as its real object the erection of a double barrier, against subversion by the French liberals and against the revolutionary machinations of such British journals as the *Morning Chronicle* and the *Edinburgh Review*.[33] Among the prospective French contributors were Chateaubriand, Vitrolles, the Duc de Levis, Montmorency, Freuilly, and Michaud, all of whom Stoddart had personally recruited; and among the English, in addition to Southey, such eminent writers as Stephen, Jacob, Professor Walker of Glasgow, Dr Inglis of Edinburgh. The journal was to appear every two months from 1 January 1817 but collapsed from unrelieved tedium after three numbers. Six days after it was launched Hazlitt ended an *Examiner* article with an apology for indiscreetly naming Coleridge's early political principles, saying he did not 'well understand how men's opinions on moral, political, or religious subjects [could] be kept a secret except by putting them in the *Correspondent*';[34] Southey groaned that it was constitutionally dull and 'want[ed] relief, and general matter' (he himself enlivened it with an article on Methodism).[35]

The *Correspondent*'s demise did not end the editor's connection with Chateaubriand. The journal, as well as serving Tory policy and perhaps also Stoddart's personal political ambition, was to provide him with an avenue of escape from the control of Walter and the interference of Barnes. When finally dismissed on the last day of 1816 he speedily doubled his resources by setting up (or having set up for him by the Government) a rival newspaper treacherously named *The New Times*. Like the *Courier* (now edited by Mudford), it was certainly in the pay of the Government (Stoddart at this time regularly called at the Treasury Office several times a week) and it was probably also subsidized by the French *ultras*.[36] Chateaubriand and Vitrolles, its Paris correspondents,

---

[33] EUL MS LA III 584, fo. 29.

[34] 7. 129. Hazlitt evidently mentioned Stoddart's mission to Hunt, who remarked in a leader on Chateaubriand, 29 Sept. 1816, that whenever the *Examiner* made any reference to Voltaire 'the Doctor [was] sure to fall into convulsions, or throw himself into the sympathetic arms of his correspondents'. The three numbers of the *Correspondent* appeared on 6 Jan., 22 Mar., and 28 May 1817 (*Courier*); its French counterpart, *Le Correspondant*, which began later, survived until April 1818.

[35] Southey, *New Letters*, ii. 149.

[36] Anon., *The History of The Times* (1935–52), i. 464–5, report from the French Ambassador in London to the Quai d'Orsay on the British press: 'The *New Times*, enjoys Government support, is paid by the Treasury, and is under the immediate protection of Mr Arbuthnot [one of the two joint secretaries], to whose office its editor, Mr Stoddard, goes three or four times a week. [It] does us infinite harm. It must be heavily subsidized by the Ultra-Royalist faction, for every day its long

used it to report their own reactionary speeches when these were excluded by Decaze's censorship from the French papers, an exclusion so irksome to Chateaubriand that at one time he considered coming over to London, where he had been exiled in the 1790s, to set up his own newspaper in opposition to Louis XVIII.[37] And when Southey went to Paris in May 1817 he bore a letter of introduction to Chateaubriand from Stoddart.[38]

A letter of Southey's of December 1816 suggests that Hazlitt's acquaintance with the *Correspondent* was more immediate than may be guessed from his reference to it in the *Examiner*. Something Southey here lets fall to his old friend, Grosvenor Bedford of the Exchequer (a close friend also of William Gifford), requires investigation. The whole passage must be quoted, not only on account of the *Correspondent* but also of what it hints about the Keswick misadventure.

You do Stoddart wrong. A fellow by name Barnes, quondam coadjutor to Hunt in the *Examiner*, has got into the *Times*, and puts in all the poison there, greatly to Stoddart's discomfort, who I believe would get out of it, if he could find any other means of employing himself so well for the public and himself. Sir Tarquin has married his sister — a worthy couple — they quarrel, fight, make it up over the gin bottle, and get drunk together. Stoddart paid him for a paper for his first number, and the said Tarquin has postponed the paper sine die, as might have been expected.[39]

The suggestion that Hazlitt was somehow connected with the *Correspondent* seems confirmed by a vaguer remark of Stoddart's a week later to Southey: 'I need not tell you that the first engagement in a work of this kind is a work of trouble. I believe I informed you how Hazlitt served me. My friends in France, with all the good will in the world, have no notion of exactness.'[40] Here is a puzzle. If we grudgingly accept that Hazlitt agreed to contribute we must concede the equal improbability that Stoddart in the first place invited him, his calumniator, to do so. However, Stoddart evidently assumed that some such arrangement existed, and the only plausible explanation is that, impressed in spite of his resentments by the strength and originality of Hazlitt's writings,[41] and being hard pressed to find someone competent to write on

columns are full of delirious nonsense against the French Government, the King, and personal criticism of the ministers, especially of His Grace the Duc de Richelieu, Comte Decazes, [etc]. This paper is under the immediate influence of its Paris correspondents, M.V . . . [Vitrolles] and His Grace the V. de C.B. [Chateaubriand].' The British Government's support of the pro-*ultra New Times* and *Courier* continued until the French intervention in Spain in 1823 (see NLS MS 4010, fo. 151)

[37] Mme de Boigne, *Mémoires* (Paris, 1908), ii. 242.
[38] BL Add. MS 47888, Southey to his wife, 17 May 1817.
[39] Southey, *New Letters*, ii. 144–5.
[40] NLS MS 1706, fo.197, 20 Dec. 1816.
[41] For Stoddart's later praise of Hazlitt's work (non-political work, however) see Lucas, ii 283.

philosophy, he reluctantly approached him. But only indirectly, through his sister. He asked Sarah to procure from her husband an article on 'The Scotch Metaphysicians, Reid, Stewart, etc',[42] and paid her in advance as an inducement. Hazlitt, when she passed on the commission, flatly refused (hence a quarrel between husband and wife, which Southey took to be one of many). Only an involved explanation of this kind will meet the case, for it is certain that Hazlitt with his obsessive hatred of Legitimacy would never have fired a single shot under such a banner and with such companions-in-arms. Nor, of course, did he.

As to that other dubious glimpse of domestic strain and tipsiness in the bare rooms of York Street we can only guess at the source of Southey's information, as at its truth, but it is a rumour of a strife that sounds much like a harbinger of the final break-up of the marriage in 1819. Finally, before leaving this bleak, significant letter we should note that the recipient Bedford, a person unknown to Hazlitt's biographers, was evidently well enough informed of the escapade in the Lakes to be able immediately to recognize Hazlitt under the nickname, Sir Tarquin, and that consequently the story was far more widely circulated than has hitherto been suspected. We may imagine how tongues had been set a-clacking by this delicious morsel of gossip, and how it spread from circle to circle in proportion as Hazlitt's fame grew as the author of *The Round Table* and *Characters of Shakespear's Plays*, and how it effectively perverted that fame into notoriety. This alone can explain the unelaborated remark of someone who made his acquaintance shortly after this, and who, though apparently a stranger to the Lake poets and indeed to everyone else known to Hazlitt, said that long before meeting him he had heard the most shocking rumours about his personal character and was persuaded that he was little better than an incarnate fiend.[43]

On 1 December Hazlitt opened an attack on *The Times*, which had become, he said, 'a nuisance which ought to be abated'. He had hardly mentioned Stoddart in twelve months and the suddenness and violence of the present attack on 'the Writer in *The Times*, a very headstrong man with very little understanding and no imagination' and on his championship of Legitimacy are probably explained by the affair of the *Correspondent*. And when, a few days later, a bellicose *Times* leader on the Spa Fields meeting came as a confirmation of the description, he wrote to the *Examiner* (8 December) over the pseudonym *Scrutator* making Stoddart out to be as hare-brained as Fluellen and as sinister as Dr Sacheverell.[44] On the following three Sundays he pressed home the

---

[42] The title is among the provisional, unallocated list of projected contributions drawn up by Stoddart and his collaborators in Paris in Aug. 1816, (EUL MS LA 584, fo. 31.)

[43] Patmore, ii. 260.

[44] See the writer's 'Three Additions to the Canon of Hazlitt's Political Writings', *RES* NS 38 (1987), 355–63.

assault with a series of 'Illustrations of the Times newspaper'. They succeeded in their essential object: the nuisance was abated; Barnes became editor after the year ended. (Apparently there had been a stay of execution: Walter approached Barnes two years earlier, and Hazlitt had then urged Barnes to accept.[45])

Although the immediate target was the fire-eating Stoddart, Hazlitt's theme was bound to concentrate within his sights those others who had also deserted the cause of the people and defected to the enemy. He unleashed a perfect cannonade against Coleridge, Wordsworth, and Southey. Meditating, no doubt, the implications of his brother-in-law's journey to Paris, he told himself that he did not see why he should repress legitimate anger, nor limit himself to a fowling-piece against members of an enemy force which not only used the heaviest verbal artillery but had possession of the actual sinews of war in the Treasury, and indeed of the real War Office as well. His earlier criticism of Mackintosh's debating style shows he had no illusions about the efficacy of sweet reason in politics. He seemed in these articles possessed of a demon and his sustained flights of furious rhetoric were powered by some pretty unguarded language. But the arguments themselves were not unconsidered. He adverts to his distinction between the true Jacobin and the false, or vulgar, Jacobin, the latter an egotistical leveller, the former a disinterested lover of truth and freedom. If he is asked what is the love of freedom, he answers (and it is the only practical answer) the hatred of tyrants. But such an emotion is the easier part; the harder is unremitting vigilance: he did not mean a mere vulgar hatred, but a sustained and demanding dedication. And here is the authentic personal note in these 'Illustrations': he is honest enough to confess his own inadequacy, and to admit to weakness and to the temptation to defect. To be a good hater of oppression 'is the most difficult and the least amiable of all the virtues, the most trying and the most thankless of all tasks'.[46] The flesh is weak, and often it is too hard not to lapse into acquiescence and have the comfort of bringing the world to one's side.

There is a craving after the approbation and concurrence of others natural to the mind of man . . . It exhausts both strength and patience to be always striving against the stream. Public opinion is always pressing upon the mind, and, like the air we breathe, acts unseen, unfelt. It supplies the living current of our thoughts, and infects without our knowledge.[47]

The odds are great: 'The love of truth is a passion in [the mind of the friend of liberty] as the love of power is a passion in the minds of others.

[45] Pope, iii. 640–1.

[46] 7. 151. He also recognizes its unamiability elsewhere: 'the most indispensable part of the love of liberty has unfortunately hitherto been the hatred of tyranny' (5. 91).

[47] 17. 27.

Abstract reason, unassisted by passion, is no match for power and
prejudice, armed with force and cunning. The love of liberty is the love
of others; the love of power is the love of ourselves.' Can we not hear him
grinding out these formulas and equations between his teeth as he
stumbles homewards down Whitehall at midnight through the dark
streets and past the sentries outside the Horse Guards, and concluding in
a momentary fit of despair as he catches sight of these looming symbols
of power, 'The one is real; the other often but an empty dream'? Man,
he said, is of his nature attracted to power; and he did not except himself
from mankind. Can we not hear him also condemning himself for his
human weakness, attracted to the forms of evil, in spite of himself, by
natural bias, natural inertia, and coerced into the neglect of reason by
the difficulty of the task, by the daunting effort needed to maintain the
sympathetic imagination in play? 'It is the excess of individual power
that strikes and gains over [the] imagination: the general misery and
degradation which are the necessary consequences of it, are spread too
wide, they lie too deep, their weight and import are too great, to appeal
to any but the slow, inert, speculative, imperfect faculty of reason.' And
how slow, inert, speculative, and imperfect he felt his own reason to be in
moments of discouragement, and how hopeless the odds! He said he was
no politician; neither was he. No politician ever risked his credit by such
damaging admissions of the weakness of his position. 'The cause of
liberty is lost in its own truth and magnitude; while the cause of
despotism flourishes, triumphs, and is irresistible in the gross mixture,
the *Belle Alliance*, of pride and ignorance.'[48] 'Ye are many — they are
few', said Shelley. Hazlitt, more subtly and more realistically, thought
the opposite. He gives full weight to the *caput mortuum* of the spirit of the
age: no single mind, he said, could move in direct opposition to the vast
machine of the world around it.[49]

But not external circumstances alone were to blame. We are bound to
suppose that his analysis of the weakness of human nature was also based
upon evidence plucked from his own breast, and his indignation with
others exacerbated by his own consciousness of the temptation to throw
up the struggle and cultivate a comfortable and profitable moral insensibility. Not that he lays claim to the martyr's virtuous crown: he
knows his resistance to be half-involuntary, and knows also that both
long-standing prejudice and deep-rooted obstinacy too play their part.
As he says, with the complex wisdom of experience, indeed with the
experience of a lifetime spent among minorities of one kind or another,
when he describes the hard conditions of self-consistency and the fatal
pull, at the same time as the irreversible finality, of apostasy:

[48] 7. 149.   [49] 5. 96.

It requires an effort of resolution, or at least obstinate prejudice, for a man to maintain his opinions at the expense of his interest. But it requires a much greater effort of resolution for a man to give up his interest to recover his independence; because, with the consistency of his character, he has lost the habitual energy of his mind, and the indirect aid of prejudice and obstinacy, which are sometimes as useful to virtue as they are to vice. A man, in adhering to his principles in contradiction to the decisions of the world, has many disadvantages. He has nothing to support him but the supposed sense of right; and any defect in the justice of his cause, or the force of his conviction, must prey on his mind, in proportion to the delicacy and sensitiveness of its texture: he is left alone in his opinions; and, like *Sam Sharpset* in Mr. Morton's new comedy (when he gets into solitary confinement in the sponging-house), grows nervous, melancholy, fantastical, and would be glad of *somebody* or *anybody* to sympathize with him . . .[50]

From the discouragement here expressed it becomes correspondingly plain that the familiar assertion of his critics that it was in his nature to take a perverse pleasure in malevolence and remain unshaken in obduracy had no truth in it. He was on the unpopular and unprosperous side, and could not help feeling discomfort, and often misery. He says it again:

It requires some fortitude to oppose one's opinion, however right, to that of all the world besides; none at all to agree with it, however wrong. Nothing but the strongest and clearest conviction can support a man in a losing minority; any excuse or quibble is sufficient to salve his conscience, when he has made sure of the main chance, and his understanding has become the stalking-horse of his ambition.[51]

This oppressive awareness of isolation, of having the whole world to struggle with, partly accounts for the anger vented elsewhere in the articles, but his anger is also attributable to the nervous tension of the times, as manifested in the riots of that winter. In the last weeks of 1816 the distresses that had accumulated during the previous twelve months erupted at a number of public meetings addressed by Orator Hunt in Spa Fields. After these meetings the mob spilled over like lava westwards through the streets, smashing windows and pillaging bakers' shops. A particularly spectacular disturbance, in which a gunsmith's shop was raided, took place on 2 December (after the meeting mentioned in Hazlitt's *Scrutator* letter), when the police arrested as scapegoat one John

[50] 7. 137–8.
[51] Ibid. G. O. Trevelyan, *The Early History of Charles James Fox* (1880), 509–10, has a fine passage on the predicament of these writers: 'When to speak or write one's mind on politics is to obtain the reputation, and render oneself liable to the punishment, of a criminal, social discredit, with all its attendant moral dangers, soon attaches itself to the more humble opponents of a Ministry. To be outside the law, as a publisher or a pamphleteer, is only less trying to conscience and conduct than to be outside the law as a smuggler or a poacher . . . '

Cashman, an unfortunate slow-witted sailor not long off his ship down-river and newly arrived in town that morning with no idea of what was happening, who had been drinking and was drawn into the commotion by another sailor. Details of the affair filled whole columns of the paper for weeks after. Finally, the pathetic prisoner was condemned to death on 30 January 1817, and on 12 March brutally hanged before a horrified and hostile crowd in the middle of Skinner Street (almost on Godwin's doorstep).

To return to Hazlitt's articles against Coleridge and Southey at this time, we are to remember that however angrily, brilliantly, and unfairly Hazlitt in his 'Illustrations' turns the shortcomings of the Lake poets into devastating debating points he privately made no mistake about their capabilities and the loss they were to the cause. He became impatient whenever he thought of the powers Coleridge had wasted and compared them with the feeble resources of many a successful nonentity, and as late as 1820 in a remarkable prosopopeia he besought Coleridge to rouse himself and preach once more the ideals of freedom, justice, and humanity, to 'start up in [his] promised likeness and shake the pillared rottenness of the world'.[52] Nay, as late as August 1830, when the July Revolution had again overturned the Bourbons, he had hopes that his old friend, 'still vulnerable to truth, accessible to opinion, because not sordid or mechanical', would 'recover his original liberty of speculating' and embrace the right.[53] Now in 1816 he hardly touches upon the supposed venality of the apostates. He does not dwell on their 'rewards', which, compared to the great possessions of their new patrons, were of course derisory (stamp-receipts, a hundred a year, a butt of sack, and for poor Coleridge a few shillings from newspaper editors: 'They, with the gold to give, doled [them] out silver'); he was too clear-sighted not to recognize that the fact that the rewards were trifles was a proof of the good faith of the recipients, evidence of their honesty. But he could not swallow a sense of betrayal that was bitter in direct proportion to his sense of loss. The blow, the shock of regret that Coleridge's defection was to him, reverberates painfully in his letter of 12 January 1817 to the *Examiner*, where, remembering the year 1798, he recalls the undimmed elation, the yearning towards good, the promise of spring, the spirit of youth and hope that attended his solitary twenty-mile walk through the misty landscape of winter, from Wem to Shrewsbury and back, nineteen years earlier almost to the day, when he went to hear Coleridge preach a sermon against the callous oppression and unrelenting materialism of the State.[54] But Coleridge had gone over to that enemy for whom there

[52] 7. 118; 8. 251.   [53] 17. 378 ('The Letter-Bell').
[54] 7. 128–9. See the writer's 'First Flight: Image and Theme in a Hazlitt Essay', *Prose Studies*, 8 (1985), 35–47, on his treatment of Coleridge here.

could be no quarter, and the rage that possessed Hazlitt at that enemy's injustices he must needs extend to his apostate friends also. The true Jacobin 'makes neither peace nor truce with [the oppressors]. His hatred of wrong only ceases with the wrong. The sense of it, and of the barefaced assumption of the right to inflict it, deprives him of his rest. It stagnates in his blood. . . . It settles in his brain — it puts him beside himself.'[55]

This implied admission of the risk of losing control goes some way to excuse the violence of his attack (he omitted certain passages when reprinting these 'Illustrations' in *Political Essays*). In private talk he was probably still more savage. Haydon remembered him at about this time (it was, he said, 'when Cobbett's twopenny things were making a great noise') letting slip something that made him shudder. He does not say what, but he has just been imagining Hazlitt as a potential Robespierre.[56] This of course may be the violent Haydon's exaggeration of the lesser, but to him unwarrantable because opposed, violence of others with unwelcome views, and a like restriction might also apply to another moment when he says of Hazlitt and the Hunts, 'A revolution is often talked of by them'.[57] Here we can only invoke Hazlitt's measured deprecation of Cobbett's intemperate language at another time. 'Cobbett', he said, was 'simple and mild in his manner, deliberate and unruffled in his speech, though some of his expressions were not very qualified.'[58]

The reactions to the 'Illustrations' of the level-headed Robinson are however a different matter and, when he not only felt obliged to protest against Hazlitt's article of 22 December but also became thereby permanently alienated from one to whom he had hitherto been a sincere friend and servant, we are bound to take a closer look at the question. His admiration for Hazlitt's talents does not seem to have been affected in 1815 by Wordsworth's story of the Lakes escapade. We find him after that date recording with pleasure energetic debates with Hazlitt at Lamb's and at Alsager's. But the summer holiday he spent at Rydal Mount in 1816 was decisive: it dissipated his earlier uneasiness about Wordsworth's reactionary intransigence and confirmed his growing allegiance. During the autumn he was impressed rather than angered by Hazlitt's attacks on Coleridge (he was perhaps influenced by Wordsworth himself, who said of the friend of his youth effectively what Hazlitt was saying, that he 'neither would nor could execute anything of important benefit either to himself, his family, or mankind'), but in December when these attacks were extended to Wordsworth, his hackles began to rise. Even so, his reaction to the first 'Illustration' of

[55] 7. 152. [56] Haydon, *Corr.* ii. 96. [57] Pope, ii. 84. [58] 8. 59 n.

15 December was, on his own deliberate recognition, ambivalent. He reported to his brother Thomas the next day that his 'admiration and disgust' were excited 'in the highest degree', and indeed, said he, all of Hazlitt's 'late writings', in which he had 'heartily delighted' even though he thought them 'morally disgraceful and even infamous', were 'magnificent and daemoniacal'. Hazlitt 'deserved flogging', but 'with a golden scourge'.[59] It is worth pointing out here that Robinson, unlike Hazlitt, had not known Wordsworth during his revolutionary period, in the 1790s; he did not meet him (or Coleridge) until ten years later, in 1808, at a time when he himself was already being accused by his radical friends of wavering in his democratic sympathies. At his first meeting with Wordsworth a basis of harmony was established between the two men in politics, and in literature also, where he willingly deferred to the older man's very personal standards of judgement. Although Wordsworth 'spoke freely and praisingly of his own poems, [he] never felt [this] to be unbecoming, but the contrary'.[60] And now in 1816, confirmed in this feeling, he 'rever[ed] and love[ed]' Wordsworth, 'admire[d] and pit[ied]' Coleridge, though he 'care[d] nothing' about Stoddart, and in consequence was vexed at witnessing the laceration of 'the most glorious creatures God ever made', even though their assailant was 'a man of the very finest talents'.[61]

And yet it was perhaps because he was embarrassed by Wordsworth's egotism that on Sunday, 22 December, at Montagu's, he so strongly resented Hazlitt's attack in that morning's *Examiner*, which elaborately mocked this self-centredness. However, this incident has been so regularly misinterpreted by Hazlitt's biographers as a head-on collision between Hazlitt and Crabb Robinson over Wordsworth that it demands scrutiny. Here is Robinson's account:

After tea went to Basil Montagu's. Hazlitt was there. I could not abstain from adverting to a scandalous article in this morning's *Examiner* in which he attacks Wordsworth. Hazlitt, without confessing himself the author, spoke as if he were but did not vindicate himself boldly. He said: 'You know I am not in the habit of defending what I do. I do not say that all I have done is right.' In the same tone, and after I had said that I was indignant at certain articles I had read, and at the breach of private confidence in the detail of conversation, Hazlitt said: 'It may be indelicate, but I am forced to write an article every week, and I have not time to make one with so much delicacy as I otherwise should.'[62]

[59] Robinson Papers, 1816 vol., unpub. letter 133a.
[60] Morley, 10.
[61] As n. 59 above.
[62] Morley, 200–1.

The conversation, fatal though it was to relations between the two men, seems, if we look closely, to have been extraordinarily evasive. Robinson did not mention Hazlitt's authorship in denouncing the article, and expressed his indignation in such veiled terms that (according to him) Wordsworth was not actually named either. It must have been a curiously cryptic discussion, in which Robinson and Hazlitt talked more *at* than *to* each other. When Robinson, circling warily, spoke of the scandalous attack, Hazlitt simply said that as a journalist he saw an excuse for it. It was not always possible to gauge accurately what constituted exaggeration. He quoted his own case, in general terms which implied that that case was merely illustrative ('I am forced', etc.), and then distinguished between opinions that were notorious and those momentarily adopted in the accidents of discussion, but Robinson refused to admit the distinction. He saw nothing in the writer's words but a violation of 'confidence of friendship'. In this tangle of allusion and innuendo and appeals to confidentiality Robinson may even have been paradoxically resenting Hazlitt's having got to hear of Wordsworth's disdainful reflections upon the praise Hazlitt accorded his poetry. When Hazlitt in his article said '[Wordsworth] scorns even the admiration of himself, thinking it a presumption in any one to suppose that he has taste or sense enough to understand him' he is evidently referring to some such occasion as Wordsworth's reception of his review of *The Excursion* as reported by John Wilson: 'It was presumption in the highest degree for these cockney writers to pretend to criticize a Lake poet.'[63]

In any case Robinson's resentment seems excessive and it is hard not to suppose that something else was involved. The expression 'confidence of friendship' could hardly apply at that period to Hazlitt and Wordsworth, but it might then fit Hazlitt and Robinson, and although in general Robinson was admirably disinterested, there is here more than a hint of personal pique. As a late-comer to the Bar he was unconfident of his qualifications and dubious of his prospects. He had many briefless friends. He assiduously sharpened his debating powers in private arguments and was always proud to record a victory in his journal. In the letter of mid-December 1816 to his brother he had shrugged off an unspecified hit at himself by Hazlitt: he does not name the essay, but he evidently thought he recognized himself in Hazlitt's gallery of commonplace critics, and with his touchy pride in his conversation it rankled to be held a purveyor of clichés. And now a week later this same friend had attacked his profession. Hazlitt's reasoned denigration of lawyers in that morning's 'Illustration' as men who

[63] 9. 6.

bartered their opinion and spoke falsities in defence of wrong and to the prejudice of right, whose business it was to confound truth and lies in the minds of their hearers, and by a natural consequence confused lies and truth in their own, coming from a man whose penetration he admired, must have jolted his self-esteem.[64] Montagu, the host of that evening, although also a lawyer, was older and more securely established, and perhaps therefore less susceptible.

Robinson accused Hazlitt of untrustworthiness. It did not occur to him that in fact he showed spirit, and, with regard to another target, Southey, courage. He admitted the attack on the Laureate's *Carmen Nuptiale* to be 'unexceptionable', but did not seem to consider that it required any kind of nerve to cross a man whose recent articles in the Government organ, the *Quarterly*, had gained him the ear of the Ministry. Robinson must have known — like Hazlitt — through Rickman or Lamb, that a few months earlier the Prime Minister and the Chancellor of the Exchequer, after reading an article of Southey's in the *Quarterly*, had invited him to London to assume a more sustained role in the Government propagandist organization by editing a new journal. Southey did not accept, but the influence he gained by this official recognition remained strong. In a further article on the condition of the country in the October *Quarterly* (published 11 February 1817), which covered some recent publications, including the *Political Register*, he said that Cobbett, 'a convicted incendiary, and others of the same stamp', ought not to be 'permitted week after week to sow the seeds of rebellion, insulting the Government, and defying the laws of the country'.[65] The 'others of the same stamp' were, in the first place, the Hunts and Hazlitt. In September 1816 Southey had assured Robinson that the *Examiner* was even more dangerous than Cobbett, and had told Rickman, placing himself with characteristic inflation above the whole Cabinet, 'I know very well what I have at stake in the event of a Revolution, were the Hunts and the Hazlitts to have the upper hand. There is no man whom the Whigs and the Anarchists hate more inveterately, because there is none whom they fear so much.'[66] When the October *Quarterly* appeared Southey was furious. Gifford, he said (whose high-handed editorial methods he had sworn to resist), had pruned out everything that was most forceful, but unfortunately protest was impossible on this occasion because it had been done 'in compassion to the weakness, the embarrassment, and the fears of the Ministry', who, however, had 'express[ed] themselves much indebted to him [Southey]'.[67] In a few months' time, in

[64] 7. 139–40. See a letter on lawyers in Parliament, *Examiner*, 17 July 1814, 463.

[65] QR 31 (Oct. 1816, pub. 11 Feb. 1817 (*MC* advt.)),

[66] Morley, 89; Southey, *Corr.* iv. 212.

[67] Southey, *Corr.* iv 239, 250. Cf. his letter of 2 Mar. 1817 to Rickman (Southey, *Selections*, iii. 62–3).

his *Letter to William Smith, Esq., M.P.*, he was able to speak more freely against 'the present race of revolutionary writers'. 'These miscreants', he said, 'live by calumny and sedition; they are libellers and liars by trade.'[68]

It is a pity we cannot know whether, in the strong language Gifford was obliged to suppress, Southey actually named Hazlitt; but however weak and embarrassed and fearful the Government may have appeared to his glowering eyes he cannot but have been gratified when, within a month of his article, they suspended Habeas Corpus (4 March), an action without parallel since the early days of the French Revolution, and, before that, the Stuart rising of 1745. Solid, bluff, resolute Cobbett felt a sudden draught, took coach to Liverpool, and set sail for America. The sequence and the speed of events were of some significance, but the rapscallions who were named along with Cobbett, the punier Hunts, and Hazlitt, frequenting as they did the newspaper-offices of London, were perhaps endowed with a finer perception of the balance of danger and better able to calculate the extent of the threat than someone who lived at a distance, on a farm in Hampshire, editor of a weekly paper though he might be. (The *Political Register* was not technically a newspaper. Cobbett had at one blow evaded the stamp-tax and reduced its price to twopence by excluding all news. The *Courier* insisted on his liability, and adduced this and not the suspension of Habeas Corpus as explaining his flight, in order to deny him the martyrdom of political principle and make him out a shyster who had burnt his fingers.[69])

Eloquent in Hazlitt's mind at this time, no doubt, was the fate of many imprisoned journalists, some of whom, like Finnerty, were not merely locked up but also shamefully exposed in the pillory. During the past twenty years there had been Perry, Easton, and Drakard, as well as Finnerty, and especially the three with whom his name was being linked at this moment, Cobbett and the brothers Hunt. At all events he was in no doubt of the significance of the *Quarterly* article, which he quotes again and again in the *Political Essays*, and it is clear from a private memorandum addressed by Southey to the Prime Minister at this time that the author of that article confidently expected the suspension of Habeas Corpus to be immediately followed by the arrest of Cobbett, Hone, and the *Examiner* group, and, he even hoped, by their transportation. Every ounce of the weight Hazlitt put into his attacks on Southey that winter was justified by the tone of this confidential missive. Here is an example of its recommendations: 'If juries, either from fear or faction, . . . give their verdict in the very face of facts, I beseech you do

[68] *A Letter to Wm. Smith, Esq., M.P.* (1817), 26–7.

[69] This version was also adopted by the Whigs: Perry, who was no friend to Cobbett and, like many members of the Opposition, blamed him for the suspension of Habeas Corpus, told Sir T. Lawrence that the Government had come down on Cobbett for £80,000 (Farington, viii. 118).

not hesitate at using that vigour beyond the law which the exigence requires.'[70] The Government, however, was a little more sane and a good deal more cautious than the inflammable cloud-borne Southey, who lived in a world of abstraction far removed from the elbowing realities of the market-place, and it did not dispense with the cloak of legality. But although the threat of an *ex officio* information constantly hung over Hazlitt's head his attacks did not cease. It is easy now, with the uninvolved eye of hindsight, to suppose that he was never in any danger of prosecution, simply because he was never prosecuted, and to note only the violence and personality of these attacks, without taking account of the risks he was running. It is nevertheless the fact that what he asserted, and persisted in asserting for another two years despite the suspension of Habeas Corpus and the successive 'Gagging Acts', was of the same order as what had caused Cobbett to flee the country.

His courage, barely acknowledged by his biographers, was no doubt sustained by his memory of the Treason Trials which took place at the most susceptible period of his life, which were still annually celebrated at a dinner at the Crown and Anchor, and continued to be so until after his death. The examples of Hardy, Thelwall, and Tooke encouraged him at this time when, as Brougham said, 'Every one who rose in a meeting, or sat down at his desk, to attack the measures of his majesty's ministers, now knew that he did so with a halter about his neck'.[71] His stand is in signal contrast to the panic that seized Wordsworth in 1808 after sending his *Convention of Cintra* to the press, when, terrified that he had risked Newgate by writing too strongly, not even the scoffs of his womenfolk could prevent him from writing post-haste to effect cancellations.[72] Hazlitt defined his own brand of fortitude some years later when he declared: 'Mental courage is the only courage I pretend to. . . . In little else I have the spirit of martyrdom: but I would give up anything sooner than an abstract proposition.'[73]

His courage, I have suggested, goes far to extenuate his violence, and it was a violence directed against those whom he saw as the enemies of the people. If of the same order, it had not precisely the same tone, the same extravagance, as that of the transatlantic fugitive, who laid about

[70] C. D. Yonge, *The Life and Administration of Lord Liverpool* (1868), ii. 298. Another passage in this remarkable document of 19 Mar. 1817 is: 'You have passed laws to prevent men from tampering with the soldiers, but can such laws be effectual? Or are they not altogether nugatory while such manifestoes as those of Cobbett, Hone, and the *Examiner*, etc. are daily and weekly issued, fresh and fresh, and read aloud in every ale-house where the men are quartered . . . I did hope that the first measure after the suspension of the Habeas Corpus Act would have been to place the chief incendiary writers in safe custody . . . No means can be effectual for checking the intolerable licence of the press but that of making transportation the punishment for its abuse.'

[71] *Parl. Deb.* 35, (1817) col. 1124 (14 Mar.).

[72] Moorman, ii. 143–4.

[73] 20. 112.

him 'with a three-man beetle' against friend and foe alike. A fair example of the blundering vehemence that too often subdued Cobbett's judgement and upset the repose and digestion of the more nervous among the governing class is his article of 2 November 1816, where he makes a set at one who could hardly be called an enemy of the people, the editor of the *Champion*:

The unfortunate journeymen and labourers and their families have a *right*, they have a *just claim* to relief from the purses of the rich. . . . [How then were they to obtain it?] There is a canting Scotchman, in London, who publishes a paper called the *Champion* who is everlastingly harping upon the virtues of the 'fireside' and who inculcates the duty of quiet submission. Might we ask this Champion of the tea-pot and milk-jug, whether Magna Charta and the Bill of Rights were won by the fireside? Whether the tyrants of the House of Stuart and of Bourbon were hurled down by fireside virtues? Whether the Americans gained their independence and have preserved their freedom, by quietly sitting by the fireside?

This passage, it hardly needs saying, was singled out in Southey's *Quarterly* article.

The wretchedness and anger of the time, and the opprobrium thrown on the Government reached a climax with the abortive and pathetic June uprisings at Huddersfield and Pentrich. On 10 June news came to London of the revelations by the *Leeds Mercury* of the role of Oliver (W. J. Richards), the Government spy who had been sent as *agent provocateur* to foment these movements. This information, 'by a violent Opposition Paper', was quoted, in order to be contradicted, by the *Courier*. Four days later came a leader in the *Courier* protesting against the acquittal of Dr Watson, charged with high treason at the Spa Fields meetings, on the grounds that the jury ought not to have been weak enough to allow themselves to be so influenced by their repulsion at the odious character of the chief witness 'Cassels' ('so absolutely infamous a witness I never heard of', said Robinson, present at the trial[74]) to find Watson innocent. 'Do we want', continued the leader (echoing Stoddart's earlier ana-thematizing of Orator Hunt, and invoking 'a vigour beyond the law'), 'any evidence to convince us that the proceedings in Spa-fields were highly seditious?' The misspelling of the name Castles was a vain attempt to conceal the *Courier*'s knowledge that he also was a Govern-ment spy. The manifest determination of the Government to persist in the same course is again apparent in next day's *Courier*, which begins, 'A man by the name of Oliver is charged in a Leeds paper . . . with having ensnared ten men . . . ' and goes on to assert that if this man were questioned it would emerge that he had nothing to do with the riots, was

[74] Sadler, ii. 53.

ignorant of where the men were to meet, knew only one of the ten
'delegates', etc., in apparent unconsciousness of the contradiction
between the phrase 'a man by the name of' and the knowingness of the
claims following. It was on the same day that Sir Francis Burdett named
Oliver in the House as 'one of the authorized spies of Government'. On
27 June the *Courier* changed its tactics and acknowledged its outcast
allies, Castles and Oliver: 'We trust that some Member will dwell
strongly upon the fact that Government employs Spies, not to prompt
but to report sedition . . . Let it be avowed with boldness, and the utmost
publicity given to the avowal, that Government do employ Spies. This
will serve to destroy confidence among the disaffected . . . '. At the end of
the month came Hazlitt's article 'On the Spy System' in the *Morning
Chronicle*, describing with searing irony Castlereagh's devious defence of
the spy system:

According to his Lordship's comprehensive and liberal views, the liberty and
independence of nations are best supported abroad by the point of the bayonet;
and morality, religion, and social order, are best defended at home by spies and
informers . . . The Noble Lord in the blue ribbon took the characters of Castles
and Oliver under the protection of his blushing honours and elegant casuistry,
and lamented that by the idle clamour raised against such characters, *Gentlemen*
were deterred from entering into the honourable, useful, and profitable
profession of Government Spies.[75]

The odium incurred by the spies was as ineradicable as the self-
righteousness of their employers was unshakeable. Years later, in 1827,
when Oliver, who had ultimately been forced to emigrate under an
assumed name, died, he was dismissed in a contemptuous paragraph in
a Cape of Good Hope paper (his alias had been penetrated) but was duly
praised by the relentless and ever-watchful *Courier* in London.[76]

Shortly after the spy scandal of 1817, Hazlitt's directly polemical
political writings in the *Morning Chronicle*, to which he had carried them
from the *Examiner*, came to an end, but he continued for the rest of the
year to write more general political essays in the *Champion*, now edited by
R. D. Richards ('On the Regal Character', 'What is the People?'), and
until the middle of 1818 for John Hunt's *Yellow Dwarf* ('On Court
Influence'). A collection of his writings in all these papers, the *Political
Essays*, was published in 1819 by William Hone.[77]

He may already have been acquainted with Hone (who attended his
poetry lectures early in 1818[78]) at the time of his three trials for
blasphemy in December 1817, which were as resounding and as sig-

---

[75] 7. 208   [76] *Courier*, 10 Oct., *Examiner*, 14 Oct. 1827.
[77] 14 Aug. (*MC* advt.).   [78] Godwin Diary, 3, 10 Apr. 1818.

nificant in the struggle against repression, although not as loftily heroic, as the Treason Trials of 1794. He was a small bookseller and publisher, rather unbusinesslike but of immense industry and determination. He had a flair for the sensational: a lightning opportunist, he brought out memoirs of Sheridan and Princess Charlotte the instant they died, and regularly ran off 'true accounts' of Crim. Con. cases and hangings. But he was wholly committed to parliamentary reform and to opposition to the Government. He was a man who courted publicity and in this instance he perhaps too assiduousIy trailed his coat, but if so the stratagem was superfluous. The authorities had kept him in sight ever since the Corn Bill riots, when he stood witness in the Vyse affair, and particularly since he made light of the alleged attempt on the Regent's life in January 1817. A sharper thorn in the Government's side was his weekly *Reformist's Register* (February–October 1817) which Cobbett had recommended to his readers when his own paper was discontinued on his flight to America.

He now came up before the fearsome Lord Ellenborough, the Lord Chief Justice himself, who took over the second and third trials when Mr Justice Abbott failed to secure a conviction in the first. He defended himself with a stubborn, rude skill and courage that deeply moved the thronged court-room. The persecution was patent and disgraceful; the intention was clear: Hone was being pursued not for blasphemy but for political satire construed as sedition. Hone became the defender of the freedom of the press, the champion of the constitutional rights of the people.

Robinson did not know whether to rejoice or be angry. He was astonished at the humiliation of the domineering Ellenborough whose high-handedness had long been deeply resented by the Bar. In this scene 'without a parallel in the history of the country,' he said, 'he [could] not but think the victory gained over the Government and Lord Ellenborough a subject of alarm, though at the same time a matter of triumph . . . This illiterate man has avenged all our injuries.'[79] Hone, with his crude sling, had brought low one of the seemingly unassailable Goliaths of the Establishment. Robinson's friend Burrell (friend also of Stoddart, a regular target of Hone's 'Dr Slop' satires) reproached him angrily for his pleasure at the verdict.[80] The *Courier* saw the confrontation and 'the boisterous acclamations at every coarse attack made by Hone upon the Chief Justice of England,'[81] as a brazen indication of 'the radical corruption in the popular mind'. It complained with ominous impercipience of this 'acquittal in defiance of the mandate of the Judge'. The *Morning Post* was equally sinister: 'The verdicts were no acquittal as the

[79] Sadler, ii.78.   [80] Ibid. ii. 80   [81] 23 Dec. 1817.

Judge had decided the parodies were blasphemous.'[82] Hazlitt's own judgement came six years later: 'that he did not suffer himself to be crushed to atoms, and made a willing sacrifice to the prejudice, talent, and authority arrayed against him, is a resistance to the opinions of the world and the insolence of power, that can never be overlooked or forgiven.'[83]

[82] 17 Jan. 1818.
[83] 11. 159 n.

12

The Leopard and the Scorpion

THE image of Hazlitt as a sinister scowling misanthrope was in the main politically inspired and easily constructed of materials adapted from his own violent attacks on the supporters of the stolidly callous Government between Waterloo and Peterloo. It took additional colour, however, from the apparent corroboration of such as Lamb, who, in defending him publicly in 1823 in the *London Magazine* against the aspersions of Southey, where he paid him as fine a tribute as any man ever received, yet found it indispensable incidentally to refer to him as a 'soured' victim of 'spleen' and 'sullenness', and to 'wish he would not quarrel with the world at the rate he does'. We have seen how Patmore, before they met, thought him an incarnate fiend. Procter was less influenced by rumour, and although 'fierce and envious passions', 'coarse thoughts and habits' had been ascribed to him, and he was 'crowned by defamation', saw at their first meeting that where he might be thought sinister he was merely shy.[1] The impression conveyed by his essays is of a diffident, compliant man. When he spent an evening in company with Master Betty, years after the actor's precocious fame had vanished, he wanted to hint a complimentary allusion to the past but 'couldn't bring it in', and it was only at the last minute that he 'ventured to break the ice'. In 'The Fight' he 'ventures, with some hesitation' to draw his coach-companions' attention to the moonrise, and when he wants to ask one of them some questions about the late Mr Windham, 'had not courage'. His diffidence during the evenings at the Southampton Arms allows him to do no more than hint some obvious oversight in an argument, and easily persuades him to reckon his sketches poor and faint compared with the mimic Roger Kirkpatrick's, or prevents his escaping from a bore who '[does] not perceive he [is] tiring you to death by giving an account of the breed, education and manners of fighting-dogs for hours together'.[2]

He was so far from misanthropic that it is clear from many a page that in company he was the last and most reluctant to leave. The man who admired Burns for 'exerting all the vigour of his mind, all the happiness of his nature, in exalting the pleasures of wine, of love, and good-fellowship' was bound to have many friends. It was his great and deserved good fortune — and we cannot but suppose he appreciated it — to attract at this stage of his career, at the approach of middle age, the

---

[1] Procter, *Lamb*, 96; Procter, *Fragment*, 176.   [2] 8. 197.

sympathy and friendship of a number of young men of talent who were themselves setting out in life with the same hopes that had cheered him on the winter road from Wem to Shrewsbury in 1798. One was in fact a Shrewsbury man, John Hamilton Reynolds. Another was Keats, about whose deep respect and admiration for Hazlitt and Hazlitt's great influence upon him a good deal has been written. Charles Cowden Clarke, whom he met at the Hunts', wrote to his grandson half a century later, 'I think no one (not even Mr Procter) had a more fervid and implicit esteem and love for his heart-principle . . . *I* never wavered respecting him'.[3] Shelley he also met at Hunt's, but although they once found themselves on the same side in an argument, upholding republicanism against Coulson and Hunt, who were for monarchy, he never took to the young poet, whose feet seemed to him never quite on the ground.

Another young friend, Thomas Noon Talfourd, was of exactly the age he had been when he first heard Coleridge. The son of a Dissenter, he had come to London in 1813 to study law, bearing a letter of introduction to Robinson who, in his usual kindly way, kept an eye on him, although he found him rather bumptious. He thought him 'very disputatious and verbosely rhetorical', but adds, scrupulous as ever, 'as I was, but with far less talent, twenty years ago'.[4] Talfourd must have been conscious of this fault: he was intrigued by the directly opposed style of discourse he encountered in Hazlitt, and in describing it casts a neglected sidelight on the role that the Lambs, and particularly Mary, played in drawing Hazlitt out of his diffidence and uncertainties:

It was . . . by the fireside of the Lambs, that his tongue was gradually loosened, and his passionate thoughts found appropriate words. There, his struggles to express the fine conceptions with which his mind was filled, were encouraged by entire sympathy . . . ; there he was thoroughly understood' and dexterously cheered by Miss Lamb, whose nice discernment of his first efforts in conversation, were dwelt upon by him with affectionate gratitude, even when most out of humour with the world. When he mastered his diffidence, he did not talk for effect, to dazzle, or surprise, or annoy, but with the most simple and honest desire to make his view of the subject in hand entirely apprehended by his hearer. [And at this point it becomes clear that Talfourd is telling what he himself saw, and not merely echoing the Lambs.] There was sometimes an obvious struggle to do this to his own satisfaction; he seemed labouring to drag his thought to light from its deep lurking-place; and, with timid distrust of that power of expression which he had found so late in life, he often betrayed a fear lest he had failed to make himself understood, and recurred to the subject again and again, that he might be assured he had succeeded.[5]

---

[3] BL Add. MS 38899, fo. 223.

[4] Robinson Diary, 8 May 1813, 13 Nov. 1816, 22 Apr. 1818.

[5] Talfourd, *Memorials*, ii. 171. Talfourd's own debt to Hazlitt is apparent from an unpub. letter of 3 July 1841, where he speaks of 'the memory of one who was my great Master in the Art of

In the spring of 1817 Hunt, whose health had improved, wished to take back the 'Theatrical Examiner' column, and in June Hazlitt moved to *The Times*, where he remained as dramatic critic until the end of the year. He then resigned, 'from want of health and leisure' according to the preface to *A View of the English Stage*. His want of leisure is explained by the new career into which he was now moving. When the change-over to *The Times* took place he was about to bring out his first work of literary criticism, and at the end of his season with that paper he was about to begin the first of his lectures on literature.

That first critical work, the *Characters of Shakespear's Plays*, was his ninth book, and the indifferent fortunes of the earlier ones made its success a matter of great moment. Constable's exasperating and inexplicable tardiness in bringing out *The Round Table*, which was not advertised until the previous 14 February, and the sluggishness of its sale, had made him apprehensive and wary, and he conceived a plan to ensure the success of the new book. He sent it in sheets to Jeffrey more than two months before it finally appeared at the booksellers, begging the favour of a notice in the *Edinburgh*, the which accolade would make it, he said, 'a marketable commodity' (and the 'market' he secretly had in mind was the publishers and not the readers). A month later he wrote to Constable, referring to what he had sent Jeffrey as a work he had written 'for a friend who wishes to dispose of the edition of 1,000 copies for 200 guineas', and enclosing a puff (clearly inspired or written by himself) from *The Times*.[6] The stalking-horse was almost certainly C. H. Reynell the printer, and Hazlitt's proposals were part of a scheme he had devised to outmanœuvre the publishers, and in particular Constable, by postponing negotiations with them until the book was out ('with a title-page destitute of any indication where the book was to be found'), and even reviewed.[7] Jeffrey was perhaps too wily a bird to fall into the trap, or it may have been the inveterate delays of the *Edinburgh* that account for the lateness of the notice, which came out on 17 September, five months after Hazlitt's request, and five weeks after the book had finally been published by Hunter & Ollier on 9 July. However, Hazlitt need not have been so cautiously devious. The book was warmly welcomed and well reviewed, particularly by the *Examiner* in an extended three-part notice.

His satisfaction at seeing the *Round Table* in print and the *Characters* in

---

Thinking, and the recollection of whose society is dearer to me than the enjoyment of that of my dearest living friends . . . [my *Ion* being published after he died] it has often been to me an affecting consideration that I could never hope for the praise of one whose lightest commendation I should have prized above the most elaborate eulogies.' (Buffalo.)

[6] *Letters*, 171, 173.

[7] NNHL, 272–5. I am gratified to see that my hypothesis has found acceptance: see Wardle, 197; C. E. Robinson, in *Shelley Revalued*, ed. K. Everest (Leicester, 1983), 189.

prospect may account for the unusually large supper-party he gave at
York Street, on 14 June, attended by the Godwins, the Lambs, John
Black (now editor of the *Morning Chronicle*, where his articles were once
more about to appear), Coulson, Alsager, H. White (unknown to us),
and James Ogilvie. Ogilvie was a schoolmaster who had given up his job
in America to try his luck as lecturer in England.[8] He had used Godwin's
Political Justice in his teaching, and when he came to London he attached
himself to its eminent author. His 'exhibitions' succeeded in Edinburgh,
but not so well in London: he lacked persuasiveness, his 'orations' did
not carry conviction, and his health and spirits were undermined by
laudanum. Hazlitt attended his performances, perhaps out of curiosity,
probably out of friendship for a man whom he found sufficiently likeable
to invite to his table (and who later attended his own lectures).[9] Ogilvie
was a man of no great mark or likelihood, and Robinson considered a
lecture of his on oratory delivered at the Surrey Institution to be 'a
ridiculous display of affectation and bad taste'.[10] Godwin saw well
enough that he had small talent, but could not abandon so keen a
disciple. Hazlitt never mentions him in either writings or letters,
although it is suggested in a letter of Mr Blackwood's that they were
much in each other's company. Mr Blackwood, who admired him,
deplored his having fallen into the clutches of such radicals as Hazlitt and
Robert Stodart (bookseller in the Strand, and subscription-agent for
both Ogilvie's lectures and Hazlitt's).[11] In a few years' time, when his
exhibitions had foundered, he wearied Godwin with complaints of
neglect and after a period of obscurity put an end to his sad life in
September 1820.

It was perhaps seeing Ogilvie get so far on so slender a talent that
encouraged Hazlitt to venture again on to the rostrum himself at that
same Surrey Institution. Having turned his passion for Shakespeare to
good account, he saw the possibility of extending his critical survey to
other equally familiar areas of English literature, but he determined first
of all to reach a more immediate audience and gain useful publicity by
giving as lectures the critical studies he intended to publish later as

[8] See B. R. Pollin, 'Godwin's Letter to Ogilvie, friend of Jefferson, and the Federalist
Propaganda', *Journal of the History of Ideas*, 28 (1967), 432–44; and *MC*, 12 July 1817, 12 Jan. 1819
(on his 'extraordinary powers').

[9] Godwin Diary: Hazlitt attended Ogilvie's lectures on 2 Aug. 1817 (and Sarah Hazlitt on
12 July); Ogilvie heard Hazlitt on the Comic Writers on 3 Nov. and 22 Dec. 1818. Ogilvie and
Hazlitt both called on Godwin, and probably met, on 25, 31 May, 28 June, and 11 Aug. 1817.
Ogilvie supped at Hazlitt's on 19 Dec. 1818 with Godwin, Washington Irving, and one of the
Kirkpatrick brothers; and on 4 June 1817 supped at Godwin's with Hazlitt and Coulson, and
dined there on 12 Feb. 1819 with Hazlitt, Washington Irving, Joseph Hume, and Booth.

[10] Robinson Diary, 13 Nov. 1818.

[11] Murray Papers, Blackwood to Murray, 3 Oct. 1818 ('I knew Ogilvie when he lectured here.
He is a respectable man but a great friend of Hazlitt's').

books. On the recommendation of Alsager, a member of the Institution, he came to an agreement with P. G. Patmore, now Assistant to the Secretary there, to deliver a series of lectures on the English poets, from Chaucer to Cowper, in early January 1818, and this was announced in the programme for the 1817–18 season on 30 September.[12] His preliminary interview with Patmore was unpromising; he was awkward, silent, diffident, and inspired no confidence in the showing he would make;[13] but in the event the lectures were a considerable success. He was better fitted to the task than in the year 1812. The conversational, direct style he had adopted in his essays, his five years of vigorous journalism, and the cautionary example of muffled, stumbling, or interminable parliamentary speakers, as well as the having a less abstruse subject, gave his delivery energy and variety, and after a moment of stage fright on the first night he carried his course through to a triumphant conclusion.

The praise began with the earliest lectures, which Robinson says were 'full of beautiful and striking observations'.[14] We have some record of the names of Hazlitt's friends who attended: John Hunt and his son Henry Leigh, Charles Jeremiah Wells, John Landseer and his sons, Keats, William Bewick[15] to whom Hazlitt gave a ticket and who was evidently, from a letter to his brother, rather at sea, but we get an echo of the public view when he reports that 'they are said to be the finest lectures that ever were delivered'.[16] The young Landseer brothers, as well as Bewick, were pupils of Haydon's, but of Haydon himself there is no sound. His eyes were bad in the second half of January. He was hardly able to leave the house and it was a long way to the Surrey Institution. But Hazlitt dined at his house in Lisson Grove with Keats and Bewick on 18 January, the Sunday following his first lecture.[17] This dinner is not mentioned in the diary, nor are the lectures. In view of Haydon's subsequent very bitter attack on Hazlitt in his diary for not serving *his* career it is striking how little interest he took in Hazlitt's. This was in the month following Haydon's 'immortal' dinner, to which Hazlitt was not invited, although Lamb and Wordsworth were. In fact Hazlitt is not mentioned in Haydon's diary in the whole of the year 1818, and this is unfortunate because it is a year in which there are few Lamb letters. In February Haydon was better, but he had much ground to recover, and he immediately threw himself into his work on *Christ's Entry*.

The need to cross the river by Blackfriars Bridge to the Borough seems

---

[12] *Courier* advt.; *MC* advt. 1 Oct. [13] Patmore, ii. 250–3. [14] Morley, 218.
[15] Keats, i. 214. [16] Bewick, i. 41. [17] Keats, i. 206.

to have deterred the Lambs, and the Godwins as well, but not Robinson. With the third lecture, on 27 January, a rival entered into competition with Hazlitt. Coleridge began a series of lectures of wide-ranging scope, to be delivered in Flower-de-Luce Court, Fetter Lane, every Tuesday and Friday for seven weeks, so that every Tuesday, from 27 January to 3 March, both Hazlitt and Coleridge, on either side of the river, were at the rostrum. But not at the same time. Hazlitt's lectures were at seven in the evening, and Coleridge's at eight. Robinson with remarkable agility managed to hear both men. Unfortunately he makes no comparison between the two. Lamb disliked lectures because he either felt it superfluous to be read at or was in pain for the lecturer's confidence when venturing to speak without a script. He visited neither the Surrey Institution nor Flower-de-Luce Court, but he did hear Hazlitt on Dryden and Pope[18] when the series was repeated from 23 March, Easter Monday, at the Crown and Anchor in the Strand.[19] This was a short walk from Russell Street and with the beginning of April the fine weather was approaching. We find also that whereas Godwin only attended the last lecture at the Surrey, that on the living poets (he expected no doubt to learn Hazlitt's opinion of his son-in-law), he and his wife and son heard nearly all of the repeated series. Among other hearers at the Crown and Anchor on various evenings were Charles Kemble (whose transition from bold Marlow to bashful Marlow in *She Stoops to Conquer* Hazlitt had described some months earlier as representing for him 'one of the finest and most delicate touches of real acting [he] ever witnessed'[20]), Alsager, Martin Burney, Basil Montagu and his wife, William Hone, J. P. Curran's daughter Amelia, Dr John Reid (a schoolfellow of Hazlitt's at Hackney), Thomas Hill, and John Taylor of the firm of Taylor & Hessey, who were to publish the lectures in a few months.[21]

In the audience on 13 April when Hazlitt lectured on 'Burns and the old English ballads' was Tom Moore, and named next to him in the list in Godwin's diary is one Rees who must be Owen Rees of the firm of Longman, Hurst, Rees and Orme, who were at that moment getting out a book of Moore's. The lectures seemed to be attracting publishers' attention. This is the point in the careers of the poet and of the lecturer at which they were most in sympathy. On the 20th Moore's book, *The Fudge Family in Paris*, came out, and on the 25th Hazlitt reviewed it favourably in John Hunt's paper the *Yellow Dwarf*. Two days later a copy arrived in York Street inscribed 'To William Hazlitt Esq. as a small mark of respect for his literary talents and political principles, from the

[18] Godwin Diary, 3 Apr.   [19] *MC* advt., 23 Mar.   [20] 18. 260
[21] D. Reiman (ed.), *Shelley and His Circle* (Cambridge, Mass., 1961–   ), vi. 540–2, quoting Godwin Diary.

Author'.[22] Later on, at the time of the *Liberal*, Hazlitt conceived Moore to have done him an ill turn, and said so, and shortly after took offence at Moore's sneers at Rousseau and Mme de Warens in *Rhymes on the Road*. Nevertheless, Moore is one of the few contemporaries who ever said an explicit word in recognition of his political writings. A name surprisingly absent from the accounts of the lectures is Sarah Hazlitt. Is it conceivable that she did not attend? In 1812 she seems not to have been in London at all. But now that Hazlitt was making a mark it is unlikely that she was not on hand to enjoy his prosperity, and get what share she could of it. From what we hear of her she was anxious at all times to 'see all she could', whether in Scotland, Ireland, or Paris. We can hardly suppose her curiosity deserted her in London. If she had been at the Crown and Anchor she would undoubtedly have been mentioned by Godwin. The probability is that she was at the original series. Robinson does not mention any companions, nor does he recognize any acquaintants, and this again indicates that the hall was crowded.

There is no doubt that the lectures drew a great deal of attention. The *Examiner's* last report (Blunden believed by John Hunt[23]) of the Surrey Institution lecture of 3 March said: 'It is not a little striking that though the *subjects* . . . necessarily decreased in interest [with] every lecture (he began with Shakespeare and ended with Southey!) his audiences continued increasing to the last, and on Tuesday the ample hall in which he read was crowded to the very ceiling.'[24] This may seem partial in the *Examiner* but there is the inadvertent corroboration in Robinson, who was so incensed at Hazlitt's criticism of Wordsworth on Burns that he lost his temper and hissed; 'but', he says, 'I was on the outside of the room': he was evidently unable to get a seat. From the Surrey Institution he hurried away to Coleridge's lecture, in which he found 'much obscurity . . . and not a little cant and commonplace'. If the scrupulous and discriminating Robinson, whose diary contains excellent criticism, said so, it must have been so. 'I fear', he added, 'that Coleridge will not on the whole add to his reputation by these lectures.'[25] Whereas it is clear that Hazlitt greatly added to his.

The reports by Procter and by Talfourd of the occasions, of Hazlitt's manner, of the audience, and of the reception given to such passages as those relating to Dr Johnson as Good Samaritan and to Voltaire have been quoted many times and are familiar. I propose therefore to introduce a new witness who offers a new perspective on the last lecture in particular (the unscheduled one on the living poets, which is largely

[22] In the possession of Mr Michael Foot, MP. A *Courier* advt., 20 Apr., supplies date of pub.

[23] E. Blunden, *Leigh Hunt's Examiner Examined* (1928), 83.

[24] *Examiner*, 8 Mar., 154.

[25] Morley, 220.

neglected in the earlier accounts) and whose presence at the lectures has
hitherto not been suspected. This is a very ardent admirer of Hazlitt,
Mary Russell Mitford. A fanatical admirer of Napoleon also, she had
published a few thin volumes of verse some years since and continued to
manifest her Whig sympathies in *Morning Chronicle* verses addressed to
Leigh Hunt in Prison, to Erin, and so forth. She had not yet acquired the
momentary celebrity as tragic dramatist that was to be hers in the 1820s,
nor was she yet the author of *Our Village* that she permanently remains.
Her enthusiasm for Hazlitt emerges in a letter of 13 September 1817,
when she tells her correspondent Sir William Elford that she regrets
having sent him a critique of *Biographia Literaria* since her comments were
about to be overshadowed: 'I am told that the article on it in the
forthcoming *Edinburgh Review* is to be at once the severest and the best
ever inserted in that malicious but delightful journal. It is written by Mr
Hazlitt . . . the author of those enchanting *Characters of Shakespear's Plays*
which you have not of course waited my recommendation to read.
Schlegel is nothing to them.'[26] On 3 March she wrote home, 'Tonight we
have been to Hazlitt's lecture again — his last — quite as good as the one
we heard — I think better — certainly much longer — it was on the
Modern Poets and most charmingly he trimmed the whole set of them.
Nothing was ever so amusing.'[27] A few days later, in a letter to Sir
William Elford, she repeats and amplifies her impression that the
modern poets were being trimmed. 'Mr Hazlitt is really the most
delightful lecturer I ever heard — his last, on modern poetry was
amusing past all description to everybody but the parties concerned —
them to be sure he spared as little as a mower spares the flowers in a
hayfield. I never so thoroughly thanked Heaven for the double blessing
of being nobody and a woman as at this lecture.'[28]

 The following autumn when she had read the published lectures she
gives a yet more precise assessment of the destruction.

I heard 2 or 3 of these lectures in Town last Spring and was delighted to meet
with them in print. He is a very entertaining person, that Mr Hazlitt, the best
demolisher of a bloated unwieldy overblown fame that ever existed. He sweeps
it away as easily as an east wind brushes the leaves off a faded peony. He is a
literary Warwick — 'a puller down of kings'. I am not so sure that he deserves
the other half of Warwick's title. He is no 'setter-up' of anything. He praises
indeed pretty often but his praise has an unlucky air of insincerity which
whether intentional or not spoils the effect. And yet he can speak well of some
poets. Of Milton for instance because Johnson has abused him — Of
Shakespeare because he himself has written a book on him — Of Thompson
from contradiction — Of Pope from fellow-feeling — but of Swift and Voltaire
only with hearty goodwill. He is very like Voltaire himself — just the same

---

[26] Mitford, Letters, iii. 311.   [27] Ibid 324.   [28] Ibid. 325, 8 Mar. 1818.

shaker to pieces of the great and the old and the severest — yet he has the same light and decisive spirit — the same delicate and subtle wit — the same calmness — the same tact. As a critic he is too cold, too uncatchable — he has no enthusiasm — right or wrong — there is nothing like fire about him, neither flame nor smoke — he is a provoking unadmiring critic, but a most delightful lecturer — more delightful viva voce than in a printed book. When I read his lecture on the living poets it seemed, impudent as it is, so much civiller than my recollections that I at first thought he had softened and sweetened it from a well-grounded fear of pistol or poison, but upon reconsidering the matter I am convinced that it is unaltered and that it owed its superior effect to Mr Hazlitt's fine delivery — to certain slight inflections in his very calm and gentlemanly voice — to certain almost imperceptible motions of his graceful person and above all to a certain momentary upward look full of malice French and not quite free from malice English by which he contrives to turn the grandest compliment into the bitterest sarcasm. In short the man, mind and body, has a genius for contempt and I am afraid, very much afraid, that I like him the better for it.[29]

The Tory journals, to whom the *Examiner* and its writers had always been an object of acute interest, had not let either the *Round Table* or *Characters of Shakespear's Plays* pass unscathed. William Jerdan's newly founded *Literary Gazette* devoted three 'letters', in May and June 1817, signed 'A New Examiner', to attacking the 'poison of infidelity' and the 'spirit of irreligion and scepticism' of the 'incendiary writers of the *Round Table*', and the April *Quarterly* (published 8 September) addressed itself to the ill breeding and bad English of this 'sour Jacobin'. Prolix, laboured, and naïve, the 'letters' were relatively mild, and the *Quarterly* notice was not couched in much stronger language than it commonly used towards the *Examiner*. But the two-pronged Church-and-Throne form of attack was established. The violently contemptuous tone which that attack was increasingly to adopt was initiated by the *New Monthly Magazine*, a journal founded in 1814 to counteract the 'Jacobin' influence of the *Monthly Magazine*. Among the contributors to the latter were Godwin, Dyer, and Lamb, and it had briefly in April commended the *Round Table* to the discerning. As a natural consequence the *New Monthly* in July vilified the *Round Table* in a brisk paragraph of concentrated and ignoble abuse, and so provided a compact model and example for the future amplifications of *Blackwood's Magazine*:

A palpable misnomer distinguishes the front of these volumes against all the laws of delicacy and good manners. The title should have been 'The Dunghill', or something still more characteristically vile; for such an offensive heap of pestilential jargon has seldom come in our way . . . When, however, the reader is told that the articles here raked together have been gathered from the

---

[29] Ibid. 350, 1 Nov. 1818. After 'overblown fame' comes 'such as Johnson's', struck out.

common sewer of a weekly paper called the *Examiner*, all wonder will cease; and they who after that information can have any relish for the feculent garbage of blasphemy and scurrility, may sit down at the Round Table, and enjoy their meal with the same appetite as the negroes in the West Indies eat dirt and filth.[30]

This 'brief mention', a cloud no bigger than a man's hand, monstrous but seemingly aberrant, gave small warning of the long period of foul weather in the offing.

Hazlitt always maintained that the *Quarterly Review* effectively killed the sale of *Characters of Shakespear's Plays*, which had no sooner achieved a second edition on 30 May 1818 than the *Quarterly* notice of the first edition belatedly appeared (in the January number) on 10 June.[31] But already on about 20 May Hazlitt had published the *Lectures on the English Poets*, a month after *A View of the English Stage*, and Gifford once more found himself outflanked and behindhand in his campaign.[32] Hazlitt was producing books faster than they could be reviewed, and his rapidly increasing reputation made him increasingly troublesome. Help was at hand, however, and the task of stifling this upstart radical was taken over by a younger and more energetic journal in the north, in the stronghold of that *Edinburgh Review* which the *Quarterly* had been created to combat.

Blackwood's Magazine had violently assailed Leigh Hunt as long ago as October 1817, and had then promised to attend to other members of 'the Cockney School'. When Patmore, who was 'A.Z.', the magazine's London dramatic correspondent, suggested to Hazlitt that he should report his poetry lectures Hazlitt could hardly believe that the flattering notices Patmore showed him would be accepted without difficulty.[33] He was right. Although they did appear, in February, March, and April, their effect was partly neutralized in the second number by a calculated sneer on another page, where, while praising Patmore's drama notices (he was an over-valued contributor), Lockhart made a first cut at Hazlitt.

> Of pimpled Hazlitt's coxcomb lectures writing
> Our friend with moderate pleasure we peruse.
> A.Z. when Kean's or Shakespeare's praise indicting
> Seems to have caught the flame of either's muse.[34]

---

[30] *NMM* 7 (July 1817), 540.

[31] *MC* advts. (the 2nd edn. was still being advertised on 9 Nov. 1823 (*Examiner*, 736)).

[32] The publication date of *Lectures on the English Poets* is difficult to determine. See the writer's rev. of Keynes, *Bibliography of William Hazlitt*, in *Analytical and Enumerative Bibliography*, 6 (1982), 276.

[33] *Letters*, 181. The date (March 18) is wrong: it could fall anywhere between 13 Jan. and 20 Feb. 1818, since the article is evidently the first of Patmore's three notices.

[34] *Blackwood's*, 2 (1818), March, preliminary 'Notices', unpaginated.

No great hostility was shown for some months. The June number carried a side-sneer at Hazlitt in an abusive 'letter' to Hunt, but also a sober article on 'Jeffrey and Hazlitt', critical but decent. And then Hazlitt in July began to write for Constable's *Edinburgh Magazine*. The response was immediate: 'Hazlitt Cross-questioned' in the August number was intended to destroy his reputation, ruin his credit, and blast his career, all at one blow.

William Blackwood was driven by two passions — ambition and hatred. He was determined to make his firm the most important and successful in Scotland, and he hated and envied the man who stood in his way, Archibald Constable, his rival of twenty years, whose career had been spectacular and whose publishing empire was so extensive and personal manner so autocratic that he was known (in Edinburgh, the city of nicknames) as the Czar. Of the three valuable properties Constable controlled, the *Edinburgh Review*, the *Encyclopaedia Britannica*, and the novels of the Wizard of the North, the last was a particularly sore point with Blackwood. In 1817, after having been for more than a decade out-manoeuvred and outdistanced by his fellow-bookseller, he was entrusted almost by accident with the publication of *The Black Dwarf*. He seemed within an ace of wresting the author of *Waverley's* future books from Constable's grasp, but he lost his chance by attempting to teach the author how his books should be written, and the custodianship of the works of the Great Unknown reverted to Constable.

Shrewd, coarse-grained, blunt ('ungracious' was Scott's glaring euphemism), he was cunning rather than tactful, and covered up his lack of subtlety with lavish applications of flattery and barely disguised appeals to the self-interest of those he manipulated. Of his own interest he never for a moment lost sight. When he launched his magazine in April 1817 one such Edinburgh monthly was already in existence, the old *Scots Magazine*, which had been bought by Constable some years before, and now, renamed the *Edinburgh Magazine*, led a twilight existence, played down by its owner in order not to prejudice the sales of the *Edinburgh Review*. Its chief function was to advertise Constable's other publications, while acting as a dog in the manger. But he was hoist with his own petard. Blackwood saw that its very ineffectiveness disabled it from blocking the establishment of a rival monthly. Unfortunately his shrewdness failed him in his choice of editor, or perhaps his illiberality was to blame. The first number of *Blackwood's* was not only feeble and colourless but he found, to his fury, that while his back was turned his editors (he had chosen two, the inept pair Cleghorn and Pringle) had with exasperating honesty praised their rival, the *Edinburgh Review*. He bundled them out as soon as he could and installed in their place two frequenters of his front-shop saloon who were already contributors. But he took care not to appoint them formally as editors; henceforth he kept the reins in his own hands. However much they might write, the final decisions were his.[35] In the years to come he alone was the unavowed editor, whatever play was made with the pen-name Christopher North. But, for the rest, he encouraged them to take the widest scope. These lively reckless scoffers, John Wilson and John Gibson Lockhart, new-fledged briefless advocates with time on their hands and mischief in their minds, unhampered by respect or admiration for any man, having at their command, and capable of supplementing by invention, all the tittle-tattle of the gossiping Scottish capital, Oxford men to boot, with acquaintance in the south, admirably suited his purpose. Their status as gentlemen and barristers was also privately to his advantage, patronize him though they might. Cleghorn and Pringle were salaried professional editors; Wilson and Lockhart were amateurs who disdained the paid hacks of journalism (such as Hazlitt and Hunt). They also were of course paid, but not on a regular basis; theirs was an occasional piecemeal arrangement that offered large opportunities, on the proprietor's part, for forgetfulness and parsimony. His meanness (compared with the open-handedness of both Constable and Jeffrey) peeps through in his whining apologies in early letters for not being as yet in a position to 'remunerate' them as they deserved, apologies balanced by their exasperated but genteelly circuitous efforts to extract what they had earned.

Lockhart euphemistically conveys in his life of Scott what Blackwood was at this time: 'He owed his first introduction to the upper world of literature and of society in general . . . to his Magazine.' A self-taught vulgarian with a strong business sense, the bookseller had the instincts of the small-town bully, ready to force a passage and crush competition with his fists if necessary, thinking it glory rather than embarrassment. More often than not in the early years of the *Magazine* he was dragged through the courts and had to pay hefty damages,[36] but sometimes the attacks were so outrageous that the libelled took to violence themselves. There is a vignette of him pushing his way through the Princes Street crowds, brandishing a cudgel, in search of a victim of his magazine who had come all the way from Glasgow and had just horsewhipped him in his shop.[37] His attitude in telling of these frequent incidents suggests that he preferred the second way of meeting protests against attacks as being more economical, if more painful.

The bully was also the poltroon, and stealthiness went hand in hand with brutality. He encouraged Lockhart and Wilson to push the

---

[35] See Oliphant, i. 302–3.

[36] NLS MS 2245, fos. 50, 230. T. Besterman estimated the damages paid by W. Blackwood in the early days at nearly £3,000 (*TLS*, 25 Apr. 1936, 356).

[37] *Glasgow Chronicle*, 12 and 13 May 1818; Gordon, i. 278. He was still being horsewhipped in 1828: see *Examiner*, 29 June 1828, 420.

campaign of libel and scurrility for all it was worth, when after the dismissal of Pringle and Cleghorn he relaunched his *Maga*, as he called it, in October 1817. Although unfamiliar with society he had a shrewd knowledge of the machinery of Edinburgh life and its personalities, and he saw that a short way to bring it into notice would be a lampoon of all the principal figures in the capital. This he was provided with, in the biblical parody the 'Chaldee Manuscript', which made part of the first new number. The cruder the satire, the coarser the naming of physical peculiarities or disabilities, the greater the recognizability of the characters and the greater the impact. A contributor to a later number (probably Lockhart) said, 'the public curiosity is always stimulated to an astonishing degree by clever blackguardism'.[38] This was Blackwood's own calculation in 1817, and this it was that caused him to enlist the support of the clever blackguards Lockhart and Wilson.

Wilson was tall, burly, fair-haired, florid-complexioned, gregarious; Lockhart small, dark, atrabilious, conspiratorial. The moral differences answered to the physical. Wilson's moods were all on the surface, his reactions erratic and impulsive, his behaviour animated and wild; Lockhart's thoughts were hidden behind an arrogant forbidding exterior, and his actions cold-blooded and deliberate. Wilson was no doubt flattered by his description in the 'Chaldee Manuscript' — 'the beautiful leopard from the valley of the palm trees', but it is less comprehensible that Lockhart should have allowed himself to be called 'the scorpion which delighteth to sting the faces of men'. Wilson had a certain whirling boisterous talent which he exercised most successfully in invective, or more exactly in elaborate abuse. This verbal energy attracted many of his friends and readers, and it accounts for the vogue of the 'Noctes Ambrosianae', the 'celebrated papers' which he contributed to *Blackwood's* for thirteen years. It also excused the excruciating facetiousness which makes them impossible to read today. But it lacked a fixed centre. This is apparent in his criticism, which shows flashes of insight and occasional justness, but which he always sacrificed to surface animation and effect, almost immediately descending into rodomontade and coarseness. Wilson was by nature an overbearing bully, and when the hidden violence that stoked the nerve of his prose erupted he did not care who the victim was or what the bystanders might feel. The sadism he then betrays would be horrifying if it were not recognizably adolescent. The humour of the 'Noctes' is the rude aggressive humour of the fourth form, and of the fourth form also is the scatological alliterative vilification he delighted to develop when the least opportunity offered, as for example in 1823 when he reviled those who criticized *Blackwood's*:

Let execrations gurgle in your gullets, distended with the rising gorge of your

[38] *Blackwood's*, 12 (1822), 710.

blackest bile; belch out your bitter blackguardism lest you burst; clench your fists till your fretted palms are pierced with the jagged edge of nails bitten in impotent desperation; stamp, unclean beasts with cloven feet, on the fetid flags of your sty till the mire mounts to your mouths . . .[39]

The cruelty, in the most charitable interpretation, is the mindless cruelty of wanton boys. He delighted in bull-baiting and cock-fighting, and adapted these delights to criticism. His moral insensibility was inveterate, and his incapacity to understand or recognize carelessly inflicted suffering permanent. It is not possible at times to believe that he knew what he was saying. 'You deserve, Sir,' he would remark, 'to be hung up by the little finger till you are dead.'[40] Nearly a decade later this 'boy at heart' (his friends' excuse), who had been for ten years Professor of Moral Philosophy in the University of Edinburgh (the most remarkable appointment, said the *Scotsman*, since Caligula made his horse a consul), describing, as Christopher North, how he would deal with someone of differing views, said for the edification of his readers that he would 'smash his nose flat with the other features till his face was one mass of blood', and at another time that he would 'cut with a blunt knife the throat of any man who yawned while he was speaking to him'.[41]

Mr Blackwood made no attempt to conceal his own blackguardism from his associate John Murray; indeed it is a question how far he was conscious of it. He had been Murray's Edinburgh agent since 1811, and now Murray had £1,000 invested in his magazine. Parallel with the series of his office-letters to Murray is another, date for date, written at night from his home and often sent through Croker.[42] When Wilson anonymously attacked his friend and former Lake District neighbour Wordsworth in June 1817 Blackwood was delighted, and in one of these confidential night letters hastened to share the good news with Murray:

This number of the magazine is making a great noise. The Wordswortheans are quite indignant at the terrible thrashing their idol has received. I expect a great deal of fun from this article . . . I have sent the Magazine to Wordsworth himself and written him in such a way that he will be obliged to send me some reply. I should like much to see him when he receives my packet.[43]

Evidently Blackwood's covering letter to Wordsworth apologized for an article his editor had printed without his knowledge and hoped his sincere regrets, accompanied by the article, would reach Wordsworth

[39] Ibid. 14 (1823), 341. Cf. the writer's 'Hazlitt in Edinburgh: An Evening with Mr Ritchie of the *Scotsman*', *EA* 17 (1964), 123–4.

[40] *Blackwood's*, 112 (1822), 779.   [41] Ibid. 28 (1830), 596; 29 (1831), 853.

[42] Among those referring to Croker are: Murry papers, 23 May, 22 July, 19 Aug., 13 and 19 Sept. 1817; 18 July, 17 Sept., 16 Oct. 1818.

[43] Ibid., 21 June 1817.

before he saw it elsewhere. The device became a stereotype to be used in all cases where retaliation was to be feared — or provoked. Murray however proved unexpectedly squeamish, and before long terminated his association with Blackwood.

Wilson, having attacked Wordsworth in June, defended him with equal insouciance in October. His real opinion (jealousy apart) is difficult to determine, and the truth probably is that he yielded to the mood or impulse of the moment. He was a dangerous, treacherous sentimentalist, capable of maudlin comradeship at one moment and of the most violent antagonism at the next, of braggadocio succeeded by cowardice. In moments of crisis, when exposure threatened, he fell into a shivering panic, took to his bed, threatened suicide.[44] Lockhart at such times thought he was mad. Eight years later, when Wordsworth's growing fame, if nothing else, should have given Wilson pause, even Lockhart was shocked, on returning from seeing Wilson and Wordsworth on the warmest terms together in the Lake District, to find the latest *Maga* on his desk and in it a stinging attack by Wilson.[45]

When the October number was published, with the 'Chaldee Manuscript' which was bound to set the whole town by the ears, Lockhart and Wilson decamped and lay low, leaving Blackwood to face the music. His magazine was the self-appointed champion of Church and State against the seditious infidels of the *Edinburgh Review*, and he had some slight difficulty, which however he shrugged off, in defending a biblical parody of the same kind as was shortly to bring William Hone to trial for blasphemy at the Guildhall. He had little confidence in his colleagues' ability to outface their aggrieved critics. J. G. Dalyell, an Edinburgh advocate lampooned in the 'Chaldee', began an action in Court of Session. In London John Hunt called on Baldwin, Cradock & Joy, agents for *Blackwood's*, to demand the identity of 'Z', the writer of the infinitely more odious attack on his brother. Baldwin applied to Blackwood, who answered that the contributor in question was a Londoner 'of great ability', that his paper was included without careful scrutiny, that he, Blackwood, disapproved of the aspersions on the character of Mr Hunt, and that the editor had now warned the contributor that his submissions would not be accepted unless they were 'free from this defect'. Every word was a lie. And Lockhart proceeded recklessly to compound the mystification and attempt to cover his tracks by forging a letter to Leigh Hunt, purporting to be an admission of authorship of the attack, from Dalyell.

Meanwhile Blackwood craftily invited Hunt to get his friends to write in the magazine in his defence, and proposed to Wordsworth also to

---

[44] Oliphant, i. 282.　　　　[45] Ibid. 284.

become a contributor. Failing them, he enlisted the services of P. G. Patmore, promising as usual liberal payment when the journal succeeded, as it was bound to, with such sterling assistance. Not that 'gentlemen of the first literary reputation' had not come forward in Edinburgh, but Patmore's dramatic criticism, said this Sir Pertinax MacSycophant, was 'written with that elegance and simplicity which Scotchmen can admire without being able to imitate'.[46]

However, he was not getting the publicity he wanted. The London press almost entirely ignored him. In March 1818 he complained to Patmore that although he had sent complimentary copies to a number of editors they took no notice and he had given it up.[47] He would have thought, said he, that the *Courier* would have applauded his strictures on the *Edinburgh Review*. Perhaps he was right, but he added *The Times*, and in this showed how little he knew the London scene. Despite the sensational first number he had so far no more to show for his venture than a string of actions for slander. He concluded that he was not going far enough and that more sensationalism and more legal actions were needed. For this it was essential to penetrate the secrets of the enemy camp and the characters and private lives of its editors and contributors. To fuel future assaults he was henceforth constantly on the look-out for suitable and willing informants, or informers. Whenever a new contributor wrote from London he set about cajoling him into acting as spy, initially by a system of flattery. The interest, he would declare, that his deprived, uninformed Scottish readers took in the literary gossip of the Metropolis was as great as the pleasure they would find in the writer's distinguished contributions.

Patmore was one of the earliest of the London contributors, beginning in December 1817. He was in consequence made much of, and Blackwood made a point of writing to him at extraordinary length. At first, adopting the house style, Patmore kept his name secret, but Blackwood, who assured him that he recognized in him 'a gentleman and a scholar', discovered that he had written for the *Examiner*, and invited his confidences regarding that paper (that is, regarding the Hunts, Hazlitt, etc.) and also the *Champion*, confidences which of course would be respected. This was in December 1817.[48] Patmore, as one of the managers of a literary institution, was an obvious prospective mole, and Blackwood made his hints as plain as he could without entirely revealing his hand. In January he tried again, in almost the same words. 'Be assured', he said, 'that you will have no occasion to regret any freedom of communication with me . . . '[49] Patmore refused to be drawn

---

[46] Champneys, ii. 434.   [47] Ibid. 435.
[48] Ibid. 431 (20 Dec.).   [49] Ibid. 434 (29 Jan. 1818).

into the clandestine role subsequently filled by Alaric Watts, Eyre Evans Crowe, and others, in what the *Examiner* later defined as 'the *Blackwood* system of literary information and *espionage*'.[50] But if Blackwood was unsuccessful in acquiring an English secret agent in London he was luckier in Edinburgh, where in 1818 he enlisted a willing accomplice, a visitor to London. This man became a friend of Hazlitt, and it is worth taking a look at his story.

Alexander Henderson, a plausible, bustling young Scot, a bibliophile of literary leanings, had approached Hazlitt either directly or through Godwin, and had so insinuated himself into his good graces that during one visit he spent a week or more at his house in York Street.[51] He now treacherously passed reports on him which Hazlitt, in his guilelessness, never suspected and continued not to suspect. Four years later, when he was in Edinburgh, he saw a good deal of him, and two years or so after that he was still referring to him as 'my old friend', and sent him a presentation copy of his *Notes of a Journey through France and Italy*.[52] He appears for the first time in Godwin's diary in July 1818, but he was evidently already known to both him and Hazlitt. Tall, well-built, bluff, he had much in common with Wilson, and could, like him, please when he set out to do so, but lacking Wilson's advantages in education and social standing he was not in a position to inflict as much damage.[53] He affected a carelessness in dress which bespoke a defensively assertive vulgarity, and was equally crude and blunt in his manners except where he had need to be cannily circumspect. By some he was immediately disliked, but with others his buoyancy and animal spirits (traits always attractive to the melancholy Hazlitt) palliated his coarseness. To them his volubility, fired to exaggeration and fantasy by his enthusiasm, and his boisterous laughter (all typical also of Wilson) made him an agreeable companion. Sarah Hazlitt did not care for him (that he was able nevertheless to stay at York Street points up the widening gap between Hazlitt and his wife). He was very watchful.

Born near Edinburgh 23 February 1791 of poor parents and educated, not well, at the expense of a local patron, he entered the Post Office at Edinburgh as a letter-sorter. Ambitious and shrewd, he worked perseveringly to ingratiate himself with the eminent. This, and his literary interests, account for his becoming a subscriber to Coleridge's *Friend*,

---

[50] NLS MS 4009, fos. 198–237; MS 4008, fo. 107; *Examiner*, 25 Feb. 1827: 'The *Blackwood* system of literary information and *espionage* has, we understand, been set on foot in Paris. Dr Maginn was over there in the summer to organize it . . . '

[51] Le Gallienne, 331; Bonner, 249.

[52] 10. 178; NLS MS 155, p. 76 (the book was sent through Henry Leigh Hunt).

[53] This account of Henderson is partly based on a privately printed pamphlet, [Thomas Murray], *Notices of Alexander Henderson, Esq*, (Edinburgh, 1849), 4 pp. (in an edn. of 50 copies, according to a manuscript note in the Mitchell Library, Glasgow, copy).

when, with characteristic calculation he took immediate steps to recover the expense by boldly offering himself to Coleridge as contributor as well as purchaser.[54] In 1810 he published a brief turgid life of Dr Alexander Adam, the Whig Rector of Edinburgh High School, recently dead, a dominie of local celebrity.[55] Henderson had not been a pupil at the school, which was beyond his reach, but he had contrived to scrape an introduction and thereafter 'assiduously cultivated his acquaintance'. He dedicated his book to Francis Horner, MP, a former pupil of Adam but a stranger to himself, and did so 'as a very humble testimony of the author's unfeigned respect and esteem'. This is the first example of an ingenious leap-frog technique he designed for his ascent. When Horner in turn died in 1817, Henderson, although he still only 'knew' him, as he said, 'at a distance' (Horner having proved wary), pieced together an obituary, in the magniloquent style of Mr Micawber, got it into an Edinburgh paper, and then sent it to John Allen at Holland House, with, he said, the hope that it might 'be fortunate enough to obtain in any degree the approbation of one who knew so well the object of its praise',[56] but Allen no more rose to this absurd bait than Horner had done. He must have scraped acquaintance with Godwin and Hazlitt in somewhat the same way. It was certainly how he approached Leigh Hunt. He was a man of complex deviousness. Ambition taught him duplicity, how to play the courtier, and above all to flatter endlessly, but he had tact also and skill in feeling his way. Within two years of entering the Post Office at the lowliest clerical grade[57], he achieved, during a visit to London, by a combination of effrontery and calculation the remarkable feat of gaining the ear of Francis Freeling, the head of the service.[58] The secret was that he had found that Freeling was a book collector, and had put him in the way of acquiring some rare Scottish tracts.[59]

Born poor, his political views, which were always, in appropriate company, violently expressed, were reformist, or more exactly, anti-establishment. This did not prevent his corresponding with Gleig of Balliol (a move as unexpected as his approach to Coleridge) about the suppression of civil disturbances in Edinburgh;[60] an early instance of his double standards which may be explained by his literary ambitions: Gleig was a friend of Nares, editor of the *British Critic*. A letter of 1814 to Leigh Hunt illustrates the fears that often threatened to run counter to

[54] Coleridge, *Collected Works* (1969–   ), *The Friend*, ed. B. E. Rooke, ii. 434, where this is said to be Alexander Henderson, MD, author of *A Sketch of the Revolutions of Medical Science* (1806), who, however, was living in London, not Edinburgh, in 1809.

[55] *An Account of the Life and Character of Alexander Adam , LL.D.* (Edinburgh, 1810).

[56] BL Add. MS 52194, fo. 85.   [57] RHE, Post Office Records, vol. 10.

[58] NLS MS 10279, fo. 27.   [59] Ibid., fo. 24.

[60] NLS MS 3867, fo. 67. A number of letters from Bishop Gleig and his son suggest that Henderson was for a time ed. of an Edinburgh monthly magazine, but nothing confirms this.

the manœuvres of this eager young man. It was probably he who had written pseudonymously to the editor of the *Examiner* in 1810 about his Life of Adam, in an effort to publicize it.[61] But it was to the author of the *Feast of the Poets* that he now wrote, to register his admiration, uneasily conscious however that by so doing he was now communicating, on a more personal level, with the turbulent editor of the *Examiner*, the notorious radical reformer, libeller of the Regent, gaol-bird of Horsemonger Lane. His evident terror of spies befits the employee of the Post Office. He marks his letter 'Private' and directs it to *Examiner* office as being safer than trusting the Post Office to see that it got to the gaol. But in case it should be opened he takes care to insert a eulogy of his superior, Francis Freeling, the relevance of which must have puzzled Hunt.[62]

If, as seems likely, it was the fame of the author of *Political Justice* that had already before 1818 impelled him, like so many others, to seek out Godwin, it may well have been the success of *Characters of Shakespear's Plays* (by another radical) that prompted him to write to Hazlitt in 1817. But his adaptable politics and his inveterate opportunism enabled him also to join the opposite camp, to jump aboard the Tory bandwagon of *Blackwood's Magazine*, which had set off in such rattling style. He supported it from its inception, and although other early subscribers, including Thomas Campbell, James Perry, William Ritchie, an editor (there were several) of the newly established *Scotsman*, Dr Chalmers, the Glasgow minister whose published sermons that year 'ran like wild-fire through the country',[63] Henderson's friend, J. R. M'Culloch the political economist, even Dr Stoddart of the *New Times*, all dropped out in disgust after a few numbers, he never cancelled his order.[64] Why should he? The thing was a success. He already had so firm a foothold in the Princes Street shop that he was cited as a man of mark in the 'Chaldee Manuscript'. (This was the year in which he wrote, as we have seen, to John Allen of Holland House, one of the pillars of the *Edinburgh Review*.) It was important to Blackwood to have a man in the Post Office, and he would no doubt have liked to have on his side the head, the incorruptible William Kerr, Secretary at Edinburgh since before Henderson was born, but failing him, he was glad to find in Henderson someone who could supply helpful information, who could at need keep an eye on incoming and outgoing mails, who would forward urgent items (such as the *Edinburgh Review*) to him ahead of delivery time.[65] Henderson for his part, although amenable, had no intention of

[61] *Examiner*, 14 Oct. 1810, 651, letter signed 'Edinensis', or 23 Sept., 600, letter signed 'J.B', or both.

[62] BL Add. MS 38108, fo. 101.   [63] 11. 44.

[64] NLS MS Acc. 5644, H1 and H2, lists of subscribers.

[65] Murray Papers, letters of 21 Oct. and 14 Nov. 1818.

swearing exclusive allegiance to either one of the two warring camps, as the Allen letter shows. He was also in touch with Constable, although not so closely as in later years, and in May 1818 he delivered to Jeffrey a copy of the *Lectures on the English Poets* which Hazlitt had sent him for his editor (along, we may assume, with a copy for Henderson himself).[66]

Henderson now in July and August 1818 paid a long visit to London. He and Blackwood had discussed the ways in which this could serve *Maga* and on 12 July, on the eve of his departure, he wrote promising to get 'the jobbers of *The Times*' to insert a notice, and to follow up a letter he had written to the antagonistic editor of the ultra-liberal *News*: 'being on the spot myself I shall be enabled to mark how the land lies and to detect any shuffling or evasion to which, in epistolary intercourse, your jesuitical gentlemen have frequent recourse.'[67] In the event the editor turned the tables on him by making public what he very probably let slip, the secret of Wilson's connection with *Blackwood's*; it was the first English paper to do so (20 September).

However uncertain he may have been of success with these editors he felt sure of turning his friendly relations with Godwin and Hazlitt to account. He goes on, 'I can say I shall get some curious news for you respecting the operations of the enemy from Hazlitt and Godwin.' Arrived in London, he dined with Godwin on the 17th, and called on him again on 16 and 19 August[68] and, although we cannot be certain that it was on this visit that he stayed at York Street, he had ample time to see Hazlitt before Hazlitt set out on that memorable walk through Farnham and Alton to Winterslow in the glorious weather of that exceptional summer (he went down confident and light-hearted, and came back to the vexation of 'Hazlitt Cross-questioned').

The details of the gossip Henderson brought or sent back to Edinburgh are impossible to identify in the flood of cunningly disguised, often generalized abuse in the magazine. The wiliness with which this was commonly done can be seen from Robinson's puzzlement over an allusion to himself as 'the correspondent and caricaturist of Wordsworth': 'This article', he hazarded, startled by the word 'caricaturist', 'must be written by a Londoner, an acquaintance of mine probably, though I know not whom'. The author was Wilson.[69] Whatever Henderson got out of Godwin, Godwin himself was hardly a target. There is no clearer confirmation of the eclipse of his reputation at this time ('no one', said Hazlitt, 'thinks it worth his while to traduce and

---

[66] *Letters*, 182.

[67] NLS MS 4003, fo. 80, n.d., headed 'Sunday', endorsed 'July 1818' (Henderson dined with Godwin on the 17th).

[68] Godwin Diary.

[69] *Blackwood's*, 14 (1823), 505; Morley, 298; Gordon, ii. 66.

vilify him') than his almost complete absence from the early pages of
Blackwood's. Perhaps Wilson and Lockhart owed the item relating to the
actor Conway in 'Hazlitt Cross-questioned' to Henderson, perhaps he
heard Hazlitt remark that the *Edinburgh Magazine* was a millstone and the
editors equally ponderous; all that is certain is that he, 'a chield amang
them taking notes', played the Judas in Skinner Street and York Street
that summer.

Earlier in the year the less tractable Patmore had jumped to the
defence of Hazlitt after Lockhart's versified 'Notices', jeering at
Hazlitt's 'coxcomb lectures', had appeared, and perhaps he had thereby
increased the chances of a further attack. He addressed Blackwood:

Was it not a gratuitous piece of *imprudence* (to say the least of it) to admit that
line in the last number about 'pimpled Hazlitt'? In consequence of being one of
the managers of a Literary Institution I have been led to form a slight personal
acquaintance with Mr Hazlitt, and I have reason to know that such notices . . .
are exceedingly obnoxious to him — I suppose your editor is not ignorant how
tremendous his power is when he sets about to resent what he feels or fancies to
be an inury?[70]

Blackwood answered immediately, on 19 April, with the standard
specious explanation: 'As to the line on Mr Hazlitt, I believe it was put
in without thought, and I am sure that if the Editor had considered it he
would have altered it.'[71] Next day, rubbing his hands, he carefully
amplified this reply. Blowing hot and cold, he was conciliatory at one
moment, and at the next threatened to wield his cudgel if Hazlitt
ventured to feel himself aggrieved.

Your notices of Mr Hazlitt's lectures have been admirable, and must, I think,
have been satisfactory to himself. The joke about him in the notices I do not
understand nor much care about. The said Notices were written in a few hours
by a gentleman of real wit, and perfect good humour. Some one had told him
(it would seem erroneously) that Mr Hazlitt had a pimpled face, and he
accordingly said so, without much meaning . . . I can have no wish to offend or
irritate Mr Hazlitt. But neither have I the slightest fear of him. I am mistaken
greatly, if there be not a pen ready to be drawn in my service, by 'one as good
as he'. The Baron Lauerwinkel [Lockhart] is ready and able to enter the lists
with any antagonist. At the same time, I know the powers of Mr Hazlitt and
perfectly sympathize with your admiration of them.[72]

The assault on Hazlitt, as we have seen, was precipitated by his
embarking on a series of contributions to the languishing *Edinburgh
Magazine* in July, with the essay 'On the Ignorance of the Learned'. How

[70] NLS MS 4003, fo. 200, letter of 16 Apr. 1818.
[71] Champneys, ii. 437.
[72] Ibid.

could two Oxford men, mainstays of the rising intellectual journal of the
capital, let such a provocative, such an insolent title pass? But what was
of greater moment, how could Mr Blackwood tolerate this *Edinburgh*
reviewer's expanding into a second periodical, a monthly, which was
owned by his rival? He had banked on the continuing supineness of the
Edinburgh Magazine, and here it was being revived.

'Hazlitt Cross-questioned', which reached London in the August
number of *Blackwood's Magazine* on 1 September 1818, was not aimed
solely or even primarily at Hazlitt. It was a stone meant to bring down
two birds, and the second of the two (the first in order of importance) was
Archibald Constable. Hazlitt was to be exposed as a charlatan, an
ignoramus conscious of his own incapacity, but concealing it so as to
cheat (here is the essential word) his 'Task-masters or employers'. The
hoodwinked employer was Constable, and Hazlitt's immediate task-
masters were Jeffrey, Napier, and Cleghorn and Pringle (now in charge
of Constable's *Edinburgh Magazine*). The article, signed 'Z', and addressed, not cleverly, from 'Greenwich', consists of eight questions
formulated in true forensic terms, as befitted the Faculty of Advocates,
one of which was subdivided into fourteen detailed queries, and it is
these queries that take up the greater part of the space. They are devoted
to the essay 'On the Ignorance of the Learned' in the *Edinburgh Magazine*.
Hazlitt is represented as taking advantage of the 'good-natured ignorance and unsuspecting simplicity of the worthy Conductors of that
Miscellany' to 'injure their reputation and that of the said Miscellany'
by his blunders, falsehoods, and impositions. He is alleged to have
described the editors as 'perfect ninnies, and their work as a millstone',
but 'Z' (as with the preceding conjunction of 'ignorance' with 'worthy')
lets slip his real intention when he goes on to ask Hazlitt if he does not
despise himself 'for mixing, for the sake of a few paltry pounds, [his]
madness with their idiocy'.[73]

Writers on Hazlitt have tended to ignore the warfare against
Constable and dwell on the Keswick accusation, which forms no more
than an afterthought to Question II, as follows:

Is it, or is it not, true that you owe all your ideas about poetry or criticism to
gross misconceptions of the meaning of [Wordsworth's] conversation; and that
you once owed your personal safety, perhaps existence, to the humane and
firm interference of that virtuous man, who rescued you from the hands of an
indignant peasantry whose ideas of purity you, a Cockney visitor, had dared to
outrage?

In his Reply to 'Z' Hazlitt devotes one line to this quaintly phrased
question of strange abstraction, and begs to be excused answering it,

[73] *Blackwood's*, 3 (1818), 551.

except as it relates to his supposed debt. Why? If it was as grave as it has been made out to be, this was a lame response. No explanation of his refusal has been attempted. It looks much as if he did not consider it grave but ridiculous, and felt that to attempt to excuse such an episode would make it more ridiculous still. Indeed it looks as if he did not perceive the precise nature of the accusation. He remembered the village hoyden who had teased him and laughed at him after rousing him, his humiliation and resentment, and the fright he got at the hue and cry she raised, but he was far from realizing that the smacking he gave her had been magnified into an attempted rape (a word never used by either Wordsworth or Coleridge[74]). Years later, in writing 'On the Disadvantages of intellectual Superiority', the incident recurred to his memory in the same colours: 'the maid will laugh outright . . . send her sweetheart to ask what you mean . . . set the whole village . . . upon your back!'[75] We have seen the account Wordsworth gave Haydon, and what Haydon made of it. It hardly involved the kind of violence associated with rape. Three other people smiled over the story. Lamb refused to take it as seriously as Wordsworth wished: 'The "'scapes" of the great god Pan who appeared among your mountains some dozen years since, and his narrow chance of being submerged by the swains, afforded me much pleasure. I can conceive the water nymphs pulling for him.'[76] Patmore spoke of 'Hazlitt's alleged treatment of some pretty village jilt who . . . had led him to believe that she was not insensible to his attentions; and then, having induced him to "commit" himself to her in some ridiculous manner, turned round upon him and made him the laughing-stock of the village'.[77] The brisk and shrewd Sara Coleridge (whose daughter years later said, 'My mother never disliked him, uncouth and morbid and melancholy as he was. She found him civil . . .'[78]) was on the spot in 1803, and was not shocked, but she does hint now, in 1818, when news of Hazlitt's action against Mr Blackwood reached the Lakes on 27 September, that the poets were uneasy at the prospect of having to substantiate under oath gossip which might emerge a good deal less sinister under cross-examination. She wrote to Tom Poole: '. . . As Master Hazlitt would cut a very ridiculous figure, I wonder he chuses to make a stir in it, I think I told you the ridiculous story of Hazlitt's behaviour to a Peasant Girl when he was here 12 or 14 years ago.'[79]

A final point should be considered. The hue and cry need not imply

---

[74] It is used once by Lockhart and Wilson in a defensive letter to Murray, NLS MS 4003, fo. 137, 2 Oct. 1818, 'Wordsworth and Southey are both continually talking about the rape story', in a passage silently omitted by Oliphant, i. 167, between 'encouraging him' and 'Henceforward'.

[75] 8. 288.   [76] Marrs, iii. 125.   [77] Patmore, iii.141.

[78] Robinson Papers, letter of 7 Mar. 1845, misplaced in 1849 vol., no. 86b.

[79] S. Potter (ed.), *Minnow Among Tritors* (1934), 64.

that the girl was violently attacked. In the villagers' primitive collective jealousy of their womenfolk it would have been enough for a stranger to glance at a local girl, let alone risk Hazlitt's indiscretions. Also, it was easy to rouse a village in those days of country boredom. Roderick Random's enemy had no difficulty in alarming the parish with false rumours and having him hounded out. And in that particular year of 1803 boredom was liable to be shaken up by sudden alerts. This was the autumn of the invasion alarm and the spy scares, when every stranger, unwelcome even in normal times, was doubly suspect.[80]

No, the Keswick innuendo was incidental to the main aim, which was to stifle Hazlitt's voice and debar him from editorial offices, in particular the offices of the *Edinburgh Review* and the *Edinburgh Magazine*. Hazlitt knew this perfectly well. When he said in his letter of 19 September to Constable, 'My writings are before the public: my character I leave to my friends: [but] I conceive the law is the proper defence of my property', he meant by his property his earning capacity and future writings, and at that moment most urgently the *Lectures on the English Comic Writers*. His first feeling was to ignore the attack as beneath his dignity, just as he had ignored the abuse of the *New Monthly* in 1817. He must have patience. He had not spared others, and it was too much to expect that he himself would go entirely free. He had examples of the unjustly attacked in Dryden and Pope, and their fame to show that ultimately such attacks were ineffective. In any case he would have agreed with Playfair (another of the Edinburgh libelled) who concluded resignedly, 'With men all over filth it is impossible to grapple without being covered with dirt'.[81] He also had the example of the ineffectuality of Hunt's demands, finally relinquished, for the identity of his assailants. Hunt could afford to laugh at the hinted charge of incest, both from its evident enormity and because he was his own master. Neither did Hazlitt, on his side, feel threatened by the so much less grave Keswick gossip, but he was not an independent editor, and could not ignore the menace to his livelihood.

But it was not until he found, in the second week in September, that Taylor & Hessey were not now prepared to pay the £200 they had promised for the *Comic Writers* volume because 'Hazlitt Cross-questioned' had lessened the value of his 'literary e[state]' that he took steps to extract from Mr Blackwood the money he had lost.[82] He no doubt realized that Blackwood was aiming through him at Constable, but he also realized that it was his business to protect himself. On 18 September he wrote to Blackwood demanding the name of his

[80] For an arrest of a 'suspicious person' in that region and at that time, see *Lancaster Gazette*, 22 Oct. 1803.

[81] BL Add. MS 34612, fo. 224, 1 Oct. 1818.

[82] NNHL, 276.

libeller; Blackwood refused, and on the 25th Hazlitt commenced an action for damages before the Court of Session.

Blackwood fumed and blustered in letter after letter to Murray: there was nothing actionable in the paper; Constable was driving Hazlitt on to the prosecution; the Hazlitts and the Hunts had attempted to stir up a noise in London against the *Magazine* in the same way as Constable and his party had tried to run it down in Edinburgh; did any of the London lawyers think the cause could succeed? In the end he thought it would save trouble if Hazlitt, who would not have undertaken his action if he had not been urged on by Constable, gave it up: Wilson and Lockhart were to get Walter Scott 'to tell C[onstable] strongly that he must give up this system of urging on actions, else it will be worse for him'.[83] About the 20th an anonymous pamphlet, *Hypocrisy Unveiled and Slander Detected*, appeared in Edinburgh condemning *Blackwood's Magazine* and defending Hazlitt, making clear for the first time the secret reason for the attack, his *Edinburgh Magazine* articles. Lockhart and Wilson rashly replied, thereby tacitly admitting responsibility. A full account of the pamphlet in the *Morning Chronicle* on the 29th quoted, among other details, Mr Blackwood's truckling to Croker, 'the rickety bantling of the Admiralty', for his interest with the Cabinet, and their consequent alliance. The same issue included the two letters of Lockhart and Wilson to the author of the pamphlet, and his reply. Murray was aghast at the ineptitude of their move, but Blackwood persistently urged him 'firmly to deny as I do that either of them have any hand in the attacks on Hunt and Hazlitt, which are evidently from London, etc.'.[84] But he knew the secret must come out in court, and just before the case came on he persuaded Patmore to get Hazlitt to settle privately: Hazlitt must be made to realize that he was being sacrificed to satisfy Constable's malignity.[85] In the end Hazlitt accepted, probably from distaste, although he may have been unconfident of winning the case, which for him would have been disastrous. Wilson and Lockhart were not lawyers for nothing. They ensured that there were no positive assertions in the article, only questions, and a few cunningly phrased insinuations: they did not say outright that Hazlitt cheated his editors; only that 'it were well if he were honest in his humble trade'.

If Hazlitt had any hope that this would mean a cessation of hostilities he was mistaken. *Blackwood's* continued for years to snipe at him with glancing references and brief allusions. They abandoned full-scale attacks, and sprinkled their scurrilities over the whole journal, in articles which had no direct relation to their target. There is no doubt that he felt

[83] Smiles, i. 486.
[84] Murray Papers, 30 Oct. 1818.
[85] Champneys, ii. 440, 27 Oct. 1818 (misread as 27 Aug. by Champneys).

deeply the howl of execration that met his emergence from obscurity,
coming from the *Quarterly*, the *New Monthly*, and *Blackwood's*. Ten years
later he said to Northcote, 'All the former part of my life I was treated as
a cipher; and since I have got into notice, I have been set upon as a wild
beast'.[86]

[86] 11. 318.

13

The London Magazine

HAVING decided to abate the Scottish nuisance by legal action, Hazlitt abandoned the Reply to 'Z' he had sent to Constable and, in default of this means of voiding his resentment, directed his attention to his London enemy, the editor of the *Quarterly*. On 1 March he published his *Letter to William Gifford, Esq.*[1] Here, as in the Reply to 'Z', he rested his defence of his career, as a writer, on the *Essay on the Principles of Human Action*, and declared:

I have some love of fame, of the fame of a Pascal, a Leibnitz, or a Berkeley (none at all of popularity) and would rather that a single inquirer after truth should pronounce my name, after I am dead, with the same feelings that I have thought of theirs, than be puffed in all the newspapers, and praised in all the reviews, while I am living.[2]

Keats, writing to his brother, copied out page after page, to illustrate its 'style of genius' and the 'force and innate power with which it yeasts and works up itself'. As a gesture of political solidarity Hazlitt sent an inscribed copy to Hobhouse, but there is no more evidence of a response than there is to any of the articles he was collecting for *Political Essays*.[3]

The *Political Essays* appeared on 14 August 1819,[4] just before the nadir of post-Waterloo economic depression and political unrest was marked by the unsheathing of the Yeomanry sabres at St Peter's Fields, Manchester. It made little stir. The moment was unpropitious, and in any case collections of previously published pieces, as the reception of *A View of the English Stage* in particular had shown, had less chance of succeeding than new works. The prejudice was one Hazlitt always had to contend with after 1812: his *Life of Napoleon Buonaparte* was almost the only one of his works after that date that had not previously appeared in periodicals. Nor can the reputation of his publisher Hone as an opportunist and a printer of crude political squibs, have helped. Black seems to have been wary of him, but Place, to whom Black expressed his distrust and who was difficult to please, contributed extensively to his *Reformist's Register*. John Hunt thought him 'a coarse man', unscrupulous in the way of business, but immediately came to his aid at the time of the trials, an assistance repaid by Hone when John Hunt was himself brought to book on a similar charge in 1821. We find Hazlitt and Hone

---

dining at Hunt's on 31 January 1819.[5] The trials turned Hone into a
popular hero, but as a publisher he was, and remained for some years,
what Elliston was in the theatre — a not quite established adventurer,
associated in the public mind with broadsheets and ballad-mongering, as
Elliston was with lavish spectacle. Hazlitt himself, with his usual
prejudice-free indifference to conventional respectability, helped him in
the same line during the period of his collaboration with George
Cruikshank: W. C. Hazlitt tells of young Cruikshank at the Southampton Arms, outlining to Hone and Hazlitt on the table-top with a finger
dipped in beer a caricature Hazlitt had suggested.[6] It seems to have been
after the last of Hazlitt's major political writings in 1818 that he
proposed to Hone to publish them as a collection. Hone was unusually
slow in bringing them out: the contract was signed on 25 January 1819,
and reads: 'Mr Hazlitt agrees to furnish copy written, a week from this
time, and to supply it as wanted for a collection of his Political Essays
which he thinks anyway worth preserving, etc.'[7]

Hazlitt may have felt a repercussion of the *Blackwood's* attack in the
refusal of Taylor & Hessey to undertake this work, and perhaps even in
Constable's rejection of his offer of the *Lectures on the Dramatic Literature of
the Age of Elizabeth*.[8] But this may have been owing to Constable's
annoyance at Hazlitt's describing as 'quackery' his intention to preface a
work of Godwin's with a portrait of the philosopher.[9] It was Godwin who
injudiciously reported this comment, but the harm he there did to his
friendship with Hazlitt was as nothing compared with his perverse
silence, later, about the work of his predecessor in his book *Of Population*,
published at the end of the year. This piece of deliberate ungraciousness
(against which the *Examiner* protested, and which Hazlitt himself
'thought a tacit admission of plagiarism') led to a coolness between the
two that was never entirely dissipated, although they continued to
meet.[10]

In late August or early September he went down to Winterslow to
prepare his lectures on the Elizabethan dramatists, which were delivered
at the Surrey Institution from 5 November to 24 December. Procter,
who lent him his collection of plays, said he did all his reading and wrote
out his lectures in no more than six weeks. Howe argues from the essay

[5] BL Add. MS 37949, fo. 56 (24 Sept. 1817), and Add. MS 38523 fo. 41; F. W. Hackwood,
William Hone (1912), 212.

[6] *Memoirs*, i. 300.

[7] Maggs cat. 425 (1922), item 1289.

[8] See the writer's 'Hazlitt in Edinburgh: An Evening with Mr Ritchie of the *Scotsman*', *EA* 17
(1964), 16 n.

[9] Constable, i. 93.

[10] See the writer's 'The "Suppression" of Hazlitt's *New and Improved Grammar of the English
Tongue*: A Reconstruction of Events', *The Library*, 6/9 (1987) 42–3.

'Character of the Country People' which appeared in the *Examiner* on
18 July, that he went down to Winterslow before that date. But he called
on Godwin on 19 and 22 July, and again on 17 August (probably to
present him with a copy of *Political Essays*[11]), and on the next day he saw
Godwin again, for the last time until 16 October. It seems Procter was
right, and he was at Winterslow from the beginning of September to the
middle of October. He worked so doggedly that he had to decline a
cordial invitation from John Hunt to visit him at the cottage to which he
had retired at Upper Chaddon, near Taunton. Despite the attractive
offer, from a man he held in such high esteem, of a writing-room hung
with prints of Raphael, Titian, and Correggio, and looking on to a
grassy orchard, he felt he had better stay where he was. But he wanted
company, and instead, invited Leigh Hunt to come to stay with *him* at
the Hut, without success – Hunt was detained by his wife's imminent
confinement.[12]

The year came to a depressing close with his eviction from 19 York
Street, probably after Quarter Day, 25 December.[13] He had fallen into
arrears with his rent, and Bentham sent in the bailiffs. This finally
wrecked his marriage and severed him from his wife. It seems pretty
clear that Sarah, weary of his improvidence and his failure to meet bills
and that '*close hunks*', his brother-in-law, but the general ground was
concluded that he would never change, and went into rooms. He was
chronically improvident, had never saved. Whenever he had a sum of
money in his pocket it was either squandered or frittered away in dribs
and drabs. His defence of players against imputations of extravagance
and intemperance was perhaps heightened, aimed as it was at *The Times*
and that 'close hunks' , his brother-in-law, but the general ground was
good: 'they live from hand to mouth; they plunge from want into luxury;
they have no means of making money *breed*, and all professions that do
not live by turning money into money, or have not a certainty of
accumulating it in the end by parsimony, spend it'. He spent it.
Parsimony was alien to him. Like the actors, he thought it not unwise, in
face of an uncertain future, to make the most of the present moment. But
also involved was a moral principle, disinterested liberality.

He apparently only succeeded in keeping on an even keel when, with a
limited horizon, he lived on a small income paid weekly. This was so
from 1813 to 1817. It is striking that in the four years from 1814 to 1817
he contributed as many articles to periodicals (some three hundred) as in

[11] Inscribed, 'Mr Godwin / with the Author's be[st] respects', Quaritch cat. 1043 (1984),
no. 278 (evidently the same as was reported in 1935 to Keynes (p. 52)).

[12] *Memoirs*, i. 253, 255.

[13] Westminster Public Library, Rate Book 1819, the last in which Hazlitt appears, is dated
16 March; the house was empty the following year, at 29 Mar. 1820.

the whole of the rest of his life. The year 1817 saw his highest output, nearly a hundred. By 1819, however, the number had shrunk to a mere six, from his having devoted much of the years 1818 and 1819 to work on his three lecture series. The lump sums he got for these, and subsequently for the copyrights, do not seem to have been put to as effective a use as the small but constant trickle that kept up from 1814 to 1817. When he turned his attention elsewhere and that trickle dried up, he got into debt, fell behind with the rent, and Bentham made no bones about ejecting him.

We see painful evidence of a wider discouragement in the coda to his final lecture of 1819, which anticipates the sombre essays of his last years:

In youth we borrow patience from our future years: the spring of hope gives us courage to act and suffer. A cloud is upon our onward path, and we fancy that all is sunshine beyond it. The prospect seems endless, because we do not know the end of it. We think that life is long, because art is so, and that, because we have much to do, it is well worth doing: or that no exertions can be too great, no sacrifices too painful, to overcome the difficulties we have to encounter. Life is a continued struggle to be what we are not, and to do what we cannot. But as we approach the goal, we draw in the reins; the impulse is less, as we have not so far to go; as we see objects nearer, we become less sanguine in the pursuit: it is not the despair of not attaining, so much as knowing there is nothing worth obtaining, and the fear of having nothing left even to wish for, that damps our ardour, and relaxes our efforts . . . We stagger on the few remaining paces to the end of our journey; make perhaps one final effort; and are glad when our task is done!

The lectures were published on 3 February 1820 by Robert Stodart, Ogilvie's friend, whose author-list included Burdett and Hobhouse, and whose radical sympathies proved so inimical to his business affairs, that he became bankrupt the following year.[14]

At the beginning of 1820 Hazlitt began contributing to the newly established *London Magazine*, under the editorship of John Scott, whom he met again on 16 January for the first time since the rift of 1815.[15] The storm of politics long blown over, Scott, who had been out of the country for some years, was impressed by Hazlitt's achievements since the old days of the *Champion*, and saw, although with reservations, that he would be a valuable contributor. He was keener to engage him as dramatic critic than to accept the essays he wanted to place, but in the event he did

[14] Advts., *MC*, 3 Feb., and *Examiner*, 6 Feb. 1820, 96. 'Books published by Stodart & Steuart', advts. in the end-pages of *Lectures on the Dramatic Literature of the Age of Elizabeth*. Stodart was summoned before the House on 13 Dec. 1819, to answer for the publication of Hobhouse's *Trifling Mistake* (*House of Commons Journal*, at that date). He was posted bankrupt on 19 May 1821 (*London Gazette*).

[15] *Four Generations*, i. 135.

both, and his mistrust of Hazlitt's 'besetting errors' and 'improper subjects' was dissipated by the series of 'Table-Talks' which began in June and later joined the first collection of his mature essays with the title *Table-Talk* in April 1821.

As a monthly, the *London* could take a more leisurely view of the dramatic scene than was demanded of the weeklies and dailies, and Scott allowed him great freedom. It also happened that in the first months, having hardly begun, he was unexpectedly released from his duties by the King's death on 29 January, which closed the theatres for over a fortnight. A later allusion in the *Magazine* to a visit to the West Country in March 1820, together with the performance-dates of the plays he saw in the first quarter of the year, indicates that after 26 January he was out of town.[16] His biographers assume that he was at last taking up John Hunt's invitation, and that following this he probably went on to see his parents at Exeter. The order of motives must be reversed. The weather hardly encouraged visits to friends. It was a dreadful winter, with the Thames frozen almost solid; in January in the West Country five children perished of cold; and the snow lay deeper on Dartmoor than even in the terrible winter of 1814.[17] To have undertaken such a daunting journey so soon after these events can only imply response to a family crisis. It was his parents' first winter in their new home, to which they had moved from Bath the previous September, and the old minister was eighty-two, and frail. The urgency might have been some such catastrophe as a storm-damaged roof, or a sudden deterioration of the old man's health. What seems to have happened is that Hazlitt's presence became necessary at Crediton about the time he was temporarily released (a happy coincidence) from his duties, and that he took advantage of the journey and the improving weather to visit Hunt on the way back. It is clear from the order in which he names Ilminster and Hinton in the later allusion that he visited Hunt on the way back: if he had gone first to Taunton and then to Crediton, he would have returned direct from Exeter to London, without touching either Ilminster or Hinton.

He was back in town for the opening of Haydon's exhibition of *Christ's Entry into Jerusalem* on 25 March (which, although he rejoiced at its success after Haydon's years of labour, he thought 'a rude outline, a striking and masterly sketch',[18] but no more). In the same way he came up from Winterslow in May, having retired there in order to write 'Table-Talks', so as to witness the triumph of his young friend of former days, the 'simple-hearted, downright, and honest' James Sheridan

[16] 18. 343 and n., 368 and n.
[17] *MC*, 1, 6 Jan.; *Courier*, 6 Jan.; *Times*, 12 Jan.
[18] 16. 209.

Knowles, with his tragedy *Virginius*.[19] But he was spending more and more time at Winterslow, in part because he found he could write more easily there, but in the main (and the one cause implied the other) because he felt himself drawn again to that dreaming world. He desired solitude. As he said in the second of his 'Table-Talks', written there (quoting *The Excursion*), for the writer, 'Solitude "becomes his glittering bride, and airy thoughts his children" '.[20]

Another link with the past — they were growing fewer — was severed by the death of his father, a significant date in any man's life, turning the mind to early days and receding origins. This happened on 16 July, but the poor communications of the time and his awkward, independent habit of coming and going without a word to anyone meant that no one knew where to find him when the end came. Not until the last days of the month did he learn that the revered and loved figure had 'gone to rest, full of years, of faith, of hope, and charity'.[21]

Hope, for so long eclipsed, sprang up violently and without warning in his own breast that August.

When he returned to town, it was to Southampton Buildings, the old familiar ground of his bachelor days. He took a couple of rooms at No. 9, fourteen shillings a week, two floors up, at the back, looking west over the garden to the lawyers' offices in Chancery Lane, and beyond them to Lincoln's Inn. The rattling and grinding of coaches and wagons over the cobblestones of Holborn, the whistles, shouts, cracking whips, all the city's clamour and turmoil beyond the narrow northern exit of the street, echoed but faintly there in that secluded backwater on the edge of the ancient silences of the Inns of Court. He sought the peace and quiet he remembered in that street from former days. The irony was that he could not know that the move ushered in the most unquiet period of his life, and that before long it would seem to him as if he would never know peace again.

His landlady was the wife of a tailor named Micaiah Walker, a decent respectable man, and, as might be guessed from his Christian name, a Dissenter.[22] He was a member of Elim Baptist chapel, Fetter Lane, where lay his eighty-five year old father, who had come up to London

---

[19] F. Pollock (ed.), *Macready's Reminiscences* (1875), i. 213; 18. 348.

[20] 12. 279.

[21] 8. 13.

[22] My information about the Walker family derives from two sources. First, from a document headed 'Particulars of my Family', drawn up on 10 Apr. 1882 by Sarah's brother, Micaiah Hilditch Walker (1803–99). Secondly, from a 'Walker Family Tree' established by Reginald Field Walker (b. 1864), the son of Micaiah Hilditch Walker's son Edmund William (1832–1919). The second source brings the account up to the present century, and is less detailed about the first three generations, beginning with Anthony Walker (1734–1819), Sarah's grandfather, than the first, the 'Particulars', which it largely duplicates. I am indebted to Sarah's brother's great-grandson, Mr John Fieldwalker, for permission to consult and quote these documents.

from Lyme Regis in the previous century, and had died the year before.
The tailor does not however appear to have been a religious man. Those
of his children who died young (including the three-year-old John
Anthony who died 6 March 1821, while Hazlitt was in occupation of his
lodgings) were indeed buried at Elim, but his son Micaiah Hilditch
Walker, writing in after years, was only able to say that he 'believed' his
father's people were Baptists. The Walkers were of much the same
generation as their new lodger: the tailor was four years older, his wife
Martha little more than one.[23] Mrs Walker (her maiden name was
Hilditch) was shrewder and more determined to prosper than her milder
husband, and with six children to fend for she had need to be. She let
rooms to professional men, for the most part young lawyers and lawyers'
clerks who stayed there two or three years, for the term of their articles.
In addition to Hazlitt there were at this time a pharmacist named
Griffiths from North Wales and a married couple, the Folletts. Mrs
Walker's ambitions had received the year before an unexpected, an
unhoped for, fillip in the marriage of her eldest daughter Martha, aged
twenty, to one of the lodgers, Robert Roscoe. Roscoe, a man of thirty,
educated at Cambridge and come to town as a solicitor, was the fourth
son of no less a person than William Roscoe, the banker whose portrait
Hazlitt had painted, banker no more since the house foundered in 1816,
but still the premier citizen of Liverpool and a liberal of national repute.
Young Roscoe and his wife were living around the corner in Dyer's
Buildings, Holborn, where Roscoe senior, so far from disapproving the
match, stayed when he was in town, and even invited his friends, Fuseli
among others, to dine.[24] Their first child, Emma, was born a few days
after Hazlitt arrived.[25] In her calculations Mrs Walker saw her son-in-
law, with his distinguished name, not only as a means of providing for
her daughter but also as a life-line for the rest of the family. The Miss
Stimsons, who had kept the house before the Walkers, had married into
the legal profession, and she confidently anticipated that her second
daughter Sarah also would follow in their footsteps and in her sister
Martha's.[26] And indeed the connection seems to have enabled the son
Micaiah, then aged sixteen, and (as far as we can judge) with no
particular inclination or aptitude for the profession, to become himself in
the fullness of time a solicitor, and eventually the founder of a
respectable law-firm.[27] Whether or not it was this connection that paved

[23] Micaiah Walker, b. 25 July 1774, m. 9 Mar. 1797, d. 16 Apr. 1845; Martha Walker (née
Hilditch), b. 5 Nov. 1776, d. 22 Dec. 1835.

[24] J. Knowles, *Life and Writings of Henry Fuseli* (1831), i. 328, Fuseli to J. Knowles.

[25] Emma Jane Roscoe, b. 21 Aug. 1820.

[26] Holborn Public Library, Land Tax Assessment, Parish of St Andrew and St George,
Holborn, 2825–81 (years 1813–15). The Walkers moved to Southampton Buildings in 1816 (9. 161).

[27] Walker, Son, & Field, 61 Carey St., Lincoln's Inn.

the way for the fifth child, Leonora Elizabeth (the Betsey whom we meet in *Liber Amoris*) to marry in ten years' time Charles Augustus Nott, of the Temple, eldest son of one General Sir William Nott, cannot be known, but that is certainly what she did.[28]

Writers on Hazlitt have generally accepted too readily his own prejudiced view in *Liber Amoris* and assumed that the family were lower in the social scale than they actually were. The father had some education (like most Dissenters); the eldest daughter had enough intelligence and accomplishment to make a successful marriage with a man of professional status and cultivated background. Her husband was so far satisfied with his bargain that he declared marrying her to be the best thing he ever did. The second daughter wrote a neat hand and seems to have read a good deal — and not only almanacs, broadsheets, and newspapers, or even the titillating poems of Mr Moore and the well-publicized works of the dashing romantic Lord Byron: she had also the patience to read and discuss the mild verse of Barry Cornwall and the difficult prose of William Hazlitt (although it is true that we do not know that she would have read them without the new lodger's prompting).

It was this daughter Sarah, 'born', as family records state, 'the 11th of Nov. 1800, ¼ before 12 night — No 13 Gt. Smith St., Westr.' (Hazlitt would have been moved by the precise recording of so momentous an event), who was to change Hazlitt's life.

The course of his existence for the next three years was decided the following morning, 16 August, when Sarah brought up his breakfast tray. On her first entering the room he hardly noticed her looks. She seemed rather plain and mousy, and her expression not at all attractive. After her usual fashion, she kept her eyes demurely lowered. What did strike him, however, was the exceptional grace of her movements. He admired her gliding walk as he had admired the Miss Dennetts, the 'three kindred Graces', in the ballet at Covent Garden in 1816, and like them, no doubt, it touched the same distant chords in his memory that he had earlier spoken of in his essay 'On Beauty', the glittering air-borne thistle-down that had delighted him as a boy, or the Duchess in *Don Quixote*, or Northcote's fleeting glimpse of Marie-Antoinette and the famous description in Burke which paralleled it: 'And surely never lighted on this orb, which she hardly seemed to touch, a more delightful vision'.[29] In that essay he speaks of the poison which those charms left in the heart of the Irish orator. Little did he now suspect the far more deadly poison with which the present homely encounter with an ordinary London girl was about to leave him.

[28] 8 Aug. 1832. Her father-in-law's rank and title are as given by Micaiah Hilditch Walker. I find that his rank was a local one, East Indies.

[29] 4. 71; see also 20. 278, 'Trifles Light as Air', no.3.

The grace of her movements would not of itself have been enough, but the next moments were decisive. When she was leaving the room she turned in the doorway and gazed full at him. This was probably a meaningless trick to create an impression, imitated from someone she admired or picked up from some novelette. She evidently practised it upon all eligible lodgers, since she did not omit it even on so inappropriate an occasion as what was almost her last interview with Hazlitt, in July 1822, when she had already decided that the affair was over. But it was enough. Hazlitt's imagination was immediately fired. He was unused to such notice, convinced as he was of his being incapable of inspiring love, of having no attraction for women. This girl seemed different, for that very reason. And as she closed the door that August morning the unexpected possibility that she had taken a liking to him opened up a new and delightful vista to his imagination. He was launched on a vague sea of hope. The more he thought about her the more he was persuaded that she must like him and the more attractive she appeared. It was the beginning of what Stendhal in a famous image in *De l'amour* called crystallization. Stendhal, who at this very moment was putting the finishing touches to the manuscript of *De l'amour*, had seen the previous year the collapse of his greatest love-affair, with Métilde Dembowski, who had played the same paramount role in his life as Sarah was to play in Hazlitt's (the difference in social class — Métilde, wife of a general and a baron, moving in a brilliant circle in Milan; Sarah, the daughter of a tailor in a London side-street — only clinches the truth of *De l'amour* and of *Liber Amoris*).

Sarah, a young girl whose experience of life hardly extended beyond what she saw of the lodgers in her mother's house, was curious about this new arrival, and her curiosity in his case may have been sharper if she knew that he was a writer. She must soon have found it out. Her father happened to be tailor to one of Hazlitt's friends, John Payne Collier, but her mother would probably learn it from Hazlitt's visitors, from one or other of the contributors to one of the best-known journals in London, Procter, or Talfourd, or Lamb. Or it might be Joseph Parkes, a solicitor and sportsman, a protégé of Bentham's, whom he had known in York Street, or Haydon, Bewick, or Martin Burney.

There must have been something very disturbing in her manner, uncanny even, which explains Hazlitt's constant reference in *Liber Amoris* and in his letters to her alien, non-human attitude and behaviour. Procter, who often called at his lodgings, was equally struck:

Her face was round and small, and her eyes were motionless, glassy, and without any speculation (apparently) in them. Her movements in walking were very remarkable, for I never observed her to make a step. She went onwards in a sort of wavy, sinuous manner, like the movement of a snake. She was silent,

or uttered monosyllables only, and was very demure . . . The Germans would
have extracted a romance from her, endowing her perhaps with some diabolic
attribute.[30]

This unvarying stare and undulating movement directly inspired (just as
reminiscences of Keats and Burton indirectly confirmed) the terrible
image at the end of *Liber Amoris* where he has the illusion that Sarah is
transformed, and that she 'started up in her own likeness, a serpent in
place of a woman. She had fascinated, she had stung me, and had
returned to her proper shape, gliding from me after inflicting the mortal
wound, and instilling deadly poison into every pore; but her form lost
none of its original brightness by the change of character, but was all
glittering, beauteous, voluptuous grace'.[31] After that first meeting with
Sarah on 16 August 1820 he was never able to see her for what she was.
The invitation and promise he read into the lingering steadfast gaze she
bent on him as she went out of the room that day created the image of an
ideal Sarah, and set in train the changing, fading, recurring series of
reflections that filled the succeeding twenty-four hours — hopes,
scenarios, queries, enigmas, doubts, calculations, speculations, resolutions. Mingled with these were memories of exchanges with earlier
mistresses, half-recollections of similar encounters which had come to
nothing or had soon ended in disappointment, or had ushered in a
period of suffering, for all of which this girl was, he longed to believe,
now going to make it up to him. It was in vain that he told himself that he
was almost certainly mistaken, that studious, bookish, self-absorbed
men like him knew nothing of women, were condemned to remain
unloved, that he had long seen clearly into himself, and written it all
down years ago as a final judgement on his character and temperament;
it was of no avail. The wild hope still sprang up. The more he thought
how unlikely it was, the oftener he told himself he was old enough to be
her father, the tighter grew the bonds that he was himself at that very
moment weaving, by the very reflections that carried these objections
and reproofs, and winding around his heart, so that the more he told
himself it was impossible the more possible it came to seem. It was in
vain that he repeated to himself what he had said of Desdemona:

The idea that love has its source in moral or intellectual existence, in good
nature or good sense, or has any connection with sentiment or refinement of
any kind, is one of those preposterous and wilful errors, which ought to be
extirpated for the sake of those few persons who alone are likely to suffer by it,
whose romantic generosity and delicacy ought not to be sacrificed to the
baseness of their nature, but who, treading secure the flowery path, marked out

---

[30] Procter, *Fragment*, 181–2.
[31] 9. 153.

for them by poets and moralists, the licensed artificers of fraud and lies, are dashed to pieces down the precipice, and perish without help.

The next moment, a willing sacrifice, he advanced secure along the flowery path towards the precipice, in entire neglect of his own admonition. The words he now heard in his memory were quite other. What now welled up from the past was what he had said of Petrarch and Laura:

The smile which sank into his heart the first time he ever beheld her, played round her lips ever after: the look with which her eyes first met his, never passed away. The image of his mistress still haunted his mind, and was recalled by every object in nature. Even death could not dissolve the fine illusion: for that which exists in the imagination is alone imperishable.

He knew also, intellectually, that that which exists in the imagination is alone toxic, and might well prove to be a self-administered poison, but it was already too late. He had already begun to drink of the phial, small, but of ever increasing potency, and the long disease must take its course.

At their next meeting, if he did (as is unlikely) offer any familiarity, it was simply to confirm to himself that his absurd trust was well placed. He was only too eager to be convinced that she was not one of your common lodging-house decoys. It was a help that she was not an ordinary servant or chambermaid, but the daughter of the house (and the sister-in-law of Roscoe), but these reflections were unnecessary. The world had already narrowed to Sarah; and already it only needed the flutter of an eyelash or a pursing of the lips to dispose of his doubts for ever.

On her side she was flattered by the impression she had made on the new lodger, but she could not understand him. When he talked in his romantic passionate style she was thinking of the dresses in *Ackermann's Gallery of Fashion* for that month. She was fascinated by his attentions, but not surprised. Her sister had hooked a lawyer, the son of a prominent family; why should not she in turn make the conquest of a prominent writer? But how could she be sure he was sincere? Men were such deceivers. And he did not hide that he was a married man. Perhaps he meant to get what he could get from her and throw her over, move without warning to another part of town? And as her mother said, once a girl loses her reputation what has she to depend on? Besides his way of making love was so extravagant: all those high-flown speeches, which he must have got from books . . . (He could, in fact, have found them in *The Academy of Compliments*, and while it is true that the conversations in Part I of *Liber Amoris* are laden with irony, he must have uttered many of them.) Often she could make neither head nor tail of them, and naturally concluded he was making game of her. This led to resentments which were largely suppressed, since the rudenesses she, in her dependent situation, could risk offering in retaliation were mild enough; but they cut him to the quick.

Sarah Walker was a coquette; of that there is no doubt. Nor is there any doubt that she was interested in her mother's lodgers from an ambition to follow her sister's example. She hoped to make an equally advantageous match. Her motive in pausing at the door on that first morning to give Hazlitt that long, lingering look is clear. But she was unlucky in her choice. Instead of encountering a commonplace calculable lodger, she found herself confronted with one of the subtlest and most acute, as well as one of the most self-tormenting minds of the age. She did not know how to deal with the situation. Any other lodger she could, without any embarrassment, without any misunderstanding, have alternately fended off, and led on. The affair would have taken its predictable course, and ended either in her gliding adroitly out of difficulty, or in the one exceptional case, in marriage, and nothing more would have been heard of it.

But Hazlitt was quite different. The poor girl was out of her depth, and of course she alternately panicked and warded Hazlitt off and then calmed down and returned to her calculations, in proportion as Hazlitt became agitated, despaired, and regained hope. It was a drama, complete in all respects, of the kind adumbrated in his pages on Desdemona and on the 'literary character' years ago and which no doubt he had acted out and re-enacted time and again in earlier years.

What followed, in the remaining months of 1820 and in 1821, the ebb and flow of his hopes and disappointments, of encouragements and rebuffs, of advantages he thought he had gained and setbacks he knew himself to have suffered, all this is obscure and can only be assumed and dimly reconstructed from *Liber Amoris*. There can be no doubt that they quickly became very intimate. She spent hours at a time in his room, and he had to tear himself away from her distracting presence in order to write. Hence his more frequent and more prolonged absences at Winterslow, where he could 'work double tides'.

For the latter half of 1820 however his movements are obscure. In September Lamb wrote to convey the pleasure Mary had taken in his memories of Mitre Court evenings in his essay 'On the Conversation of Authors'. He himself was pleased at this time also by the unexpected homage of an unknown admirer. In October Haydon had read him, for a fairly evident reason of his own, a letter from Miss Mitford in which she contrasted him unfavourably with Haydon. It said:

Mr Hazlitt, delightful writer as he is, does not produce so much effect as you do, partly because he is by profession a 'setter-up and puller-down of Kings', and we are accustomed to see him tossing great names into a topsy-turvy confusion, and partly that he has a trick of escaping occasionally from some serious and earnest criticism by a piece of sudden pleasantry which gives an air of persiflage to all that he has said. He is the most delightful critic in the world, puts all his taste, his wit, his deep thinking, his matchless acuteness into his subject, but he does not put his whole heart and soul into it, and you do, and there is the difference. What charms me most in Mr Hazlitt is the beautiful candour with which he bursts forth sometimes from his own prejudices, such as the character of Dr Johnson in the Comic Writers, and the account of the Reformation in the Age of Elizabeth. I admire him so ardently that when I begin to talk of him I never know how to stop.[32]

When Miss Mitford learned of this venial breach of confidence she apparently wrote a letter of apology or explanation to Hazlitt, to which she received a reply, now lost, on the 20th.[33] We can guess at its tenor from her letter of 11 November to Sir William Elford,

You are not the only traitor among my correspondents if that be any consolation. I happened to say to a friend in town that Mr Hazlitt was 'the most delightful and most impudent of writers' or words to that effect, and what did my correspondent do but read him this curious panegyric the first time they met. Luckily Mr Hazlitt is good-humoured and took it without being astounded at my impudence.[34]

The burgeoning relationship between Miss Mitford and the writer she held to be 'out and out the most sparkling and brilliant . . . of the day'[35] went no further. In 1823 Hazlitt applied to Jeffrey to review her successful tragedy *Julian*, without effect.[36] At about the same time Haydon promised Hazlitt to invite him and Miss Mitford together, but before he could do so she had gone back to Three Mile Cross.[37] It does not seem that they ever met.

The December *London Magazine* carried the last of his articles on the drama, and towards the end of the month, able, no doubt for the first time since his father's death, to visit his mother and sister, he set out for Crediton for the second time within a year. As so often, we are obliged to deduce his movements from an isolated scrap of evidence, in this case hitherto undetected. The letter ascribed by Howe to 'the first days of 1821', and adduced as evidence of a visit to Winterslow, is in fact

[32] Mitford Letters, iv. 420, 2 Oct. 1820.

[33] BL C. 60. b. 7 (M. R. Mitford's' unpub. Diary), 'Friday 20th [October 1820.] Heard from dear Granny and Mr Hazlitt.' Other entries record her reading of the *Lectures on the English Comic Writers* (17, 18 Apr. 1819), *Lectures on the Dramatic Literature of the Age of Elizabeth* (19, 20 July 1820, and again on 22 Oct., when she 'made some extracts'), *Political Essays* (8 Mar. 1821). Her invariable, indiscriminating comment, on each of these occasions, is 'famous!'

[34] Mitford, Letters, iv. 423.

[35] Ibid. v. 487.

[36] V. Watson, *Mary Russell Mitford* (n.d.), 155.

[37] Letter of 27 Mar. 1823 sold at Sotheby's, 2 Apr. 1973.

postmarked 'Crediton 12 Jan. 1821', and was sent just before he left Devon.[38] It is likely that he spent Christmas and the New Year's Day with his mother and sister. Whether he called, as before, on John Hunt on the way back we cannot tell, but we do know that he stayed a long time at Winterslow. It was probably there, where his wife still kept her cottage, that he saw young William on this trip. Not that that was his sole object. His first duty was towards his bereaved mother, his second to his son, but he also meant to do some necessary work to complete the *Table-Talk* volume for publication by Warren in April. There is an implication that he was unable to settle to work in his mother's house, and found there distractions of another kind than in town. The correction of the proofs of 'On Reading Old Books' he could manage, but not sustained essay-writing. He may have been harrowed by the signs of her increasing weakness, and by the sad admission to himself that she was old. She and Peggy (now fifty) were unreconciled to his broken marriage, and may have reproached him with it and with neglect of the boy.[39] And finally there were winter chores he felt obliged to undertake.

Sarah Walker sent him at Crediton the January *London Magazine*, where he found, in Lamb's essay 'New Year's Eve', the very thoughts he had put into 'The Past and Future', left with the editor before he came away. The coincidence does not surprise us, in view of the consanguinity of mind and the long friendship, but it surprised him, and he asked Scott to return his essay to him at Winterslow. We cannot say what changes he made, but they must have been so drastic that he concluded it ought not to go into the *London* just yet. It is clear that he introduced that apostrophe to the woods of Tytherley which gives a glimpse of his solitary walks and his endless, cheerless musings on Sarah.

He got a letter from her at Crediton,[40] and another at the Hut. Neither helped to diminish his isolation. They were an expected but not the less sore disappointment. They were, as he said of others later, 'so short, with scarce one kind word in them'. He chided her for their matter-of-factness: she was 'such a girl for business'. I quote the second, dated 17 January, for its formality and guarded unresponsiveness, the determination not to commit herself, which he longed to see her put aside:

Sir,

Doctor Read sent the London Magazine, with compliments and thanks, no Letters or Parcels except the one which I have sent with the Magazines according to your directions. Mr Lamb sent for the things which you left in our

---

[38] Howe, 280. The original is in the Liverpool Record Office.

[39] As may be inferred from their joint letter of 22 Mar. 1822 to Sarah Hazlitt; see J. A. Houck, 'Hazlitt's Divorce: a Family Letter', *English Language Notes*, 18 (1980), 33–5.

[40] As is implied in the letter which follows.

care likewise a cravat which was not with them. — I send my thanks for your kind offer but must decline accepting it. — Baby is quite well, the first floor is occupied at present, it is very uncertain when it will be disengaged.

My Family send their best Respects to You
I hope Sir your little Son is quite well.
From yours Respectfully
S. Walker[41]

He was at Winterslow, or had just returned to London, when the tragic outcome towards which the coarse abuse, brutalities, cudgellings, and horsewhippings of the *Blackwood's* gang had been tending for years supervened. It was clearly bound to come about sooner or later, and on 16 February 1821 the vendetta in which the *London Magazine* and *Blackwood's* had been locked for months came to an end with John Scott lying mortally wounded at Chalk Farm. Years later Thomas Campbell said that Hazlitt had egged Scott on to fight this duel. Nothing could have been less in character. The allegation probably derives from the recent Crediton letter to Scott: 'Don't hold out your hand to the Blackwoods yet after having knocked those blackguards down.' This was a perfectly natural thing for him to say. Having suffered from them as he had, he would be bound to urge Scott not to spare them *in the magazine*. But it did not occur to him that Scott would be drawn into a duel, any more than he himself, or Leigh Hunt, had been.

Nothing seems to have been said about the duel when Robinson met Hazlitt at Lamb's on 21 February, but their own quarrel was there smoothed over, and the charitable and forgiving Hazlitt whom Robinson penitently describes could not have been less like a man of violence: 'Hazlitt and I now speak again but he does not omit the *Sir* when he talks to me. I think he behaves with propriety and dignity towards me; considering the severity of my attack on him, which, though warranted by my friendship with Wordsworth, was not justified according to the customs of society.'[42]

The death of Scott abruptly left the *London Magazine* hulling in mid-stream without a helmsman. Hazlitt's offer to the owner to do what he could to supply his place appears to have been meant to indicate willingness to step up his own contributions, but he was taken literally, and for a few months he helped Baldwin with the editing until the

[41] Ascribed to 17 Jan. 1822 in *Memoirs*, ii. 27, and by Le Gallienne, who reproduced it in facsimile (209), it has since been ignored by writers on Hazlitt, presumably because the dating involves an otherwise unrecorded and awkward absence from town. The difficulty is not the absence but its duration. Hazlitt certainly went down to Winterslow in Jan. 1822, but on a flying visit which would not require the careful instructions and reports here indicated, whereas they fit an absence of many weeks in January 1821, when Hazlitt's interest in the *London Magazine* was much keener.

[42] Morley, 261.

magazine was bought by Taylor & Hessey. When his responsibility ended at the beginning of June, he went down once more to Winterslow, where he remained until the end of the month writing 'Table Talks'.

His first volume of that title, comprising what he later called 'Essays on Men and Manners', had appeared on 6 April.[43] This 'series of plates of . . . that most interesting part of natural history, the history of man',[44] of which the *Round Table* had been the forerunner, was to form, with the second volume and the related two volumes of *The Plain Speaker*, the major work of his life. The essays had immediately been condemned for their egotism, which the *Monthly Censor* later declared to be 'the most palpable and unblushing that ever courted exposure'.[45] But when he speaks of himself it is rarely in a narrow, exclusive, personal manner, but of himself as a representative figure convenient for the study of man, in the way a painter would execute a self-portrait. He was less a successor to Steele and Addison than a descendant of Montaigne, who said at one moment, 'C'est moi que je peins; je suis moi-même la matière de mon livre', and at the next, 'De l'étude que je fais, le sujet c'est l'homme', and of whom he himself said, 'he began by teaching us what he himself was';[46] so that the circumstantial details we find in his essays are not dwelt upon as being peculiar to himself. He is so reticent in that regard that he never once mentions his wife Sarah in his writings; his parents become 'the friends I have left', his sister 'a person who took some notice of us'; and his brother he names only as the author of a saying he wishes to quote, once. His father he speaks of more frequently, evidently from his being at once so like and so unlike himself. Even in the *Liber Amoris* he is not so much recording his own unhappy story as elaborating a study of a man in love. The only point at which inappropriate egotism obtruded and threatened the balance of his work was in his passion for Sarah Walker. His obsession, already visible in 'The Past and Future', found its way into several other essays, notably 'On Great and Little Things', where he failed to eliminate it before publication in the *New Monthly* (so that 'every body in London', according to his wife, 'had thought it a most improper thing'[47]), and 'The Fight', where he was able to suppress it in time.[48]

---

[43] *Courier* advt., 6 Apr. 1821.

[44] 4. 75 (Hazlitt is here speaking of Hogarth).

[45] *Monthly Censor*, 1 (1822), 834.

[46] 6. 93.

[47] Le Gallienne, 330; Bonner, 248.

[48] S. C. Wilcox, *Hazlitt in the Workshop* (Baltimore, 1943), 17–19, 24–5, 48.

14

The End of Private Hopes

> It is said that he acts very foolishly, and talks very sensibly.
> There is no inconsistency in that.
>
> Hazlitt

By the end of the year 1821 his situation at Southampton Buildings had become intolerable. The miseries of frustration were wearing him out. He was persuaded, on good days, that 'little Yes and No' was not indifferent to him, but on bad days he was consumed with jealousy, jealousy of the uncouth Griffiths, of Tomkins, of anyone who came near her. She allayed his mistrust of Tomkins, a gentlemanly young solicitor who came to the lodgings in the autumn, by declaring that she 'despised looks'. But he could no longer endure the endless alarms and uncertainties and disappointments. The only solution was to offer her marriage, and to do that he needed a divorce.

The decision to seek a Scottish divorce must have come at some time between the excursion to Hungerford to see the fight on 10 December, when he was depressed at the prospect of 'the waste of years to come' and had little more to hope for than a mere 'truce with wretchedness', and, on 24 January, the 'delicious night' at Covent Garden when he saw *Romeo and Juliet* with Sarah and her mother, and knew that in a few days he would be off to Scotland.[1] In between those dates he must have gone down to Winterslow, ostensibly to bring young William up to town for the beginning of school-term on 21 January, but in reality to persuade his wife to agree to a divorce.[2] Her response, if not enthusiastic, was encouraging enough to enable him to proceed. It could signify little to her whether they were divorced or not: she had her own income of £150 a year,[3] and could manage very comfortably on that, but she may have seen the opportunity to extract a formal recognition of his responsibility for his son's education, which was soon to cost at least thirty guineas a year.[4] On the other hand she may simply have been sorry for him.

[1] 9. 114; *Letters*, 243. See *Courier*, 25 Jan. 1822 for a full account of the performance.

[2] *The Times* advts., 21, 22 Jan., reopening of boarding-schools after the month-long Christmas holiday.

[3] Le Gallienne, 294; Bonner, 222.

[4] In 1809 (*The Times*, 15 July) the fees at Dawson's Academy, for pupils under ten, were 25 gns. a year.

He probably brought William back to town with him by Saturday the 19th, and on Monday morning entered him at his boarding-school, Dawson's Academy, at the corner of Brunswick Square. It had been established by the Revd Jonathan Dawson, a Church of England clergyman, a dozen years before, and took about forty boarders. One of the former day-pupils was John Hazlitt's son William, eighteen months older than his cousin.[5]

On the Thursday night he took Sarah and her mother to Covent Garden, the happiest evening of the whole love-affair, but one which he did his best to spoil by begging her to assure him that his presence had not lessened her pleasure. This was a truce which broke down on the Saturday, when, on the eve of his journey and their long separation, he tried to prise from her the reassurances he craved, and when he inevitably failed, started once more upon his old litany of complaints and recriminations. The quarrel ended with a hollow enough reconciliation.

He left for Scotland on the 27th, but broke his journey at Stamford, partly in order to write about the pictures at Burleigh House.[6] He also found himself writing about Sarah, making a record of his conversations with her. He pondered the meaning of every action, every word and glance: an empty place on the breakfast tray, the interposition of the word 'before', something blurted out by her little sister. When he had finished, a part only, no more than a quarter, of the unfolding story had been told. He came to see his 'book of conversations' as a slight, but an effectively contrasting prologue to the slow tensions and accumulating violence that developed over the six months between Stamford and the conclusion of the divorce in July. In it the self-critical, satirical, ironical self that co-existed inside him with the enslaved, suffering hero dictated the trite, stilted form to clothe the sincere, heartfelt sentiments. It is an exercise in romantic irony, a conjuror's trick, at once conveying and criticizing the sentiment. From a literal point of view, the one most critics have adopted, it is embarrassingly naïve, descending at times to equally embarrassing badinage, but from another, as R. H. Horne was the first to point out, fourteen years after Hazlitt's death, it is ironical.[7] It is the dialogue of the lover and his coy mistress, couched on his side in language that dramatizes the familiar romantic dichotomy between the emotional and the intellectual (the dilemma of the 'Literary Character'), and on hers in commonplace words of notable economy. Our sense of his frustration and despair deepens with the unresponsive woodenness of

[5] Ibid., and BL Add. MS 38903, fo. 2.

[6] Howe, 308; Maclean, 438; Wardle, 305, all say he left before the middle of January.

[7] R. H. Horne, *A New Spirit of the Age* (Oxford, 1907 edn.), 214. See also M. Butler, 'Satire and the Images of Self in the Romantic Period: The Long Tradition of Hazlitt's *Liber Amoris*', in C. Rawson (ed.), *English Satire and the Satiric Tradition* (Oxford, 1984), 209–25.

her replies: 'Do not, I beg, talk in that manner', 'As you please', 'I must go now'. This Part I of *Liber Amoris* is intended to illustrate an axiom in the essay 'On the Conduct of Life': 'All your fine sentiments and romantic notions will (of themselves) make no more impression on one of these delicate creatures, than on a piece of marble.'

He arrived in Edinburgh on Monday, 4 February, and took lodgings in George Street.[8] Next, he consulted a solicitor, probably recommended, like the lodgings, by Alexander Henderson, who lived in the same street. Having thus established a commencing date for his residential qualification, he set about laying a trail of evidence for his wife's action when she should arrive, on the expiry of his forty days in Scotland. This was in a bawdy-house, at 21 James Street, kept by a Mrs Louisa Knight. The name of the girl there — an irony — was Mary Walker. Very likely it was again Henderson, a bachelor, who helped him there, since he knew no one else in the city who could have done him such a service. In that first week he must also have found time to complete his review of Byron's *Sardanapalus* for the *Edinburgh*, but he could not settle down to essay-writing, and after delivering it to Jeffrey (perhaps together with an inscribed copy of *Table-Talk*, although he had probably already sent him this from London) he went back out to Renton Inn, forty-two miles away, the last posting-inn his coach had stopped at on the way north, isolated in the middle of the countryside like the Winterslow Hut. This was probably on Sunday, 10 February, for on the 11th he began work on the new volume of *Table-Talk*.[9] In a few days Jeffrey wrote to propose some alterations in the *Sardanapalus* notice which probably also involved his adding some observations of his own on *Cain* and *The Two Foscari*. He also invited Hazlitt to come to see him. Hazlitt replied on the 14th:

My dear Sir,
 Do what you please with the article. I shall be content. I am obliged to you for your very kind invitation to Edinburgh, and hope to avail myself of it when I can get away from this which will be in about a month's time.
 I remain Dear Sir,
  Your ever obliged humble servant
Thursday afternoon   W. Hazlitt[10]

---

[8] The date is established by *Letters*, 243, 'I shall leave this next Sunday or Monday for Edinburgh. I shall then have been in Scotland five weeks', and p. 245, 'shall have been here three weeks next Monday'. The first stay in Edinburgh is established by p. 249, 'I am come back to Edinburgh', and doc. 9 in the Divorce records (see Ch. 8, n. 54 above), deposition of Mary Walker that she had known Hazlitt since she went to lodge at 21 James St. in 'the middle of February'.

[9] See the writer's 'Hazlitt's Missing Essay "On Individuality" ', *RES* ns 28 (1977), 421–30 (which gives a full account of the work Hazlitt did at this time), and the writer's letter to the Editor, *RES* ns 32 (1981) 197.

[10] Watt Monument Library, Greenock, printed in the writer's 'Some New Hazlitt Letters', *NQ*, July – Aug. 1977, 337.

He worked doggedly throughout the month of February. Apart from his afternoon walks along the high road overlooking the Eye Water, which he took as much to calm his nerves as for exercise, he spent all his time at his writing-table. His spirits were being sadly tried. The necessary but oppressive solitude, the sense of being cut off, of not knowing what was happening in London and what Sarah was thinking, even the violent gales which day after day in those exceptionally stormy months prowled around the house and moaned in the chimneys, put him on edge.[11] His sleep was disturbed. He usually did not dream much, but now he dreamt of loss, abandonment, and death. He dreamt he was back in the Louvre, 'and that the old scene returned — that I looked for my favourite pictures, and found them gone or erased. The dream of my youth came upon me; a glory and a vision unutterable, that comes no more but in darkness and in sleep: my heart rose up, and I fell on my knees, and lifted up my voice and wept . . .' He dreamt that he was reading *La Nouvelle Héloïse*, 'and came to the concluding passage in Julia's farewell letter [from her death-bed] . . . "Trop heureuse d'acheter au prix de ma vie le droit de t'aimer toujours sans crime et de te le dire encore une fois, avant que je meurs!" '[12]

By Sunday, 3 March he had got on so well with the essays that he was able to write to Colburn, who had acquired the *New Monthly Magazine* the previous year, and who was also to publish the second volume of *Table-Talk*, to say that the work was practically done. He asked for an advance of £30 to enable him to get back to Edinburgh.[13] On Tuesday the 5th he sent Patmore a list, to be communicated to Colburn, of the nine essays he had written since 11 February, naming also the two to be finished by Saturday.[14] That evening he wrote a long letter to Sarah, on the crest of a wave of satisfaction. Little did he suspect into what hands that letter would eventually fall. He was cheerful; it overflowed with love and hope, as well, of course (he could not help it) as the usual inept questions and ingenuous clumsinesses that always either made her suspect his sincerity or earned her contempt. But no trace of the agony and despair that were to come later. The end of his ordeal was in sight.

---

[11] See the writer's 'Hazlitt and *John Bull*: A Neglected Letter', *RES* NS 17 (1966), 163–70.

[12] 12. 24, written in Mar. 1822. If the quotation is from memory, it is notably accurate.

[13] *Letters*, 238 (n. 2 on this page is wrong: Hazlitt refers to his work).

[14] *Letters*, 237. What is there presented as Letter No. 100 (written, according to Howe, 309, who appears to have seen the original, when Hazlitt had been in Scotland 'a month yesterday') is described in *Memoirs*, ii. 68 as 'the end of' a letter. I now believe it to be the end of the unsigned Letter No. 102 (*Letters*, 239), and that, Hazlitt having arrived in Scotland Mon., 4 Feb., it was written Tues., 5 Mar., and that Letter No. 103 to Sarah Walker was written later the same day. The postmark 9 Mar. is not fatal to this hypothesis: Dunbar was 15 miles from isolated Renton Inn, and there could have been delay. The date scribbled at the end of the MS of 'On Individuality' here provides a *terminus ad quem*: they were certainly written *before* Thurs., 7 Mar. (see the following note).

His wife had agreed to join him in Edinburgh when his forty days were up, that is, after 16 March. The divorce, he thought, would then proceed apace, he would take a brief trip to the Highlands to recuperate, and then start homewards to begin arrangements to marry Sarah. His progress on the book exceeded his expectations. The imminence of his wife's arrival lent wings to his pen; his hand flew over the two essays to be done by Saturday, 'On Dreams' and 'On Individuality', and on the Thursday he signed the last page with the note 'Begun Feb. 11 / Monday / Ended Thursday March 7, 1822 / Well done, poor wretched creature. W.H.'[15] He was pleased with what he had accomplished against such odds — eleven essays in twenty-six days, 'a volume in less than a month' — and on 10 or 11 March he returned to Edinburgh in good spirits to await the 16th and his wife's arrival.

Sometime in the next few weeks his hopes were shattered, apparently by a mysterious blow, in a letter of Patmore's, which is dimly reflected in Hazlitt's tardy reply of 30 March telling of his despair. He was prostrated with weakness, and had impulses to suicide: 'It is as well', he said, 'I had finished Colburn's work before all this came upon me.' But precisely what 'all this' was, is hidden. In addition to this set-back there were his wife's delays. He cannot have failed to write from Renton Inn to remind her of the all-important date, the 16th. But the independent Sarah Hazlitt had her own ideas, and saw no need to hurry: she had first to come up to London for a break, and to look after young William during the school holidays, a fortnight or so around Easter Day, 7 April.[16] Evidently she would not even be leaving Winterslow until the end of March. He revolted at the thought of kicking his heels in Edinburgh, where he could not work. Charles Kemble and Mrs Henry Siddons were at the Theatre Royal, playing to crowded houses in some of his favourite pieces, but it does not appear that he saw them. He was invited out to Craigcrook, where Mrs Jeffrey received him very kindly, but he was wretched, and felt obliged to apologize for his want of spirits.[17]

His anxiety and agitation are implied in his febrile flitting to and fro between Edinburgh and Renton. On 7 April he wrote to Patmore from Renton. He had heard from his wife, who was sailing at the end of the following week, but not from Sarah Walker, so that he was still in low spirits. 'I cannot', he said, 'describe the weakness of mind to which she has reduced me.' On one page he says, 'By Heaven, I doat on her . . .

---

[15] Hazlitt MS in Manchester Public Library, described in the writer's 'Hazlitt's Missing Essay "On Individuality" ' (see n. 9 above).

[16] Le Gallienne, 314; Bonner, 236.

[17] *Letters*, 248; by 17 Apr. Jeffrey had been for some time in London (NLS MS 791, at that date).

When I touch her hand, I enjoy perfect happiness and contentment of soul'; and on the next, 'To think that I should feel as I have done for such a monster'. There were several reasons for her silence, over and above the plain fact that she had no mind to marry him anyway, which was something he refused to admit. She may have behaved badly to him (Haydon, a close confidant, has a perceptive comment here: 'she laughed', he says, 'at his passionate timidity');[18] she may have seemed calculating and ambitious, and also prepared, forgetting modesty, to permit extraordinary liberties, but in the end (as we shall see) she left her home for the man she preferred, and clung to him even though they were not able to marry. One reason for her present silence was the instinctive dread of her class of 'putting anything in black and white', which Hazlitt does not seem to have suspected (he was disappointed that she did not imitate Julie's copious epistles). Another was that Tomkins had left Southampton Buildings, and she was revenging herself upon Hazlitt.[19] It is also probable that she was still seeing Tomkins away from the house, as she certainly did in a few months' time, so that when Hazlitt said in this same letter, 'I am afraid of being kept some time longer in the dark — yet that is better than the Hell of detecting her in an impudent intrigue with some other fellow',[20] he was nearer the malign truth than he knew. Yet another was that she had discovered that Hazlitt, in a clumsy attempt to enlist support, had compromised her by telling her aunt what had been going on between them. Her aunt would have been shocked, or anxious, or both; she would certainly warn her against marrying a divorced middle-aged man encumbered with a troublesome son.[21]

April was a wretched month. He did what work he could. On the 8th or 9th the proofs of the second volume of *Table-Talk* arrived, with a request for an additional essay to make up the volume.[22] He also owed copy to Taylor & Hessey, and sent them another article on the Elgin marbles, but felt obliged to apologise for it (important though it is to us) as being somwhat of a rehash, and in any case insufficient (although twice as long as the first). To make up the sheet, he sent, in a few days, with equal diffidence, a notice of Hugh Williams's views of Greece, which pleased him more than the exhibition of Living Artists in Bruce's

---

[18] Pope, ii. 440.

[19] *Letters*, 264; 9. 159.

[20] Misread as 'independent intrigue' in *Letters*, 250.

[21] 9. 151. There is no corroborative document, but a comparison of other places in the text with parallel documents shows that Hazlitt's invention was confined to attempts to evade detection. It was essential to the whole 'experiment on character' that he should not invent. As to the identity of the aunt, Sarah's mother was an only child, so it was probably her father's only sister, then aged about 40, a spinster, shown, but not named, in the family tree.

[22] *Memoirs*, ii. 5.

Rooms, Waterloo Place (following upon Haydon's *Judgement of Solomon*, *Dentatus*, and *Christ's Agony*).[23] His acerb notice of these Scottish painters was commissioned by William Ritchie, editor of the *Scotsman*. This newspaper, founded to remedy a lamentable situation in which 'Scotland, . . . and Edinburgh especially, was bound hand and foot under the most ultra form of Toryism [when] the Tory papers . . . used to say that not a single "Reformer" could be found in all Scotland who had a good coat on his back',[24] had from the beginning reviewed Hazlitt's books with consistent warmth and enthusiasm. Its notice of *Lectures on the English Poets* praised its 'originality, acuteness, and force of conception and expression', and ended, 'We are not sure that a work of *greater power* . . . has appeared within these thirty years'.[25]

He was at supper at Ritchie's house in Nicholson Street on 17 April, when his fellow-guests were J. R. M'Culloch, the lawyer George Combe, a fanatical phrenologist, and Thomas Hodgskin.[26] We may guess that he found the last of these the most congenial. 'Dry, plodding, husky, stiff M'Culloch' was wont to relieve his sober study of economics with bouts of deep drinking and boisterous merry-making. Combe was a 'person with one idea', the business of whose life was the popularization of the systems of Gall and Spurzheim. Ritchie also was an enthusiast, and it is unlikely that phrenology was not mentioned, but Hazlitt, who had just written two essays on it, must have seen that dispassionate discussion was out of the question, and kept silent. Combe however, tireless worker in the cause, took careful note of his bumps, and added a description of his appearance at this time, published by Howe. Howe did not have access to one observation of Combe's, which is of great interest, and unique in accounts of Hazlitt: 'He is quite a gentleman in his manners and did not utter one sentiment which the most delicate and scrupulous female might not have listened to.'

The quiet retiring Thomas Hodgskin, ex-naval officer, traveller, teacher of languages, and newcomer to political economy, was a figure of no note at this time, but his writings later influenced Marx, and his life was written by no less eminent a historian than Elie Halévy. Francis Place's description of him as 'of rather a gloomy disposition, singularly modest and unobtrusive, easily excited, but more to mirth than anger, . . . speculative, and [harbouring] some very curious speculative might, with some slight modification, equally be applied to Hazlitt. He

---

[23] *Letters*, 252, 253; *Scotsman*, 24 Nov. 1821, 2 and 16 Feb. 1822.

[24] J. Grant, *The Newspaper Press* (1871–2), iii. 436–7.

[25] *Scotsman*, 27 June 1818.

[26] Howe places the meeting in early May; the correct date and a full account of the occasion and of the guests are given in the writer's 'Hazlitt in Edinburgh: An Evening with Mr Ritchie of the *Scotsman*', *EA* 17 (1964), 9–20, 113–27 (based on the Combe MSS in NLS).

was a disciple of Ricardo, and had reservations about the politics of
M'Culloch, who patronized him, and whom he considered to be more of
a Whig than a genuine Reformer. Hazlitt's writings, however, he knew
and admired. It is probable that when Hodgskin moved to London at the
end of the year and began to work for John Black, there was further
communication between them. This would seem to be the only reason
why Hazlitt would send to Hodgskin from abroad, as he did three years
later, a contribution intended for the *Morning Chronicle*.

Ready to seize at any straw, Hazlitt was lifted out of his depression by
a letter of Patmore's defending Sarah and urging him to take courage.
On 21 April he replied that the thought that her regard was true, 'out of
the lowest depths of despair, would at any time make me strike my head
against the stars'. But equally important was a dizzyingly unexpected
stroke of good fortune. As he told Patmore, he was just on the point of
setting out for London to plead with his wife to keep to their agreement,
and was actually in Leith Walk, making for the harbour, when at the last
minute he ran into his old acquaintance Bell, recently arrived to settle in
Edinburgh, who was providentially able to tell him that by a coincidence, when he was boarding his vessel in London he had met Hazlitt's
wife at Miller's Wharf, where she had booked her passage by the smack
Superb which was due to dock at Leith at any time.

Bell was making a tactical withdrawal from a commercially embarrassing and perhaps personally dangerous position. A brief sketch of his
career since we last saw him will explain why.[27] He became bankrupt in
the summer of 1819, protesting that he had been 'robbed by those whom
he too liberally trusted', which perhaps hints an accusation against his
partner, who himself in turn became bankrupt in June 1820. How it
came about that the bankruptcies of these partners took place separately
and at a long interval does not appear, but Bell seems to have been
caught up in some dubious transactions, even involving forgery (a
hanging matter), which obliged him to flee to America in 1819 or 1820.
After travelling through Upper Canada he was back in Britain in 1821,
and we must suppose that his father, who seemed more prosperous than
his official salary warranted, had in the mean time reached an
accommodation with his creditors. He attempted to reopen his agency
for the supply of timber to the Victualling Board, but either their needs
had changed since the war or they had no confidence in him. At the
beginning of the present year came rumours of corruption in the
Victualling Office and a *Times* leader accusing Commissioners of
venality and Bell's father in particular of fraudulently amassing nearly
£70,000 on contracts. J. R. Bell was mentioned but not named. He

[27] The information here is summarized from the writer's 'Hazlitt's Mysterious Friend Bell: A
Businessman Amateur of Letters', *EA* 28 (1975), 150–64, which specifies sources, mainly in PRO.

immediately, on 26 January, came forward in a letter to the editor of *The Times* with a pained defence of his father and himself. But the scandal spread, there were dismissals, a clerk committed suicide, and in February the *Morning Chronicle* demanded an investigation of the accounts of all agents and contractors for Naval provisions during the late war, promising the recovery thereby of hundreds of thousands of pounds. It may have been this unnerving development that impelled him to get out of town and so remove from danger his father's hoard, that '*moiety* of £60,000 or £70,000 which [as he declared in his *Times* letter] was at present involved in [his] commercial affairs'.

Hazlitt may have been aware of all this. He was in Stamford, with time on his hands, when the letter appeared, and he certainly knew, or later got to know, of Bell's flight to America. But he was chiefly glad of the help he could provide with the divorce proceedings, and at first he was not disappointed. Bell's initial zeal was extraordinary, and his later attempts at the seduction of Sarah Hazlitt prompt the thought that he was excited by the divorce and already had designs on Sarah himself. He went down to meet her off the boat on 21 April, missed her, but tenaciously followed her to the Black Bull.[28] The following day he accompanied her to the office of the highly respected advocate George Cranstoun. She was doubtful whether she might not be perjuring herself in taking the oath *De Calumnia*, which was designed to prevent collusion. Cranstoun, whose 'knowledge of the law was profound, accurate, and extensive',[29] was able to reassure her, and recommended a solicitor, John Gray, who told her the case, if undefended, would take two months and cost £50. Everything was set in motion, and at first moved fast. By the following Saturday Hazlitt had arranged to be detected *in flagrante delicto* at James Street, and was formally identified as the guilty party, at his lodgings; Gray had taken the depositions of Mrs Knight and Mary Walker; and Sarah, whose qualms had been reawakened on reading for herself the form of oath, was again reassured, this time by Ritchie. That Saturday afternoon, 27 April, Hazlitt set out for New Lanark, probably on foot, intending to write an article on Owen's model village, but he was defeated by 'all the apparatus . . . and its deadly monotony' and nothing further is heard of his visit. He almost certainly walked the thirty-four miles back the following Tuesday: when he arrived at George Street he had only a pound in his pocket. He found awaiting him the expected summons to appear on 17 and 24 May, but another hurdle

---

[28] Where details in the following pp. covering events from 14 Apr. to 18 July 1822 can be verified by reference to the relevant dates in Sarah Hazlitt's journal (Le Gallienne, 239–335; Bonner, 185–252) I give no source, except with long entries, or where for some other reason location might be difficult.

[29] Cockburn, i. 209. Cranstoun later became Lord Corehouse.

arose when his wife told him she would have to return to London unless he paid her expenses. He persuaded her to wait, saying he expected £30 from Colburn on Thursday next week, £100 for two lectures in Glasgow, and £80 for *Table-Talk*, in a fortnight. He thought her unreasonable, and grumbled at the way she 'appeared to love money', but he sent her a few pounds on Saturday, 4 May, before taking coach to Glasgow.

The lectures were probably arranged by Sheridan Knowles, who taught elocution there. He was also proprietor of the *Glasgow Free Press*, and a citizen of some prominence since the success of *Virginius*. His great admiration for 'William' (as he called Hazlitt, to Robinson's displeasure[30]) made him anxious to help. Hazlitt probably stayed at his house in Laurieston, south of the river, but the lectures were given at the Andersonian Institution in John Street. The audience numbered about a hundred at each of the two lectures, the first on Shakespeare and Milton, and the second on Thomson and Burns, although the Tory *Glasgow Sentinel* reported the second as thinly attended. The *Sentinel* correspondent, as might be expected, complained of cockneyism, defended Wordsworth, quoted Southey, and begrudged paying '5s. . . . for an old lecture read at the Surrey Institution'. The Whig *Glasgow Chronicle*, equally true to its principles, was 'especially pleased with [the lecturer's] opening remarks on Milton; for love of liberty and heroic sentiment are as much his own as they ever were the immortal poet's'.[31]

On Monday, 6 May, the day of Hazlitt's first lecture, Knowles's elocution classes changed from a winter timetable, which included two afternoon hours, to a summer, in which lessons finished at 8 a.m.[32] It seems to have been his custom to take a short break every year at the time of this change in early May, and the next morning he and Hazlitt started off betimes on a walking-tour up Loch Lomond. They reached Luss for dinner, and probably spent the night at Tarbet, after covering a distance of thirty-six miles, the latter half with Ben Lomond constantly in view across the water, 'its giant shadow clad in air and sunshine', a walk that sufficiently impressed itself on Hazlitt's imagination for him to include it in *Liber Amoris*. Crossing the loch next morning to Inversnaid, they found the weather unpropitious, with chill showers from the north-west, and their intended ascent of Ben Lomond was abandoned. Hazlitt does not say what direction they took, but it seems likely that they went on to the Trossachs and returned through Rob Roy's country and Aberfoyle

[30] Morley, 319.

[31] For further details see the writer's 'Hazlitt as Lecturer: Three Unnoticed Contemporary Accounts', *EA* 15 (1962), 15–24. Hazlitt received £56 for the two lectures (Le Gallienne, 296; Bonner, 223) , so, allowing for hire of hall etc., the combined total attendance must have been at least 224.

[32] *Glasgow Herald*, 10 Apr. 1822.

to Glasgow. On Saturday the *Glasgow Chronicle* reported the outing:

Mr Hazlitt, we understand, is employing the interval between his first and second lectures in an excursion to Loch Lomond and the Trossachs. That strong feeling of the sublime and picturesque which manifests itself in his writings, must fit him peculiarly for appreciating the beauties of the scenery. Mr Hazlitt commenced his career as an artist, and the very first picture he exhibited, a portrait of his mother, was [given an] honourable mention . . . in the report of the exhibition . . . Accident or choice gave, subsequently, a new direction to his talents. We may doubtless expect a sketch of the highland landscape, from his original and powerful pen.[33]

On Tuesday 14th, the day after his second lecture, unable to contain himself any longer, he took the morning coach to Edinburgh so as to get to Guthrie's, the agent, in time to book his passage on the *City of Edinburgh* for the following morning at eight. He left from Newhaven on the 15th, in high spirits, and the vessel docked at Blackwall on the Friday evening. But all the joyous anticipations of his long journey were shattered when he arrived at Southampton Buildings only to find that Sarah avoided him. Having with difficulty got to see her, he met with a stubborn evasion and rejection of all his pleas that put an end to his self-control, and shrieking his fury and despair he ran from the house as though pursued by all the fiends. After this scene, the most violent episode in the story, he thought to regain his calm by going into the country, his old remedy, but a misleadingly encouraging word from her thirteen-year-old sister Betsey persuaded him to stay. To no purpose. He was alone all week, except for one day when he went with Patmore to Coldbath Fields to visit John Hunt, whose imprisonment was nearly ended.

Having contrived only one interview with Sarah, he gave up the hopeless struggle, and on Wednesday 30th went back to Scotland. He wrote two letters to Patmore during his passage in the *James Watt*. The first, sent from Scarborough, where the vessel touched on the 30th, was filled with black despair. In the second, posted at Edinburgh when he landed on the 31st, he raged at Sarah, a 'monster of lust and duplicity' and a 'hardened, impudent, heartless whore', whom he suspected, now that Tomkins had left the house, to have taken up with Griffiths, the

---

[33] Together with a similar par. in the *Glasgow Courier* of the same date, 11 May, this provides the necessary confirmation, hitherto lacking, that the walk described in *Liber Amoris* actually took place. I see no reason to doubt that Knowles accompanied Hazlitt, although he is not mentioned in the newspapers: the only point at which Hazlitt alters the facts is when he wishes to speed up the action, saying, 'we returned, you [Knowles] homewards, and I to London', which has persuaded all his biographers that the Loch Lomond outing came after the two lectures, whereas it came in between (he could never have gone to the Highlands *and* travelled to London, between 14 and 17 May).

Plenipotentiary, 'a tall, stiff-backed, able bodied, half blackguard', because 'the bitch wants a *stallion*'.[34] His fellow-passengers were moved to pity by his wretchedness, and Ritchie, meeting him 'in a state near to frenzy', the day after he got back, thought he was like a person in a fever.

He had been bludgeoned into so dulled a sense of realities that he failed to perceive the fears that a letter he wrote to his son at the end of the first week in June might arouse in a lonely little boy. Nothing except the extreme tenderness and directness of his attachment to his son can explain the ingenuous, insensitive candour of his confidences. It begins, 'My dear little baby, the only comfort or tie poor Father has left!', and ends, 'Thy mother is to take the oath next Tuesday, the only thing that can save me from madness. Once more, farewel. I have got a present of a knife for thee. Every one pities poor father but the monster who has destroyed him.'[35] The incongruity of the last two sentences is a sufficient indication of his disturbed mind, but it is nevertheless clear that father and son withheld nothing from each other. Concealment was alien to Hazlitt, and he observed neither reticence nor taboo in his confidences to his son; if he could hold nothing back from chance-met strangers, how much less could he do so with someone he loved? And in any case, young William was well acquainted with Sarah.

His wife was to take the oath on 11 June, and the business would be over on the 28th. His entire absorption in Sarah and the divorce is reflected in his ignoring a political event which set all Edinburgh by the ears on 11 June, the trial of James Stuart for murder, following upon his duel with Alexander Boswell, who had libelled him in the *Beacon* (the scandalous Edinburgh equivalent of London's *John Bull* set up by a number of Tories, including Walter Scott). When Stuart, who was defended by Jeffrey, was acquitted, at four in the morning, the verdict was cheered by waiting crowds in the streets.[36] Lord Holland's son, on a visit to Edinburgh, was in court the whole day.[37] Hazlitt in normal circumstances would have rejoiced in the judge's condemnation, clearly aimed at *Blackwood's* as well as at the *Beacon*, of the 'system of libelling lately introduced', and would have learnt without surprise of Wilson's reaction: 'The said Judge is a mean idiot with his personalities.'[38]

On Thursday evening, 13 June, he went out to Renton Inn, but he could neither work nor endure the solitude, and in less than forty-eight

[34] *Letters*, 263, 264. The unglossed nickname Plenipotentiary (*Letters*, 270) is an allusion to a ribald song of that title (composed by Captain Morris, a favourite at Carlton House) whose hero was remarkable for sexual gigantism: it is given in *The Merry Muses of Caledonia* (privately printed, Edinburgh, 1959), 162–5.

[35] *Letters*, 262. For dating, see the writer's rev. of *Letters*, in *The Library*, 6/2 (1980), 359.

[36] BL Add. MS 52173, John Allen to Lady Holland, 11 June 1822.

[37] BL Add. MS 52086, 10 June 1822.

[38] As n. 36 above; NLS MS 4009, fo. 278.

hours was back in Edinburgh. On the 27th, stupefied and worn out with the emotions of the past month, he wrote a numbed, subdued letter to Patmore from which it was clear that in his present mood if he could not have her he would kill her, and himself afterwards. His jealousy is no longer exclusively of Griffiths; he even thinks the favoured rival might be a brother of Roscoe; but he never suspects the departed Tomkins. He again suggests putting her virtue to the test, as the only way of finally deciding whether she has been making a fool of him or not.[39]

On the following day, the long-awaited 28th, the Commissaries did not grant the divorce but required specific charges to be laid. This impelled Prentice, Hazlitt's Solicitor, to counsel contesting the charges, to avoid suspicion of collusion, so Hazlitt was obliged to 'deny that he [had] been guilty of the crime imputed to him', initiating a further delay.[40]

On 6 July he and Alexander Henderson drove out to Dalkeith Palace, and we may here take a brief respite from Hazlitt's tribulations to say something of Henderson's career since 1818. At the end of that year he saw his chance of a big promotion, upon the death of one of the two Post Office Surveyors (inspectors) for Scotland, and burst into a frenzy of activity, directed into two channels. He got Blackwood to ask John Murray to intercede with Freeling, and at the same time persuaded Constable, then in London, to approach Freeling independently (but enjoined upon him to avoid any suggestion that his candidate had any 'Whiggish leanings' — the slightest hint of that sort would be fatal).[41] His dissimulation went undetected; his boldness and pertinacity succeeded; and he was appointed, at a salary of £300 a year, an increase of £230. The letters of thanks he thereupon wrote to Murray and Freeling are models of abject sycophancy.[42] He continued thereafter impartially to cultivate both Edinburgh publishers. In 1821 he acted as intermediary between Blackwood and William Maginn, a formidable Irish recruit to *Maga* and a self-appointed scourge to Hazlitt, and one who, in reckless, gratuitous offensiveness, soon outstripped both Wilson and Lockhart. In February 1822 a letter to Henderson from Constable in London greeted him as 'My dear Friend'.[43]

In March 1822, however, Henderson met with tribulations of his

[39] *Letters*, 278–80. For the dating, see the note to this letter in the writer's rev. of *Letters*, in *The Library* (see n. 35 above), 360. The name of the person he proposes on four other occasions for this task (*Letters*, 257, 265, 273, 284) is Elder, about whom we know nothing (see the writer's 'Some Notes on the Letters of William Hazlitt', *The Library*, 6/5 (1983), 272).

[40] Divorce records, doc. 7 (this is not given in Houck's article (see Ch. 8, n. 54 above)).

[41] NLS MS 7200, fos. 146–9, Henderson to Constable, 6 Nov. 1818; Murray Papers, Blackwood to Murray, 'Saturday night' [14 Nov. 1818].

[42] Murray Papers, Henderson to Murray, 14 Nov. 1818; NLS MS 10279, 19 Nov. 1818.

[43] NLS Acc. 5643, B1, 26 Feb. 1821; ibid. MS 7200, 18 Feb. 1822.

own. We do not know how much he and Hazlitt had been seeing of each other in the months up to July, but he is unlikely, given his secretive nature, to have confessed to Hazlitt that he had been forced to resign his surveyorship on suspicion of fraud. The talent for duplicity he had shown at the time of the *Blackwood's* affair in 1818 had long been put to more consistent and profitable service in his place of work. Reports of the 'deplorable turpitude and villainy that had obtained in the Post Office' (in the words of Kerr, the head) had briefly appeared in the press, but Hazlitt was preoccupied.[44] Within the office, systematic and long-standing malversation had come to light. Henderson, with others, had been feathering his nest for years past. He had amassed considerable property, much more than was possible on his annual salary. After he had been turned off, he lived comfortably and never needed to work again. (He did however undertake commissions for Constable, but it is significant that he disappears from the list of Blackwood's correspondents: his usefulness had ceased when he left the Post Office.)

During the investigation he denied everything, although his guilt was abundantly established by the evidence of witnesses on oath. To the dismay of his friends he accepted his dismissal without a word, and never publicly protested his innocence. He no doubt saw it as an unfortunate accident preventing him from continuing to appropriate what his ill-rewarded abilities entitled him to, and thus rationalized the part he had played for over a decade, in milking the Post Office of an estimated annual £6,000. His superiors forbore to prosecute, because, they said, they feared the bad effect on public confidence, but they also no doubt feared the exposure of the hidden function of the postmasters, which was to report through the surveyors to the Postmaster General any threat of political agitation in their districts. Since the surveyors travelled the length and breadth of the country they could quickly identify trouble spots. Humble postmen who stole from letters were hanged as a matter of course; crooked surveyors, who were in effect unavowed spies, required more circumspect handling. Finally, the evidence was deemed insufficient for an indictment, in Scots law; and Henderson knew it.[45]

To return to his excursion with Hazlitt on 6 July to Dalkeith, a favourite beauty-spot in the environs of Edinburgh. Hazlitt's heaped abuse of Sarah Walker in his recent letters to Patmore meant nothing: he could not shake off an obsession more debasing than a prison treadmill; and indeed, as if by a fatality, he saw at Dalkeith Palace a painting entitled *Truth Finding Fortune in the Sea*, by Luca Giordano, in which the naked figure of Fortune conjured up so overwhelmingly the image he

---

[44] *Edinburgh Advertiser*, 19 Mar.; *Caledonian Mercury*, 21 Mar.

[45] *Parliamentary Papers, 1822: Post Office Revenue, United Kingdom, Part III, Scotland*, at 19 Mar., 20 Apr., 24 Apr., 1 May, and esp., as regards Henderson, 15 May (pp. 60–2).

cherished of her that on the following Monday, haunted by the memory of it. He hurried down to Bell's house in Pilrig Street to get his friend to accompany him on a second visit to Dalkeith.[46] Out they went together, six miles through heavy showers, so that he could gaze again on that figure, the very pattern of Sarah, all delicacy, all softness, and all grace.

On 16 July he wrote to Patmore that the Commissaries would pronounce on the following day. He affected little interest in the affair, now that it seemed Sarah wanted nothing to do with him. He glumly congratulated Patmore on his impending marriage and his honeymoon abroad; it was, he said, just what he had proposed to Sarah: to cross the Alps to Italy, to visit Vevey and the rocks of Meillerie, and tell her in that hallowed setting the pathetic story of Julie and Saint-Preux. The following morning the divorce was granted. He may have told Patmore he had no hope, but all the same he left Edinburgh as swiftly as he had left Glasgow in May, and without saying goodbye to either Bell or Ritchie. Getting out of Scotland, he said, seemed 'like the road to paradise'.

He saw Sarah the day he got back, 20 or 21 July, and was relieved to find her 'kind and comfortable, though still cold and distant'. He did not suspect that she was 'comfortable' because she was seeing Tomkins again, and 'kind' because her mother had charged her not to offend Hazlitt, who was about to take the whole lodgings at a hundred guineas a year to lighten Sarah's work. She could not deal with his fatal, compulsive persistence, however, and when he asked if 'there was a bar from a dearer attachment to his being admitted to her friendship' she took refuge in tears.

He also was on edge. He went up to Lisson Grove and told Haydon about the divorce, obliquely betraying the madness and anguish of the previous months by inventing outrageous details, partly no doubt from perverse bravado, partly to revenge the stares and whispers that had so humiliated him in staid, respectable Edinburgh. The story he told sprang from the exasperated disgust with which he now regarded the whole useless, sordid mess of his Edinburgh expedition, and from a perverse impulse to exaggerate his degradation.[47] It is as demonstrably

---

[46] RHE. Dalkeith Estate Docs., GD 224, Box 420/1, Visitors' Book (signed by Hazlitt for Bell and himself). The picture is identified in *A Catalogue of the Pictures at Dalkeith House* (privately printed, 1911), No. 122, 'Fortune, a nude female figure, lies on the sea, on a raft on which are objects emblematical of riches. Truth, a flying figure, slightly draped, raises her. In the right hand of the painting are fairy and satyr children, some swimming.'

[47] Pope, ii 373–4 (cf. J. Forster, *Walter Savage Landor* (1869), 48). The story, nowhere corroborated, should have taxed Haydon's credulity. It answers rather strangely to an anecdote in the *Examiner*, 20 Jan. 1822, 46, of one Arthur Haslewood (a striking name) of Norwich, who picked up a woman in the street, and when he got her into a tavern found she only had one eye and was ugly; he said to her 'Drink and go!', which became a catchword in the town.

untrue as the less preposterously embellished version (equally designed
to punish Sarah by punishing himself) with which he regaled Landor in a
few years' time in Florence.

Monday, 29 July dealt the last, fatal blow to his happiness. All day he
wanted to ask her to go with him to Scotland to be married, but
something held him back. In the evening, Betsey having told him she
had gone out, he set off in the same direction, suspecting an assignation,
and almost immediately ran into her.[48] She was with Tomkins. He went
home. When she returned they talked. He felt deep grief, but no enmity.
It was (as it had been with so many other women who had turned away
from him) his bad luck, no more. Stifled, he went out to roam the streets,
and again encountered Tomkins. After a long conversation in which he
learned that for months before his departure for Scotland she had been
playing the same game at Southampton Buildings with Tomkins as with
him, while denying it to both, he was finally convinced that the creature
on whom he had thrown away his heart and soul was a hypocrite, a
lodging-house decoy, low, sensual, without human feelings. He wrote
his sad farewell to her in the last lines of *Liber Amoris*:

I am afraid she will soon grow common to my imagination, as well as worthless
in herself. Her image seems fast 'going into the wastes of time', like a weed that
the wave bears farther and farther from me. Alas! thou poor hapless weed,
when I entirely lose sight of thee, and for ever, no flower will ever bloom on
earth to glad my heart again!

At the end of the summer it no doubt gave him a pang to see his two
young confidants, Patmore and Talfourd, achieve the happiness that
had eluded himself by both getting married (by a coincidence, on the
same day).[49] Ever since his return to London he had been unable to resist
the melancholy temptation to pour out his troubles to anyone who lent a
sympathetic ear; in his loneliness and misery he responded to the
slightest encouraging word or look, from friend, from acquaintance,
from stranger. His story was bound to be widely known. It became even
more widely known when the *Examiner* clumsily quoted, by way of a puff
of the recently published *Table-Talk*, the lines on love at first sight (with
that phrase as title) unmistakably describing his first view of Sarah on 16
August 1820: 'Oh! thou, who, the first time I ever beheld thee, didst
draw my soul into the circle of thy heavenly looks . . . '[50] The title could

<hr>

[48] 9. 156–7, ' "Where is she gone?" — "To my grandmother's, Sir". "Where does your
grandmother live now?" — "At Somers' Town".' This illustrates how closely Hazlitt adhered to
fact in *Liber Amoris*: Sarah's paternal grandmother died in 1817, but her mother's mother, Sarah
Plasted (2nd married name), who died on 12 Feb. 1823, was buried in Old St Pancras Churchyard,
the burying-ground for Somers Town.

[49] 30 Aug. (*Examiner*, 1 (560); 8 (576) Sept. 1822).

[50] *Examiner*, 25 Aug., 541, under 'Newspaper Chat'.

hardly fail to catch Sarah's eye, and even if she missed it someone would certainly draw her attention to Hazlitt's name at the end. It is even possible that she had already found in the volume this passage which was just as sure to do her an injury as 'On Great and Little Things'. The first journal to review the second volume of *Table-Talk*, the *Literary Register*, all but named his Infelice, proclaiming in plain English that she was his landlady's daughter.[51] From this, his enemies' steps were easily directed to Southampton Buildings, and it was not long before they were able to get into their hands one of his most unguarded letters to Sarah, with dire consequences.

The clean break, the decisive departure from Southampton Buildings which seems to be implied, but is nowhere stated, in the last page of *Liber Amoris*, did not take place. The miserable obsession dragged on. At the end of August he was still there, and he does not seem to have moved to his new address in Chapel Street West until late September.[52] But the last glimmer of hope was not quenched even then. In another attempt to break out of the treadmill, he stooped, in March 1823, to an entirely uncharacteristic and unworthy act. His conclusion, reached after his interview with his young rival the previous July, that she was a practised, callous jilt, was in the nature of things a verdict that could not be definitive, and now in the late winter of 1823, ill, depressed, dogged by misfortune, he badly needed to know whether he might not after all have judged wrong. The remedy involved a degrading act on his part, but that could not be helped. His motives are never explicit but it seems that the friend he sent to lodge at Southampton Buildings (probably the Elder he had earlier named) was intended to attempt the seduction of Sarah.[53] He did not wish her to be seduced by elaborate means, but if she gave up as a matter of course[54] it would be a way out of the impasse, a means of finally obliterating the recurrent image of a benign, tender Sarah, and stifling the resurgence of a hope he knew in cold reason to be impossible. His motives were mixed and vacillating, if not confused. He wanted to know for certain whether she allowed other lodgers the liberties she had allowed Tomkins and himself; he wanted revenge; but he also wanted the opposite outcome — to be proved wrong. And yet perhaps he wanted more than anything else simply to dull the ache of separation. Any news of her, however trivial, was a relief to his tortured imagination.

[51] 3 Aug., p. 68.

[52] *Letters*, 291, 322.

[53] The 'journal' of his friend's stay was printed in *LH* (113-27), and then in Bonner, 269-77. The text, very imperfect in both, is corrected in the writer's 'Hazlitt's Journal of 1823: Some Notes and Emendations', *The Library*, 5/26 (1971), 325-36.

[54] His expressions, *Letters*, 265.

When shall I burn her out of my thoughts? — Yet I like to hear about her — that she had her bedgown or her ruff on, that she stood or sat, or made some insipid remark, is to me a visitation from Heaven — know that she is a whore or an ideot is better than nothing. Were I in Hell, my only consolation would be to learn of her. In Heaven to see her would be my only reward.[55]

But above all there was the enigma of her real nature. He had to try once more to determine what it was. His journal of his friend's stay at 9 Southampton Buildings in 1823 extends from Tuesday, 4 March to Sunday, 16th. At first it seems to confirm that she was coquettish, but soon the flirtation began to get out of control, and on the 15th Elder suddenly laid such a boldly crude hand on her that she could not doubt either what he intended or what he thought of her. She was deeply offended, and the next day resolutely avoided him. When that night they met by chance in Lincoln's Inn Fields, and he attempted to force his company on her, she stood stock still, immovable as a statue, until he was shamed into leaving her alone. There was a significant finality about her resistance that befits its position as the last recorded incident in the journal. The attempted seduction ended there: the lodger's grossness the previous day was so shockingly contemptuous of Sarah as a person that she could no longer have anything to say to him. It is also possible that just then her affair with Tomkins had taken a more favourable turn. She seems to have cajoled him out of believing what Hazlitt had told him. She was seen with him opposite the house on the first Saturday night,[56] but there seemed to be some kind of impediment, perhaps connected with his family.[57] He was articled to a solicitor, and they may have considered a 'tradesman's daughter', despite the Roscoe alliance which he was able to cite, as not good enough a match for him.[58] This could not however have been an insuperable bar, and what seems more likely is that he had already committed himself elsewhere, was perhaps even married. An insurmountable difficulty of that order is implied in what followed. At all events before the year 1824 was out Sarah had borne a son to John Tomkins, and although the only evidence that they lived together comes eighteen years later, in June 1841, when they were at 31 Gloucester Street, they must be supposed to have continued together

---

[55] Bonner, 272, and the writer's 'Hazlitt's Journal of 1823' (see n. 53 above), p. 331.

[56] Bonner, 270 (9 Mar.), and 'Hazlitt's Journal of 1823' (see preceding notes) p. 329.

[57] A similar suggestion repeatedly made in *Liber Amoris* may be based on her relations with Tomkins (9. 103, 105, 111, etc.).

[58] He first appears as a solicitor in the 1825 *Law List*, and annually thereafter until 1847, when some professional or financial disaster seems to have supervened. He is not in the 1848 *Law List*, and when he died ten years later his occupation is given as 'lawyer's clerk' (other details on the certificate: d. 5 Feb. 1858, King's College Hospital, St Clement Dane's aged 60; cause of death, paralysis, erysipelas).

until his death at the age of sixty in 1858. But they never married.[59] There is no hint in the 'journal' that Hazlitt suspected she was about to leave Southampton Buildings, and there is no further allusion to them in his writings after the appearance of *Liber Amoris* in May 1823.

His motives in publishing the story to the world in this volume are obscure. Haydon saw them simply as cathartic:

He has written down all the conversations without colour, literal as they happened; he has preserved all the love-letters, many of which are equal to anything of the sort, and really affecting; and I believe, in order to ease his soul of this burden, means, with certain arrangements, to publish it as a tale of character. He will sink into idiotcy if he does not get rid of it.[60]

But Hazlitt knew that Sarah could not fail to recognize the hero and heroine, and perhaps intended that the publication should succeed where the 'experiment' of March had failed, and would burn her out of his thoughts by destroying any hope of renewed contact. Perhaps he meant that it should finally alienate her from him, and that her resentment should make even the thought of reconciliation impossible. His last hope lay in taking this irretrievable step. Perhaps, on the other hand, or even at the same time, he intended the very opposite, and the *Liber Amoris* was a last attempt to touch her heart by this overwhelming evidence of the strength of his passion.

Whatever his intentions, when the book was published on Friday, 9 May it exploded a mine under his feet.[61] He seems to have assumed that he had effectually covered his tracks. It took only eight days for the secret to be detected, and exposed to the whole world in the *Literary Register*, which was published, like the *John Bull*, by Edward Shackell. The reviewer was evidently the same who on 3 May in the *Register* savagely attacked Hazlitt's *Liberal* essays. He made no bones about his political motivation: he was reviewing *Liber Amoris*, he said, to give the public 'every opportunity of seeing what materials go to the composition of these liberal and radical rapscallions, who take upon them the airs of

[59] Micaiah Walker in his 'Particulars of my Family' records that she died unmarried and childless, but his great-grandson, Mr John Fieldwalker, who was unacquainted with *Liber Amoris* and Hazlitt biography, informed me (letter of 13 May 1969) that his own father, Reginald Field Walker, b. 1864, 14 years before Sarah died, 'insisted that she was in fact married to a man named Tomkins and had a son by him'. I have not been able to find any record of a marriage, but other documents establish that a son, Frederick, was born late in 1823 or in 1824. The 1841 census returns show (PRO, HO 107/671, Enumeration Schedule 9, p. 11) John Tomkins, solicitor, aged 45, and Sarah Tomkins, aged 40 (this census gave approximate ages, to the nearest five years), living at 31 Gloucester St. (Queen Sq., half-way between Southampton Buildings and Somers Town). Their son does not appear, and may have been from home on census night. I have not been able to discover where they were living in 1851.

[60] Haydon, *Corr.* ii. 75, letter to Miss Mitford, who, when *Liber Amoris* was published, thought it showed 'fine passion' (Chorley, i. 126).

[61] *Examiner*, 4 May, 304 ('to be published on Friday next'); *MC*, 9 May.

philosophers, poets, and politicians, disseminators of truth, improvers of
taste, reformers of abuses, and ameliators of mankind, as they call
themselves'. Inured as he was to press abuse, Hazlitt might have borne
this exposure if he alone had been its victim, but Sarah was identified (as
in the review of *Table-Talk*) as the daughter of William Hazlitt's
landlord, a tailor; and, what was worse, was branded, without reprieve,
and as undoubted fact, as 'an artful, shameless, trumpery, common
strumpet — the common servant in a common lodging-house — the
creature and the property of every inmate who might chance to become
. . . the object of her lascivious fancy'. Not only did the review give the
book wide notoriety, it also made it hardly necessary to lay down money
to read it. The two numbers of the *Register*, 17 and 24 May, might almost
be said to have pirated the text in eighteen closely printed columns of
'selections' (each with a brief derisive comment at the end).

There was little consolation for Hazlitt in the praise of *The Times* or in
the judicious remarks of the *Globe* a fortnight later, 7 June, where he
read:

It is a strange story and most courageously told. The locality wants something
of dignity, but what has dignity to do with love? . . . The *Liber Amoris* is unique
in the English language; and as, possibly, the first book in its fervour, its
vehemency, and its careless exposure of passion and weakness — of sentiments
and sensations which the common race of mankind seek most studiously to
mystify or conceal — that exhibits a portion of the most distinguishing
characteristics of Rousseau, it ought to be generally praised.

Cheering though it was to know that some (there was also the *Examiner*
critic, who made the same analogy with Rousseau) had seen what he was
aiming at, it was unpleasantly apparent that the jeers about the book's
foundation in real life might start a hue and cry. And so they did. The
murder was out; and worse was to come. A week later, 15 June, the *John
Bull* gave his name, and Patmore's, and mentioned, as well as Sarah and
her father, a 'very respectable married sister'. He awaited with anxiety
the second part of this 'review', promised for 22 June.

When it came, he must have felt the very walls of the room sway about
him as he saw staring up at him from the paper the words of his own
letter to Sarah, written at Renton in the March of the previous year. It
was incredible. But a second glance showed there was no mistake: it was
his own letter, with all its intimate terms of endearment and outpourings
of love and loneliness. An inexplicable profanation. Who could have
betrayed him? Not Sarah surely, she was just as much a victim as he.
Nor could it have been Mrs Walker, or Micaiah, for the same reason.
The letter must have been stolen by one of the lodgers, either for spite or
gain, and sold to Theodore Hook ('Theodore Flash'), editor of *John Bull*.
The malice and ingenuity of Hook were notorious. One of his 'favourite

maxims, constantly occurring in his novels, [was] that there exists some weak point, some secret cancer in every family, . . . the lightest touch on which is torture'. He had access to 'sources of information peculiar and inexplicable'. His income as editor was between £2,000 and £3,000 a year, and he was known to have fraudulently amassed £13,000 during his Secretaryship in Mauritius in 1817, so that he would have no trouble in offering a bribe too big to be refused.[62]

Pondering this act of treachery, Hazlitt seems to have concluded that the real culprit was not the editor but the man behind the *John Bull*, namely John Wilson Croker. Croker had the power, the will, the malevolence. And his intermediary might have been the mysterious figure who got Sarah's letters franked.[63] Croker and Hook were so much hand in glove that when Hook was finally arrested for defalcation Croker went bail for him.[64] The editor, though mischievous, even dangerous, was a kind of mountebank, a merry-andrew, who had no special reason to seek Hazlitt out. There had been only two previous references to Hazlitt in the *John Bull*, both stressing his obscurity and nothing else. And Hazlitt had mentioned Hook only once — as author of an indifferent play he had had to notice. But when Hazlitt had spoken the previous January of the editor of the *John Bull* as an Irishman, it was not Hook he had in mind. Lady Mackintosh observed as soon as the *John Bull* was founded in 1820 that Croker had a hand in it, and when in 1822 he claimed not to have written in the newspapers for two years, the *Examiner* asked 'how much his closest friend the Public Defaulter [had] written for him, and at his desire, during that period; and how often [had] he himself contributed to certain reptile Reviews and Magazines'.[65]

Ever since his Irish days, when he had published some anonymous lampoons on Dublin society, Croker had prided himself on his facile and rapier-like pen, which others however saw more as a blackjack employed by stealth.[66] He had at this very moment in June 1823 cause for uneasiness in those early squibs. He was startled to find that when Blackwood, who was in London, introduced his new contributor Maginn, this wild and disreputable Irishman, a complete stranger, had the effrontery to tell him he had forwarded to Lockhart a bundle of old, unacknowledged, half-forgotten lampoons which he had brought from Ireland, and which were the work of the Secretary of the Admiralty.

[62] R.H.D. Barham, *The Remains of Theodore Hook* (1849), i. 205, 196, 174; *The Times*, 11 Apr. 1822. For many of the details relating to Hook, Croker, and the *John Bull*, see the writer's 'Hazlitt and *John Bull*' (n. 11 above).

[63] *Letters*, 239.

[64] *British Press*, 16 Dec. 1823; *Examiner*, 21 Dec. 1823, 822.

[65] BL Add. MS 52453, letter to Mrs Graham, c.Jan. 1821; *Examiner*, 10 Mar. 1822, 156.

[66] Lady Morgan, *Memoirs* (1862), i. 337.

Croker hastily wrote to warn Lockhart against imposition, but admitted that when he was twenty-three he had published anonymously some satires on Dublin notables (in exactly the same way, although he did not say so, as Lockhart in Edinburgh in recent years, and with as little compunction).[67]

Croker's *Quarterly* notice of Keats's *Endymion* is notorious, but it was merely an early example of a venom which showed no abatement in thirty years of reviewing, so that we find Carlyle in 1851 reading in the *Quarterly* some 'very beggarly Crokerism, all of copperas and gall and human *baseness* . . . No viler mortal calls himself man than old Croker at this time.'[68] Lady Charleville was more urbane: 'He deserves all the reprobation of candid, honourable men.'[69] But others felt compelled to resort to hyperbole: 'He had the dagger and the poison ever ready for friend or foe', and, 'He was an adventurer whose path from obscurity to greatness was paved with dead men's skulls.' The daughter of an old friend of his declared that '[his] chief pleasure in life [was] to cause mental suffering to his fellows', and Macaulay said that he 'would go a hundred miles through sleet and snow, on the top of a coach, in a December night, to search a parish register for the sake of showing that a man is illegitimate, or a woman older than she says she is'.[70]

He had for years been a target of the satire of the *Morning Chronicle* and the *Examiner*, and we remember Hazlitt's jibe at 'the poetical pen of the Admiralty Croker' as far back as April 1813. There are many incidental, disobliging references to him in Hazlitt's essays and articles after 1819, but one of them is particularly significant, and may well have pointed out to him the way to stop Hazlitt's mouth, because it disclosed a state of strained relations between Hazlitt and his landlord which argued a certain vulnerability. In 'The Disadvantages of Intellectual Superiority', published in the second *Table-Talk* in 1822, Hazlitt remarked that the attacks on him in a 'government engine' (evidently the *Quarterly*) lost him the consideration of people of whose goodwill he had need, because they were unaware of the role of political cabals in journalism and accepted the imputations at their face value.[71] Among such, he said, were his landlord and his landlord's daughter, and here he is evidently recalling the effect on Micaiah Walker and Sarah of the contemptuous review of the first *Table-Talk* in the *Quarterly* in late December 1821 which had been shown them by a fellow-lodger (the precedent was sinister). It

---

[67] NLS MS 927, no. 8.

[68] *New Letters of Thomas Carlyle*, ed. A. Carlyle (1904), ii. 114.

[69] Lady Morgan (n. 66 above), ii. 72.

[70] S. C. Hall, *Retrospect of a Long Life* (1883), i. 211 (quoting R. R. Madden); i. 324; L. J. Jennings, *The Croker Papers* (1884), i. 259; H. Martineau, *Biographical Sketches* (1869), 62.

[71] 8. 284.

was useless, he adds, for him to explain that it was not so much a review as a reprisal for political writings of his own, to which 'Mr Croker [had] peremptory instructions to retaliate'.

On the Sunday in June when his letter was published in the *John Bull* Hazlitt attended the Caledonian Chapel in Hatton Gardens to hear Edward Irving preach. In his account of the sermon the following month he names the notabilities who had flocked to listen to Irving, and for no clear reason adds that 'the "Talking Potato" and Mr Theodore Flash have not yet been'.[72] An inexplicable aside which is the pretext for a long, irrelevant note on the antecedents and opportunist career of Croker, ending with his election to Parliament and the observation that 'the case of the Duke of York and Mrs Clarke soon after came on, and Mr Croker, who is a dabbler in dirt, and an adept at love-letters, rose from the affair Secretary to the Admiralty'. The innuendo can hardly have been aimed solely at this fourteen-year-old scandal. What reason was there to refer to it, or even to mention Croker at all, in a paper on pulpit oratory? It is a little odd.

Blackwood's also at about this time decided to renew its attacks on Hazlitt, and roused him to threaten a legal action against Cadell, Blackwood's current London agent, Murray having long since withdrawn in disgust. Blackwood consulted Maginn, who replied (13 May); 'If [Cadell] be bullied by that vagabond Hazlitt, would it be impossible for you to heal the old wounds between you and Murray? Believe me, it would be worth trying, and Croker is a fine channel.'[73] Here, no doubt, in the summer or autumn of 1823, we may place the following undated letter to an unidentified editor, probably Henry Leigh Hunt (who had succeeded his uncle on his departure for Italy in November 1821, and who published in the *Literary Examiner*, in the last four months of the year, Hazlitt's 'Common Places').

Dear Sir,

Will you fly a kite on Saturday of a quarter or half a sheet to be named — *Mr Blackwood in London* or the *New Alley Croker*? I have an article ready cut & dry for that purpose, & will follow it up weekly, if you are disposed to get the thing done, which I understand from Mr P. you are. You couldn't step over, or I will come to you, if you like.

Yours truly,
W. Hazlitt.[74]

[72] 20. 113 and n.

[73] Howe, 325. Maginn then offers to call on Cadell, and the unpub. part of the letter continues (NLS MS 4011, fo. 29); 'I'd tell him, after swearing him not to disclose a word of it, that Galt was the man principally engaged, and then Hogg — and that W. and L. were the most innocent people in the world. Tell me what you think of this idea.'

[74] Unpub., in the writer's collection. The watermark is 1822. The expression 'alley croaker' appears in Pope, ii. 70.

Whether or not Croker was responsible for the publication of the *John Bull* letter, it seems fairly clear that Hazlitt thought he was. It may be that his suspicions were later confirmed. There seems to have been no other reason, in August 1824, for his contributing a savage portrait of Croker to the *Examiner*, in which each of his isolated strictures on his model over the previous decade is revived and adds its deepening brush-stroke to the picture, and in which the treacherous and malodorous blazoning abroad of this private letter is hinted at. His sitter, he says, in this indictment of the anonymous *Quarterly* reviewer and contribution to the *John Bull*, is a man 'who conceals his own writings and publishes those of other people, which he procures from his relations at a lodging-house . . . '.[75] However obscure the allusion may have been to the readers of the *Examiner*, it now seems to be plain enough, when taken in conjunction with the earlier allusion to dirt and love-letters.

[75] 'A Half-length', *Examiner*, 1 Aug. 1824, 484, will not be found in the *Complete Works of William Hazlitt*. It is given in full in the writer's 'Three Additions to the Canon of Hazlitt's Political Writings', *RES* NS 38 (1987), 361–2.

15

Second Marriage

What shadows we are, and what
shadows we pursue!

Burke

THE nauseating blow dealt by the *John Bull* letter had put an end, for the
time being, to his openness about his affairs. He could not bear to
mention it, even to people to whom he had already told the story, like
Tom Hood, then aged twenty-four and sub-editor of the *London
Magazine*, in which he had since November 1822 been describing, one
after another, the great art collections of the country. The latest of these
articles, on Lord Grosvenor's Gallery, was due to appear in the July
number, and he had promised next to go down to Petworth to describe
Lord Egremont's. He failed to do so, and asked Hood to make his
excuses to John Taylor, saying no more than that 'something happened
which hurt my mind and . . . I had only the heart to come down here
[Winterslow], and see my little boy, who is gone from hence'. He
proposed however to devote his next article to Blenheim, which he could
conveniently visit from Winterslow. Equally accessible was another
gallery he had agreed to write up, but not in the *London*, and that was at
Fonthill. Shocked and depressed as he was, in a dreadful state of health
and spirits, he still had to earn his living and support his son. He had
been arrested for debt in February, only briefly, but it was a humiliation
he never forgot, and one which was brought home to him afresh by the
fate of Haydon, who was in too deep to escape, who was sold up on
11 June, and who now found himself in King's Bench Prison.

He made a point of attending his publishers' monthly dinner on the
29th,[1] where he met and liked George Darley, and he no doubt stayed on
in town until 5 July to see the publication of his *Characteristics*, a little
volume filled with the distilled essence of his comfortless reflections of the
past few years.[2] But he felt the need of quiet and affection, and turned, as
in the year before, to 'the only comfort he had left'. Young William, now
nearly twelve, who was spending the summer holidays with his mother
at Winterslow, was to come up about 15 July for the start of the school

---

[1] *Keats Circle*, ii.444.
[2] *Examiner* advt., 29 June 1823.

year, but the few evening hours Hazlitt could expect in his company in town were not enough. He wanted the companionship of long summer days in the meadows and along the lanes, so he stole a march on him and went down. The country and his son's company restored him, and it may even be that his unconventional ex-wife, who still had a great regard for him and who had predicted that his enemies would seize upon his affair with Sarah, was able to help. When the boy had left for London, Hazlitt also left Winterslow and went down to Exeter to see his mother and sister, and his brother John who had recently moved there. After their disapproval of the divorce he wished to win them over, and the wording of a letter from the grandmother to young William the next year implies success.[3]

We have seen his palpable devotion to his father, his affection for his brother, his visits to Addlestone, Bath, and Exeter. There are hints of even wider family feeling in his early affection for his Loftus cousins, and in his tender account in 'The Pictures at Burleigh House' of his pilgrimage to his mother's birthplace. The spring of 1822 brought an unexpected sign of life from his father's side of the family, lost to sight these many years, in a letter from Tipperary. The writer, this rediscovered cousin, Kilner Hazlitt, had somehow learnt that this William Hazlitt, whose fame had reached Ireland, was connected with the *Examiner*, so when he visited London in 1822 he called at Catherine Street. John Hunt greeted him warily and told him little, suspecting a dun or a spy. Kilner then wrote, but Hazlitt, submerged in the difficulties of the divorce, did not answer. Now in June 1823 the prospect of seeing his family in the West Country put him in mind of his correspondent. The warmth with which he wrote to Tipperary on 14 June shows how deeply, revering his dead father as he did, he was touched at the thought of restoring the connection with the old minister's Irish relations.

Dear Sir,

I have felt myself much gratified by your kind inquiries & nothing but a dreadful state of health & spirits can excuse my not having answered your letters before. I have delayed it so long & apparently so inexcusably, that I am ashamed almost of setting about it now. The person of the name of Hazlitt who wrote to your relation from Wem in Shropshire about 26 years ago was my father & was born in Tipperary. He had a brother named James, a clergyman of the Church of England in Dublin, & another brother John who died in America. There can therefore be no doubt of the relationship & indeed I never knew anyone else of the same name. My father died about 2 years ago near Exeter at the age of 82. My mother is still living in the West of England & I shall be down that way this summer. If you like to make a trip of the sort, I would be

[3] *Memoirs*, ii.107–8.

very glad to give you the meeting either at Bristol or on the coast of Wales about a month hence. As to my books I will send you any you want to see if you will please to give directions to some bookseller in Dublin. I was in Scotland at the time you were in London, or I should have seen you then. I shall be happy at any time to hear from you & hope I shall prove in future a more punctual correspondent. If there was anything extraordinary in Mr John Hunt's behaviour to you it must have been occasioned by your anxiety to see so very insignificant a person as I am.

> Believe [me] to be, Dear Sir, very gratefully &
> [affec]tionately, yours
> W. Hazlitt

Please direct to me at the Examiner Office, Strand, London.[4]

Kilner Hazlitt, this new and unknown correspondent to whom in his first letter Hazlitt signed himself 'affectionately yours', was the son of James Hazlitt, trader and tanner of Fethard, Tipperary, three miles from Shronell, the birthplace of the Revd William Hazlitt, whose cousin he appears to have been.[5] James was a successful merchant, and the money he left evidently gave his son leisure and the means to travel. Whether Hazlitt meant to meet Kilner Hazlitt at the proposed rendez-vous before or after visiting his mother is not clear, but in the event they met at Fonthill. No sooner had Kilner received Hazlitt's letter at Fethard than he wrote again to the *Examiner* office, and on hearing that Hazlitt was now about to go down to Fonthill proved his eagerness by offering to meet him there rather than at Bristol or South Wales, both of which were more convenient for him. At Fonthill he produced the letter from the Revd Wm. Hazlitt which he had mentioned. Hazlitt, we are told, showed no surprise at seeing this carefully preserved paper dating from 1792, which had been sent to inform James of the minister's plans for his son's education at Hackney. This was evidently no ordinary letter of family news. Now, thirty years on, it was 'at once recognized' by Hazlitt, who had 'a clear recollection of this correspondence', so that he, who was only fifteen when it was written, must have read it before it was

[4] Unpub., in the collection of Mr Michael Foot, MP, addressed to Kilner Hazlitt, Esq., Fethard, Tipperary, Ireland. A seal-tear has suppressed 'me', and part of 'affectionately', and also the year in Hazlitt's dating, but the last is supplied by the postmark. He began to write his own address as 4 Chapel Street West, and then crossed it out.

[5] James's occupation is thus described by his son Samuel, Kilner's brother, in a letter to W. C. Hazlitt, 9 Aug. 1866 (BL Add. MS 38899, fo. 79), where there is no suggestion that he was a university graduate or had been a clergyman. Nor on the other hand does Hazlitt in his letter suggest that his uncle James, his father's brother, was ever a tanner, still less that he was Kilner's father. There is evidently confusion between two James Hazlitts (see Moyne, 33 and 131, nn. 10–13). The renewed contact was maintained. William Hazlitt the younger, at some date, probably after his father's death, even visited the Fethard Hazlitts. A further letter by Samuel Hazlitt to Hazlitt's grandson is worth quoting here: 'I should have a bad memory if I did not recollect your father's visit . . . I have felt that he was tired of us else he would have sent a line to say he got home safe and sound' — perhaps like all the Hazlitts he is an oddity.' (BL Add. MS 38899, fo. 71.)

dispatched.[6] Hazlitt's biographers have wondered how his expensive school fees of £60 a year — twice his father's income — were met. A plausible conjecture would be, by an appeal to a more fortunately placed relation, and W. C. Hazlitt tells us that James died worth £17,000.[7]

Hazlitt had already been to Fonthill the previous year and described it, with no great enthusiasm, in the *London Magazine*. This collection, inaccessible to the public for years, and then thrown open only because Beckford was forced to sell, had been a great disappointment to him. The auction then arranged did not take place: the Abbey was bought by a speculator who engaged a new auctioneer, the enterprising but dubious Phillips, to arrange his own sale. As time went on, rumours of strange activities had reached town, and by this summer of 1823, when the new auction was announced, it was being said that although it might be a Fonthill sale it was no longer a Beckford, since Phillips had crammed the rooms with the scourings of London antique shops and bookstalls. But he had a thick skin, and connoisseurs apparently assumed that a sufficient number of Beckford items would remain, and the sale proceeded. However, to ensure success, Phillips invited the foremost art critic of the day to give it the stamp of his approval, at a fee, according to Haydon, of fifty guineas.[8] A ticklish commission. Hazlitt had dismissed the collection the previous year for reasons grounded in the opinions and predilections he had held all his life, because (as he later said in *Sketches of the Principal Picture Galleries in England*) it 'exhibit[ed] no picture of remarkable eminence that [could] be ranked as an heirloom of the imagination', and held no example of 'high art', merely of the 'mechanical'.[9] His present way of dealing with it was to keep his tongue in his cheek. His account is patently ironical, so much so that Phillips must have been remarkably impercipient not to have seen merely from the title that his heart was not in it. The article he published, in a series of 'paragraphs' in the *Morning Chronicle* on 20, 22, 25 August and 1 September, was headed, 'Notices of Curious and Highly Finished Cabinet Pictures at Fonthill Abbey'.[10] He did his best to give Phillips his money's worth by wrapping up his reservations in ambiguities, but at one point his mask falls and he addresses himself briefly but explicitly to the essential question of what constitutes the 'strange difference between

---

[6] BL Add. MS 38899, fo. 79, Samuel Hazlitt to W. C. Hazlitt, 9 Aug. 1866.

[7] *The Hazlitts*, i. 10.

[8] Haydon, *Corr.* ii. 79 (the letter, sold at Sotheby's, Apr. 1973, is dated 26 Sept. 1823).

[9] 10. 58, 59.

[10] Repr. in the writer's 'The Fonthill Abbey Pictures: Two Additions to the Hazlitt Canon', *Journal of the Warburg and Courtauld Institutes*, 41 (1978), 278–96. The undated letter to John Black listed in the writer's 'Some New Hazlitt Letters', *NQ*, July-Aug. 1977, 339–40, probably refers to these 'paragraphs' (see also 'The Fonthill Abbey Pictures' p. 296, where Hazlitt mentions 'Mr Phillips' as the bearer of the letter).

the spirit that breathes from Dutch and from Italian art', and finally, when his work was done and his article safely in print, he transmuted his growing exasperation into a dramatic skit on triviality of taste, centred on the Metsu which was supposed to be the gem of the collection. It was called 'The Science of a Connoisseur' and appeared in the *Morning Chronicle* on 30 September.[11]

Happily, the country air, as always, did him good. Patmore, who happened to be at Fonthill on a like errand, remarks upon the extraordinary physical and moral effect on him of the sight and feel of the country: 'his step, firm, vigorous and rapid — his look eager and onward, as if devouring the way before it — and his whole air buoyant and triumphant.'[12] They were together eight or ten days, and made long excursions on foot to other collections in the region, notably at Stourhead and Wilton. These Hazlitt knew of old, and he enthusiastically and generously shared his knowledge with his friend and (at that moment) rival. Patmore dwells on the evident pleasure he took in showing him these galleries but says nothing of the pictures at Fonthill, nor of Hazlitt's opinion of them; but certain echoes of Hazlitt's 'Notices' in Patmore's subsequent article on Fonthill show that they must have seen and discussed them together, and that he adopted Hazlitt's deprecation of high finishing.[13]

The help he gave Patmore he extended also at this time to another of his young friends. At Fonthill he received an appeal from Bewick, who was involved in Haydon's difficulties and in consequence obliged to withdraw to Scotland. Hazlitt's reply of 25 August, just before he left for Winterslow, suggests that Bewick was looking for letters of introduction to prospective sitters.

My dear Sir,
     This is all I could do — When I learn how I stand with Jeffrey by the next Edinburgh Review, I will see if I can do anything in that quarter: but I can't till then, without exposing both you & myself to a rebuff. I have a favour to ask of you. Is the British Institution open yet for copiers? Will you learn in that case if little Wroughton is there & send me word? Or inform me when it is to open? But be cautious in your enquiries. Why don't you marry some girl with a small independence & let the Fine Arts go to Hell — the best place for them? Is Haydon getting out soon, or what is he about?
                                        I am, my dear Bewick,
                                              ever yours truly, W. Hazlitt[14]

---

[11] Also repr. in the first article named in the preceding note.
[12] Patmore, iii. 366.
[13] 'A Day at Fonthill Abbey', *NMM* 17 (1823), 368–80.
[14] Printed in 'Some New Hazlitt Letters' (n. 10 above), 338. A Miss Wroughton exhibited at the British Institution, 1826–9; see A. Graves, *The British Institution 1806–1867* (1908). 'But be cautious in your enquiries' is an interlinear afterthought.

Bewick's subsequent portraits of Combe, M'Culloch, and Jeffrey indicate that the introduction Hazlitt sent him was to Ritchie, and that the promise to approach his editor was later fulfilled.[15] The unexpectedly cynical 'let the Fine Arts go to Hell' is a reaction to the bad canvases he had just endured, and the other piece of advice is a pointer to Hazlitt's own present state of mind. He was still smarting at Sarah's rejection, and wanted to prove to himself that he was not entirely despicable. He was casting about for a woman of property as a way of mending his finances: according to Kenney, he was, at this time, 'the humble servant of all marriageable young ladies', and 'little Wroughton' may have been one of these.

But he was still not quite free of Sarah. Sometime in the week beginning 15 September he went up to town for a few days, and spent the nights watching her door. What the object of his journey was we do not know, but Haydon, to whom we owe the information, adds to it that he also went down to Hampton to ask 'another flame' to marry him.[16] One of his aims in seeking a woman with money was to be able to make the pilgrimage to Italy which he had long had in mind, and particularly since 1821. His expedition to Hampton does not appear to have answered, and he returned to Winterslow to continue with his series of Picture-Galleries for the *London*. The November number carried the paper on Oxford and Blenheim, and since it contained the exasperated aside (omitted in the volume — he evidently thought better of it) that 'There is not (thank God!) a single Dutch picture in the whole collection ', we may take it that he visited Woodstock after he left Fonthill, and probably from Winterslow (as he had done with Lamb years before).

At this time also he was cheered by Lamb's celebrated letter of reproof to Southey, published in the *London* in October, the warmest and most forthright tribute ever paid to Hazlitt by a contemporary. He had complained in 1821 of Lamb's unaccountable reluctance to give him a helping hand. He could not complain now. But the full significance of the oft-quoted commendation has not been spelled out. The legion of Lamb's friends, and their worth, and the certainty that Lamb knew they would all read these lines, have to be taken into account before we can justly gauge the import of the parenthesis in his declaration, 'I stood well with him for fifteen years (the proudest of my life)'. Not the Salutation and Cat evenings with Coleridge, not Wordsworth's visits, not the long frequentation of the celebrated author of *Political Justice*, nor the years of warm friendship with the faithful and zealous Robinson. And as if this was not clear enough, he goes on to say, 'I think I shall go to my grave without finding, or expecting to find, such another companion'. Small

---

[15] Bewick, i. 156.
[16] Note 8 above.

wonder that Robinson was taken aback. He was also significantly solicitous: Lamb, he said, must have been 'aware that he would expose himself to obloquy by such [a] declaration'.[17]

Hazlitt wrote with pleasure to his mother and sister of Lamb's letter, and Sarah, who happened to be with them, mentioned this to the Lambs. He resolved to heal the breach between himself and Lamb when he got back to town, but in the mean time the visits to the galleries opened another train of thought, turning his attention to another old friend from whom he had rather fallen away, the old painter Northcote. He now realized that Northcote's conversation differed from Lamb's in one remarkable respect: it was entirely concerned with people; but however much it seemed to resemble idle gossip, it was in fact an endless, fascinated rumination on human character. He remarked that in his sessions with Northcote 'there was none of the cant of candour' (a thrust at Godwin), but 'our mutual acquaintance were considered merely as subjects of conversation and knowledge, not at all of affection'. His determined pursuit of these 'demonstration[s] . . . as beautiful as [they were] new' is reflected, with distortion, in Haydon's remarking at this time, '[Hazlitt] is [a] man who exists only on the pleasure of dissecting the weaknesses of others', and adding, 'Yet he reconciles all by his candour, which spares neither himself or others'.[18] It may have been now that Hazlitt decided to record Northcote's conversations in the same way as he had recently recaptured two quite different conversations of former years at Lamb's.

These remarks of Haydon's, on 3 December 1823, suggest that Hazlitt visited him just at this time, and told him of his intention to write a series of 'characters' of men who were typical of the age, a genre in which his experiments went as far back as 1806, to some pages on Pitt in *Free Thoughts on Public Affairs*. Haydon's recurring resentment of Hazlitt's independent views on painting, which he ascribed to jealousy, persuaded him that the 'characters' he outlined to him were malevolently inspired, and he in his turn tried his hand that same day on a 'character' of Hazlitt filled with gall. Two days later he spoke of the contemporary figures who visited his studio as 'the Spirits of the Time' The first of Hazlitt's series of 'Spirits of the Age' appeared in the *New Monthly* in January.

In the interval, during his absence from town, *Blackwood's* had not been silent, although, for once, it was slower than its London allies. Lockhart's review of *Liber Amoris* in the June number hinted no further at the identity of the author than that he was a cockney, a liberal, and a

[17] Morley, 298. The letter was quoted at length in *MC*, 4 Oct. 1823, and in *Examiner*, 5 Oct., 643–4; 12 Oct., 660–1; 19 Oct. 1823, 675.

[18] 12. 133; Pope, ii. 440.

friend of the enemy of all Tories, Francis Jeffrey. The July number struck nearer home, but still no more than a glancing blow at 'Hazlitt, the gallant of Southampton Street, Holborn', but thenceforward the references to 'Mr H. of the *Liber Amoris*, who, for his drivelling, was despised, even by the daughter of a tailor', and to 'Southampton Row', and the Southampton Arms, are constant and are insinuated into the most unexpected pages of the journal.[19] When the September number appeared he felt that he was no more able to stem the abuse and protect his name than he had been five years earlier at the time of 'Hazlitt Cross-questioned'. An unsuccessful parody of his sketches of the picture-galleries could be shrugged off, but an ominous 'Letter of Timothy Tickler' showed him that Mr Blackwood's spies, more persistent than ever, were close on his heels: 'A friend of mine wrote me t'other day, that he had seen "Billy Hazlitt and Count Tims at Fonthill, busy writing puffs for Harry Phillips of Bond Street" '.[20] In the next month their campaign reached a daring climax of obscenity in a 'Noctes' where Wilson judges that 'Hazlitt is the most loathsome, Hunt the most ludicrous. Pygmalion is so brutified and besotted now, that he walks out into the public street, enters a bookseller's shop, mounts a stool, and represents Priapus in Ludgate Hill.' But, hardened as he had become to the nastiness of this journal, perhaps even this shocked him less than the naming of his unoffending rival, the 'gentlemanly' (as he admitted) Tomkins, identified (again, by Wilson) in capital letters in a review of Hogg's *Three Perils of Woman*, where Hazlitt is then ridiculed in his own name as 'the unsuccessful rival of Mr Tomkins for the favours of a tailor's daughter'.[21] However wretched the couple had made him, their public exposure in this abominable rag must have given him a pang.

Daunted by such extremities of filth, he concluded that his first impulse in 1818 had been right after all — to let the abuse take its course. For five years he had been assailed almost without intermission, hounded almost to death, by the Tory press. They had, as he said, almost put him underground, but it now seemed likely that before long he would be able to turn his back on them and enjoy a peaceful sojourn in the chosen land of high art.

How much the woman who was soon to become his second wife knew or understood of all this, it is difficult to guess. She was attracted to him because he was a well-known writer, and it is improbable that she remained ignorant of what was by now well known about him. She is bound to have been a reader of some, if not all, of the literary journals

[19] *Blackwood's*, 14 (1823), 82, 183, 219–35, 241, 313, 338, 472, 502, 514.

[20] Ibid. 342–3, 'Bits. By the Director-General'; 313 (Patmore, who used the pen-name of Victoire, Vicomte de Soligny, appears as 'Tims' in Hazlitt's 'The Fight').

[21] Ibid. 488, 428.

and reviews. And as Hazlitt ruefully remarked at the time, his character 'smacked'.[22] She at least knew he was divorced, but as a widow and a Scotswoman, she was less likely to be put off by this than if she were a spinster and English. She must have been very emancipated, or very independent, or both. It seems likely that they met this summer or autumn; if so, it was six months or more before she agreed to marry him.

Some account of her origins and family is required at this point.[23] She was the daughter of James Shaw of Inverness, great-grandson, on the distaff side, of a Sir Robert Gordon, and his wife Jean, daughter of Alexander Mackenzie, merchant and banker of Inverness, who claimed descent from the Earls of Athole and the Barons of Kintail.[24] This, no doubt, explains Huntly Gordon's statement that Isabella was 'of a very good family'.[25] The Shaws' first child Margaret was born in 1780, and their first son, Alexander Mackenzie, in 1781.[26] Between him and the second son, William, baptized in 1790, there were six daughters, including Mary (1785), Mitchell (1786), and Agnes (1788). The baptismal entry of the ninth child, Isabella, reads, '24th August 1791, James Shaw, Esq., of Muirtown and his spouse Jean Mackenzie had a child baptized by Mr Downie called (Isoble). Capt. Alpin Grant and Dr Robertson witnesses.' Thereafter there were four more daughters, of whom we need name only Barbara (1794). There is little indication of the kind of man James Shaw was, except that he seems to have been fond of the company of the military, and that, like his father before him, he was a freemason.[27] But even in his military friendships there is no particular significance: officers in Scotland were ten a penny in the century after the '45, when those sons of the middle class who did not go abroad as planters (notably to the West Indies) went into the army, and the Highlands became one vast recruiting ground. James's two sons were officers and so were at least two of his sons-in-law. But it is certain that he had the land-hunger. He embarked on his career as landowner with an early major purchase in 1782 of the lands of Wester Fairburn, or Muirtown of Fairburn, in the Black Isle, the origin of the style 'James

[22] See n. 81 below.

[23] I described how I succeeded in identifying Isabella in the W. D. Thomas Memorial Lecture, delivered at University College, Swansea, 17 Nov. 1981 (pub. as *The Second Mrs Hazlitt: A Problem in Literary Biography* (Swansea, 1982), which contains much of the information presented here).

[24] Alexander Mackenzie, *A History of the Mackenzies*, 2nd rev. edn. (Inverness, 1894), 582–4.

[25] Howe, 353.

[26] Inverness Parish Register. The births and/or baptisms of all the Shaw children except Margaret are recorded. Her year of birth may be deduced from her age, 85, on her death certificate, 16 Nov. 1865 (RHE). Isabella's date of birth is not given, but since all the older children, without exception, were baptized four to nine days after birth, we may assume that it was about the middle of Aug. 1791. The names of some of the children, including hers, are crudely spelled, and placed, unaccountably, within brackets.

[27] Old Inverness Kilwinning St. John's Lodge of Freemasons, Minute Books.

Shaw of Muirtown' usually found in papers relating to him. He later acquired the vast estate of Vaternish in the Isle of Skye, lands in Ross-shire and Inverness-shire, a house in Edinburgh, and shares worth at least £5,500 in businesses in Inverness and Aberdeen.[28]

Disaster struck the family when Isabella was nine. In January 1801 the father suddenly died, in his mid-forties, and this was the beginning of the break-up of the large Shaw household. The abrupt inheritance of the estate went to the head of the young heir, Alexander Mackenzie, and at his majority in 1802 he sold the greater part of Vaternish for £16,000.[29] It is true that he had the consent of his trustees, his brother-in-law Lieutenant-Colonel William McCaskill of the 92nd Foot, who had married his sister Margaret two years before, and his uncle Captain Alpin Grant, and that he used some of the money to pay debts of his father's,[30] but the excess of the amount raised over the amounts to be repaid indicates a rash prodigality.[31] The sale of this property seems to have galvanized the widowed mother into insisting upon the recognition by Alexander Mackenzie of a Bond of Provision drafted by her husband just before his death guaranteeing each of the daughters £400 on marriage as well as an annual sum for their education, and also for the education of the second son.[32] The careful, business-like habits of the father appear in the precise terms of the draft, more detailed than here given, and if his daughter Isabella inherited his temperament they go a long way to explaining her later alienation from the unbusiness-like, or rather, not to use too grotesque a euphemism, the improvident Hazlitt.

Despite his signing of the bond, Alexander Mackenzie plunged into a wild and extravagant life and squandered his patrimony like so many officers of the time in taverns, on clothes, on horses, and at the billiard-table. He became bankrupt, and his estate was wound up in July 1806. Among his creditors in Inverness, Edinburgh, and London, were a horse-coper, a spirit dealer, a jeweller, a glover, two saddlers, two bootmakers, and no fewer than six tailors.[33] By this time he was a cornet in the 7th Dragoon Guards, transferred from the 17th Light Dragoons, and stationed in Ireland.[34] Isabella was nine when her brother came into his inheritance, and fourteen when he became bankrupt. It may be that

---

[28] The preceding transactions are recorded in RHE: Ross-shire Sasines, 85 (at 29 Aug. 1782); Inverness-shire Sasines, 607 (at 28 May 1798); GR 406, fo. 100; PR 16, fo. 373; Register of Deeds, Dur., vol. 317, fos. 400–25; GR 911, fo. 7; and Com. Inverness Testaments, vol. vi, fo. 507.

[29] RHE: Register of Deeds, Dur., vol. 297, fos.158–67; GR 1004, fo. 6l.

[30] Notably a mortgage of £5,000 raised in 1795 (RHE: GR 631, fo. 61; GR 673, fo. 159; and Register of Deeds, Dal., vol. 295, fo. 729).

[31] The version of the inheritance given in *A History of the Mackenzies* (n. 24 above, p. 582) does not accord with the documents.

[32] RHE: Register of Deeds, Dur., vol. 319, fos. 1180–96.   [33] Ibid., vol. 317, fos. 400–25.

[34] *London Gazette*, 5 Jan., 30 July 1805; PRO, WO 31/173, 187.

years later her memories of the whispered scandal she overheard in her childhood about her spendthrift brother, or even family quarrels at which she had been an unnoticed wide-eyed witness, darkened her attitude to the careless habits with money she discovered in her second husband, and confirmed her in a decision that put her income beyond his reach, and ultimately even helped to part her from him.

In 1805 the widowed Mrs Shaw took refuge in the arms of a new husband, one William Murison, an Edinburgh brewer.[35] Six of her children were still under fifteen, and some were still with her in 1810 when she and her husband moved to Clockmiln in Berwickshire, but others appear never to have left the North of Scotland. These, and they included lsabella, were under the guardianship of the Revd John McDonell, of Forres.[36] He was almost certainly a boyhood friend of James Shaw's from the days when James's father ran a textile factory at lnvermoriston, and had been minister of Forres since 1792. It is curious to speculate on his influence on his ward, who was much attached to him. He seems to have been an energetic man with strong practical sense, jealous of his rights, rather aggressive, litigious even. A freemason and a JP, he was also a captain in the Forres Volunteer Rifles, and a steward of the patriotic Trafalgar Club founded in 1807.[37] Documents are lacking to establish with certainty that Isabella did live in Forres, but her devotion to her guardian argues at least some period of residence, unmistakably implied by the long journey we shall see her undertake in a few years' time to be married by him in the church at Forres.

The next clear date is in papers relating to the sale of part of her father's estate arranged by McDonell and Alexander Mackenzie's trustee to pay the widow and children the money they were entitled to. Isabella's share amounted to £700, a sum promptly re-invested for her by McDonell. She was declared to be residing at this time, 7 February 1811, in London.[38] She was only nineteen, and although a very self-reliant person, would hardly have gone to live there on her own; it is likely that she was staying with her sister Mary, twenty-five at the time, and who, two years before, had married the London merchant John Downie.[39] Downie must have been adventurous: it looks as if he did not

[35] RHE: Edinburgh Parish Registers, St Cuthbert's, 4 Mar. 1805.

[36] RHE: Minute Book, Court of Session, 1807, petition of 21 May. The Scottish term is 'Tutorship' or 'Curatorship'. As frequently happens, the name is variously spelt: this appears to be the correct form.

[37] R. Douglas, *Annals of the Royal Burgh of Forres* (Elgin, 1934), 108, 113, 315.

[38] RHE: Minute Book, Court of Session, 1807, pp. 230–1; I. McNeill M15/6 (21 May 1807); General Register of Sasines, vols. 906, fo. 181; 834, fos.167, 174, 181; 961, fo. 116; 889, fos. 7–14, respectively (her residence in London is mentioned in the last).

[39] RHE: Forres Parish Register, 17 Mar. 1809; *Edinburgh Advertiser*, 7 Apr. 1809. Downie was probably the son of the minister who baptized Isabella.

remain long in London but betook himself to those latitudes that in those
days built up a merchant's fortune as rapidly as they broke down his
health.

The career of Isabella's brother had in the mean time suffered an
interruption. He had sold his commission in 1807, evidently because he
found his position in the expensive Dragoon Guards untenable after his
bankruptcy.[40] He lowered his sights, and the following year was
appointed ensign without purchase in a regiment of the line, the 92nd
Foot (Gordon Highlanders).[41] In April 1808 they embarked with the
expeditionary force of Sir John Moore.[42] After Gothenburg, they were
already in poor case when they landed in Portugal, but the subsequent
miseries of the abortive autumn campaign, when they toiled through the
bitter cold and driving rain over hundreds of miles of desolate mountain
roads, often little better than goat-tracks, to the Escorial, and of the
chaotic winter retreat that seemed more like a shameful flight, beginning
as it did in drunkenness and indiscipline and ending in nakedness and
starvation, nearly ruined them. When, five months later, they came out
of the mountains again to the sea at Corunna they were a wild and
scarecrow army that terrified the inhabitants of the little town. They
were not engaged in the famous battle of 16 January 1809, but when they
disembarked at Portsmouth on 26 January, they were not much better
off than the rest of that ragged, verminous, and typhus-ridden army.
The following July the 92nd again embarked at Deal, amid scenes of
popular enthusiasm, for the Walcheren expedition, from which they
returned with emaciated figures shaken by the ague, having left behind
four thousand of their fellows, only to encounter the crowning misery of
a cold reception. After these trials, however, Shaw had some respite.
Finally promoted lieutenant at Bree-sand,[43] he was put on garrison duty
in Ireland until October 1811, when his battalion was transferred to
Banff in north-east Scotland, where they remained until June 1812.[44]
George Huntly Gordon, whose testimony comes into play at this point,
was the nephew of Abercromby Gordon, minister of Banff, and it was
there in the early summer of 1812 that he met Isabella. 'One of my
earliest recollections', he said fifty-four years later, 'when I was just at

[40] *London Gazette*, 8 Aug. 1807.

[41] PRO, WO 31/246, Commander-in-Chief's Memoranda; and *IND* 5499, annotated *Army List*
for 1809, correcting the published *Army List* where Alexander Mackenzie Shaw is shown as Andrew
Mackenzie Shaw. This mistake occurs over many years, and in all cases I have checked the pub.
Army Lists against the annotated *Army Lists* and the original WO documents.

[42] R. Cannon, *Historical Record of the 92nd Regiment* (1851), *passim*, and F. H. Neish, *Historical
Diary of the Gordon Highlanders* (Dundee, 1914), *passim*.

[43] PRO, WO 31/280, Commander-in Chief's Memoranda.

[44] PRO, WO 12/9329, muster rolls of the 92nd Foot, and WO 17/213, monthly returns, 2nd
battalion, 92nd Foot.

the age when one feels the full force of female loveliness, was a day passed in her charming presence, at an uncle's of mine in Scotland, when she was about nineteen, and on her way to some relations in the island of Granada . . . I still think she was one of the loveliest girls I ever saw.'[45] Evidently, before setting off on her long transatlantic journey, Isabella returned from London to Scotland to say goodbye to her mother at Clockmiln, her guardian at Forres, and her brother at Banff.

There are no documents to establish the date of her embarkation, but it is not difficult to imagine when it is likely to have been. In June 1812 America's declaration of war made it dangerous for merchant vessels to sail alone. The admiralty organized convoys, with the rendezvous for westbound ships at Cork harbour.[46] In all likelihood Isabella boarded the vessel which was to take her across the Atlantic at Belfast, having taken leave of her sister Margaret, whose husband was Inspecting Field Officer there,[47] and was probably in the convoy that sailed on 15 November under the protection of the *Circe*.[48] If so, they made Barbados on 5 January 1813, and no doubt within a few days Isabella was reunited with her relations on the quay-side at St George's, Grenada. These were probably her sister Mary and her husband. There is no trace of John Downie in the London trade directories, and this may mean that they went abroad after Isabella's visit to London in 1811. If we ask why the Downies (or whatever relations it may have been) chose to emigrate to Grenada rather than elsewhere, then the intervention of McDonell is again adumbrated. The town treasurer of Forres, one William Hoyes, had two sons, John and Lewis, who had gone out in the early 1800s to Grenada.[49] According to one source, John, a merchant at St George's, became Speaker of the House of Assembly and married 'Louisa, daughter of Judge Bridgwater'. John's brother, who married his wife's sister, Almeria Bridgwater, was a partner in the firm of Messrs. Ker & Hoyes, merchants.[50] Both were later executors of Henry Bridgwater's will. What more likely than that the Treasurer of Forres, aware of McDonell's responsibilities to the Shaw girls, should have sounded out his son on the chances of Mary's husband opening a business in Grenada? Arrived there, Mary may have recognized in Henry Bridgwater, brother of Louise and Almeria, a comfortably situated bachelor on the verge of middle age, the means of providing for one of her

[45] NLS MS 925, no. 4; BL Add. MS 38899, fo. 81.

[46] PRO, Adm. 2/1107, fos. 57 ff. (Admiralty out-letters, Convoys, 1812–13).

[47] RHE: GR 806, fo. 188; PRO, WO 31/373, 4 June 1813.

[48] Captain's log, *Circe* (PRO, Adm. 51/2175).

[49] NLS MS 6714, fo. 9; RHE, Minute Book of the Presbytery of Forres.

[50] *Aberdeen Journal Notes and Queries*, 1908, 160; R. Douglas, *Annals of the Royal Burgh of Forres* (Elgin, 1934), 214–15, 253. John's wife, Bridgwater's sister, died in 1820 (*The Times*, 31 Mar.). He returned to Forres and died there 19 June 1839 (*St. George's Chronicle* (Grenada), 3 Aug. 1839).

numerous unmarried sisters. Such manœuvres, common in the social history of the colonies, were vital to Scots girls stranded in increasingly depopulated Highland towns that had grown emptier of eligible young men as the eighteenth century drew towards its close. In Inverness-shire the percentage by which women outnumbered men had become the highest in the Kingdom, and continued so for a hundred years.

To turn to Bridgwater, he came of a family long established in the Leeward Islands.[51] There is no record of his birth, but since he is stated to be the eldest son of Thomas Bridgwater of St Kitts, and another, Edward, was born in 1776,[52] he could not have been born later than 1775, and so must have been at least sixteen years older than Isabella. Entered at Lincoln's Inn in 1796,[53] and probably called to the Bar in 1800, he returned to practise in Grenada, where he prospered and in September 1810 was appointed Chief Justice.[54] We are now in a position to resolve a difficulty that has puzzled Hazlitt's biographers, namely, the apparent contradiction between different references to him as a planter, as a barrister, and as a lieutenant-colonel. The explanation is that he was a lieutenant-colonel in the local militia, in which every able-bodied man from sixteen to sixty-five was liable for service.[55] A field return of November 1814 shows six regiments, each commanded by a lieutenant-colonel (since it was simply a statement of strength there are no names, except one, accidentally entered, which turns out to be Lieutenant-Colonel Hoyes).[56]

On 15 April 1813, soon after Isabella arrived, Bridgwater became involved in a more serious legal question than he usually had to do with.[57] As Chief Justice he had to pronounce the letters of appointment presented by the new Governor, General Charles Shipley, to be insufficient.[58] Shipley seems to have been disliked by the whole island, but when, upon appealing to the Colonial Secretary, he finally prevailed, the position of the Chief Justice who had opposed him must have been uncomfortable. Bridgwater resigned on 1 December 1813, in a letter

[51] See 'Bridgwater of Nevis', *Caribbeana*, ed. V. L. Oliver, v (1919), 209–12.

[52] BL Add. MS 43866 records the births of Mary, 30 July 1765; Edward, 30 June 1776; and Almeria, 3 Dec. 1780.

[53] *Records of the Honourable Society of Lincoln's Inn* (1896), i. 557, 'Admitted March 1st 1796, Henry Bridgwater, gent., eldest son of Thomas Bridgwater of the island of Grenada.'

[54] PRO, CO 101/53, at 1 Dec. 1813.

[55] G. Smith, *The Laws of Grenada* (1808), Law no. XCI of 1801, enacted following the native insurrection of 1795.

[56] PRO, CO 101/54, dispatch no. 82.

[57] Several cases in which he pled are described in three pamphlets by Alexander Lumsden, an aggrieved islander who came to England to seek redress in 1817: *The Case of Alexander Lumsden* (1820), PRO, CO 101/60; and two versions of his *Petition to Parliament*, 1829 (BL 1414.b.ll(16)), and 1830 (BL 1414.k.4).

[58] For this and what follows see PRO, CO, 101/53, *passim*.

addressed to Lord Bathurst from 4 Sackville Place, Piccadilly, London, giving as his grounds the poor financial rewards.[59] He adds that he had been 'compelled by sickness to leave the Colony and seek the recovery of [his] health in Europe'.

Certainly ill health was an ever-present threat in the Caribbean, but he had other business in the United Kingdom, as we see from the following entry in the parish register of Forres:

5th October 1813, This day Henry Bridgwater, Esq., of Grenada and Isabella Shaw, lawful daughter of the late James Shaw, Esq., of Muirtown were married.

The bride's affection for McDonell, almost a father to her, is one reason for their marrying at Forres, where two of her sisters were also married, Mary, as we have seen, and Agnes, to Lieutenant David Anderson of the Scots Fusiliers on 26 March 1815. Another may have been Bridgwater's curiosity to see the home town of his two brothers-in-law.

The Bridgwaters did not remain long in Britain. Henry, who was present at the Grenada Assembly on 29 April 1814,[60] resumed his law practice, refusing all further official appointments, including the Attorney-Generalship.[61] His former office however prompts the reflection that Hazlitt's second wife had, through her first husband and her brothers and brothers-in-law, a much more direct, although minor, involvement in public life than he did, who always remained on the sidelines, a free lance, an outsider. It is stranger still that in Isabella he married the widow of a slave-owner. It is true that paternalistic slave-owning was considered less heinous than slave-trading, and Bridgwater seems in any case to have owned so few that the emotive term is misleading. Since Isabella remained eight years in the West Indies we may perhaps assume that she shared the views of the majority of her fellow-whites. In an age so brutal that in Isabella's own native town a woman could in 1817 be flogged through the streets,[62] it is likely that only excessive and persistent cruelty (as in the case of Hodge of Tortola) would momentarily disturb complacency. We need not suppose that when she and Hazlitt met they wrangled over the slave-trade and the 'enormities to which no words can do adequate justice' of this system (as he had written two years before) which was 'rotten at the core'.[63] He always spoke of it with abhorrence, and of its abolition as 'that great chapter in the history of the world'.[64] The merest reference she may have

---

[59] PRO, CO 101/53. A further letter of 13 Dec. shows him still at the same address.
[60] PRO, CO 104/10, at that date.
[61] PRO, CO 101/54, dispatch n. 59.
[62] *The Times*, 26 Mar. 1817.
[63] 12. 47, 48.   [64] 11. 149.

made to the 'ingratitude of [the] negro slaves'[65] would convince him of
the uselessness of attempting to root out her prejudices. We may
however consider as a possibility that what before marriage seemed to
him simple ignorance (perhaps from her never having actually witnessed
cruelty), or lack of imagination, could easily have later become an
additional ground of incompatibility among others that brought about
their separation.

The next date we find relating to the Bridgwaters is in the summer of
1817. In the early months a terrible outbreak of yellow fever had struck
Grenada, and Bridgwater in June went 'to America for a change of
climate'.[66] The American war was over and it was a less wearisome and
less expensive voyage than to Europe. We may assume that Isabella
accompanied her husband, a sick man, on this visit, which lasted into the
new year.

From this point on we have the evidence of the annual returns in the
newly opened General Register of Slaves, Grenada. The return for 1817
of 'Henry Bridgwater of the parish of St. George' was made in his
absence by Lewis Hoyes, who had his power of attorney.[67] At that time
Bridgwater owned five male and one female slaves. The couple were
back by 9 March 1818.[68] Bridgwater signed his own returns in the
January of 1819 and 1820. Putting his name to the last he must have
been aware of his approaching dissolution, for in the previous month he
had made his will, leaving lsabella £300 a year.[69] The next return refers
to 'the estate of the late Henry Bridgwater deceased', [70] so that Isabella
was widowed, not 'shortly after' her marriage, as Gordon asserted, but
in 1820, after a lapse of seven years.

As to the fortunes during this time of the other members of her family,
her brother-in-law William McCaskill had had, since 1783, after five
years' service in the American War of Independence, an unsatisfactory
career, spending long periods on half-pay. He did attain the brevet rank
of Lieutenant-Colonel in 1801,[71] and occasional postings. But in the
summer of 1813 he contrived a staff appointment to the island of
St Martin, four hundred miles from Grenada, and was concurrently
promoted Major-General.[72] This was an unhealthy move, but he was
now able to give a helping hand to his brother-in-law Alexander

[65] The expression appears in a 'Common place' pub. 8 Nov. 1823 (20. 133).

[66] PRO, CO 101/57, at that date.

[67] PRO, T.71/264.

[68] PRO, 101/58, at that date.

[69] *Four Generations*, i. 182.

[70] PRO, T.71/273.

[71] J. MacInnes, *The Brave Sons of Skye* (1899), 156, gives a brief account of his career. The dates
are given in the relevant *Army Lists*.

[72] PRO, WO 31/373, and RHE, GR, 806, fo. 188.

Mackenzie Shaw, also in a backwater, and, at thirty, much the oldest but far from the most senior subaltern in his regiment.[73] McCaskill succeeded in getting him as his ADC, but further difficulties arose and the Major-General did not finally take up his post in St Martin until January 1815.[74] Poor man, he had not long to enjoy his new dignity: at the beginning of April he died.[75] Shaw, attempting to look after McCaskill's young family as well as his own five children, asked to be placed on half-pay and retired to live in Kirkcudbrightshire.[76] He named one of his sons Henry Bridgwater Shaw.[77]

Bereavement came also in these years to another sister, Agnes. Her husband Lieut. David Anderson was posted in 1815 to the West Indies, and in 1820, when Isabella's husband died in Grenada, Agnes was living no more than a hundred and sixty miles away in Barbados.[78] Anderson in turn died of a fever in August 1821, leaving his wife and four children 'in necessitous circumstances'; Agnes was compelled to apply for the King's Bounty.[79]

We now approach the meeting between Isabella and Hazlitt and the question of when and where it took place. It is said that they met in a stage-coach, and that she 'fell in love with him on account of his writings',[80] and this may have been during one of the journeys he undertook in 1823 in visiting the picture-galleries. That summer he met Kenney in Hampstead Fields, and the following conversation took place: Hazlitt: 'Well, sir, I was just going to Mr ——'s, there's a young lady there I don't know.' Kenney: 'But there was another young lady of colour you were about to marry — has she jilted you like [Sarah Walker]?' Hazlitt: 'No, sir, but you see, sir, she had relations — the kind of people who ask after character, and as mine smacks, sir, why, it was broken off.' The impertinent Kenney, who was notoriously prone to hyperbole, may have turned the word 'creole' (wrongly used), or West Indian, into 'a young lady of colour'.[81] Haydon, in his letter of

---

[73] PRO, WO, 27/107 (where his age is wrongly given as 33).

[74] PRO, WO 31/379, a letter dated Belfast, 25 July 1813, from McCaskill to Col. John Hope, Shaw's CO, is significant of the setbacks and stagnation in the careers of the two officers. In it is enclosed a memorial from Shaw to the Commander-in-Chief, dated Dumbarton, 20 July 1813, setting out his services. Also, PRO, WO, 17/2504, 25 Jan., and WO 25/696, 25 Feb.

[75] WO 17/2504, 25 Apr., giving his date of death as 5 April. See also *Edinburgh Advertiser*, 4 July 1815.

[76] He set out his case in another memorial to the Commander-in-Chief, dated Kirkcudbright, 23 Aug. 1816, PRO, WO 31/444.

[77] *A History of the Mackenzies* (n. 24 above), 583 where eight children of Alexander Mackenzie Shaw are named.

[78] PRO, WO 17/343, 347, and 356.

[79] PRO, WO 42/1(93).

[80] *Memoirs*, ii. 106; Pope, ii. 496; Morley, 387.

[81] *The Letters of Mary Wollstonecraft Shelley*, ed. B. T. Bennett (Baltimore, 1980), i. 375.

26 September to Miss Mitford, unwittingly confirms the drift of
Kenney's story: '[Hazlitt] had another flame who is at Hampton; down
he went to tempt her for Gretna, but her brother, an officer in the Navy,
happened to be with her; and "officers," said Hazlitt, "you know, are
awkward fellows to deal with!" Oh, the gallant Lothario! Is not this
divine?'[82] It is not impossible that this was Isabella. Hazlitt was not the
man to make a fresh conquest every few months, even under the novel
colours Kenney and Haydon ascribed to him. It is tempting to see a
connection between the naming of Hampton and the 'Pictures at
Hampton Court' of June 1823. The brother might have been Alexander
Mackenzie Shaw, who disappears from Kirkcudbrightshire records in
1818, and whose old regiment, the 17th Light Dragoons, had its HQ at
Hampton Court in that year. He would be in mufti, and Hazlitt might
have mistaken his branch of the service, or perhaps Haydon did. One of
the 'Common Places' Hazlitt published that autumn is unexpected in a
man of such sparse military acquaintance: 'An officer in a Scotch
marching regiment has always a number of very edifying anecdotes to
communicate; but unless you are of the same mess or the same clan, you
are necessarily *sent to Coventry*.'[83] Here we get an unmistakable glimpse of
a wainscoted inn-parlour at Hampton, and a glowering mumchance
Hazlitt shuffling his feet under the table while Isabella, all unconscious,
smiles from him to the loquacious and peremptory Alexander. Peremptory Alexander would be, even irascible, with a history like his. His
temper is not likely to have been improved by his early bankruptcy, by
his wretched experiences in his two campaigns, by the frustrations of the
service, or by the heat of the Caribbean, followed by responsibility for
two young families, not to speak of the crabbed existence that was the lot
of thousands of half-pay officers in the years after Waterloo. And in any
case, having seen his orphaned sister, whose livelihood he himself had
come close to imperilling, make a good match and gain a respectable
income, he would not be best pleased to find her, as a vulnerable widow,
pestered by some damned writer-fellow obviously after her money. Poor
Hazlitt was unlucky in his brothers-in-law.

An intriguing hint of their having met in the summer rather than the
autumn of 1823 appears to be playfully concealed in the essay 'On the
Old Age of Artists', which Hazlitt published on 1 September, and in
which there is a kind of charade made up of both the surname and the
Christian name of his future wife. If the passage is a coincidence, it is a
remarkable one; and if he had not already met Isabella it is a strange
presentiment.[84]

---

[82] See n. 8 above.   [83] 20. 135.

[84] See the writer's 'Isabella Bridgwater: A Charade by Hazlitt?', *A Review of English Literature*, 8
(1967), 91–5.

However closely all this may correspond to fact, there is yet another pointer of undeniable significance. It is the phrase 'tempt her for Gretna'. We recall how the hero of *Liber Amoris* proposes after the divorce to take Sarah to Scotland to marry her. Howe follows Cowden Clarke in supposing they were married in Edinburgh. But in quoting Bewick's account of the honeymoon at Melrose he neglects to say that they came there by accident, through Hazlitt's absent-mindedness. When, said Hazlitt, they got into the post-chaise (evidently after the ceremony) and the driver asked, 'Where to, Sir?', he realized he had not decided where to go. 'Looking out before me,' he told Bewick, 'I observed two pointed hills, and asked, "Where are those hills?" "Melrose, Sir." "Then drive there", and to Melrose we came.'[85] The Eildon Hills are not visible from Gretna in the west, nor from Lamberton in the east, but they can be seen from the only other place through which a London coach passed and where smithy marriages were performed, namely, Coldstream.

Hazlitt's letter of 16 April from Melrose to Taylor & Hessey shows he had had time to send them his address, as well as to receive and correct the proofs of *Sketches of the Principal Picture Galleries*. The wedding could not therefore have taken place later than 9 April. It seems that he and his wife intended to spend at least a month there. He was confident that his publishers could not only produce the book but get a copy to him before he left.

From Edinburgh Bewick had moved on to Glasgow, where he rented a studio, held an exhibition in February, and advertised for sitters.[86] It seems likely that Hazlitt had followed up his letter of introduction to Ritchie (or perhaps Henderson) with another to Knowles. At all events, Bewick's portrait of *James Sheridan Knowles, Esq.* was shown at a second exhibition, in Finlay's Gallery in April, [87] and Hazlitt's making his two young friends known to each other would be a likely prelude to inviting them both, as he did, to come and see him and Isabella at Melrose.

Although Bewick, encouraged by Hazlitt, called on Scott at Abbotsford (a short walk from Melrose) and was well received, Hazlitt himself did not go near the place. He admired the private man and the writer as much as he detested the public man's politics, his ostentatious loyalty to the throne, his siding with prejudice and power even to injustice, and his known connection with the *Quarterly Review* and the gutter-press Edinburgh *Beacon*. He had delivered a sharp attack on the reactionary laird in the very middle of generous praise of *Peveril of the Peak* in 1823, but the attack was suppressed and did not meet Scott's eye. He was just about to

85 Bewick, i. 170–1.
86 *Glasgow Free Press*, 3 Feb., 5 Mar. 1824.
87 *Glasgow Herald*, 2 Apr. 1824.

add a devastating rider in preparing, for the volume *The Spirit of the Age*, his 'character' of Scott which had just appeared in the *New Monthly*.[88] But he could not say enough in praise of the novels, that 'new edition of human nature'. So he kept his distance, at Melrose. Elsewhere he summed up his ambivalent attitude in two remarks: the first, that he admired him 'on this side of idolatry and Toryism'; the second, 'that he [would be] willing to kneel to him, but . . . could not take him by the hand'.[89] But he spoke warmly of Scott to the young artist, in these terms:

Ah! he is indeed a fine piece of Nature's handiwork! I was convinced of that when I went to see him at the Court of Session . . . Scott's large heart rises above his party prejudices . . . As a man, I am told, he is frank, free, and open-hearted, simple and natural in his manners, and ready to grant everyone his meed of praise and justice.[90]

Scott had heard that Hazlitt was in the neighbourhood, and tactfully questioned Bewick, who told him of his marriage. He abhorred Hazlitt's political views as much as Hazlitt did his, but instead of echoing some of Blackwood's milder disparagements he unwittingly traded homage for homage with the absent writer, saying that, 'Mr. Hazlitt was one of our most eloquent authors, and a man, as far as he could be allowed to judge, of great natural and original genius; that it was a pity such great powers were not concentrated upon some important work, valuable to his country, to literature, and lasting to his fame.'[91] It is strange to reflect that these two writers, opposed in politics but attracted to each other by a mutual admiration in literature, lived for over a month within a mile's distance without ever making the slightest move to meet.

Bewick said nothing to his host of Hazlitt's recent tribute, but he was keen to apprise Hazlitt of Scott's high opinion. It was a disappointment, on getting back, to find that Hazlitt had departed for the south, either that morning, Thursday, 6 May, or the day before, leaving a word of explanation, and as a gift the copy of Lady Morgan's *Salvator Rosa*. For Bewick the whole episode, evidently a delightful moment in his life, ended with his return at the weekend to Glasgow, where he closed his exhibition and packed up his pictures for Dublin.[92] After Ireland he went to Italy. He did not see Hazlitt again for five years, and probably never told him of his conversation with Scott.

Between the time when Hazlitt and his wife left for London and their departure for France at the end of August we get only two direct glimpses

---

[88] For the suppression, 19. 343; for the addition, 11. 335.
[89] 11. 64, 276; Patmore, iii. 22.
[90] Bewick, i. 172.
[91] Ibid. 197.
[92] *Glasgow Free Press*, 4 May 1824.

of him. The first is at one of his publishers' dinners at Fleet Street in early July. Coleridge established himself as lion of the evening with a three-hour monologue. John Clare, who was present, his bashfulness even further intimidated by this bravura feat, was drawn to the silent figure of Hazlitt, who appeared equally low-spirited and ill at ease, and who became animated only after the departure of the performer.[93] The second we owe to Haydon who encountered young William in the street, and wrote to Miss Mitford: 'I have found out Hazlitt at last. I met his little boy in Piccadilly, better-dressed, cleaner, more modest and more respectable than I had ever met him before. He lives at Down St., Piccadilly, No. 10. I have not called yet . . . I dare say his marriage may contribute to his happiness but it will embarrass his friends.'[94] It does not appear that Haydon did call, since it was not until after Hazlitt's return from abroad that he was able to give any account of Isabella. The embarrassed friends certainly did not include John Hunt and his wife, who sent Isabella their 'best respects' in a letter soon after the Hazlitts' departure.[95] Hunt describes her ('*this* Mrs Hazlitt') as 'pleasant and ladylike', the opposite, he implies, of Sarah Hazlitt. By early July the odd, inconsequential, self-reliant Sarah had crossed the Channel, bound for Paris, a worthy objective for her curiosity and energy, determined as she was not only to see the Louvre and the masterpieces her husband had so often and so glowingly described, but everything else within and without the city walls. She was alone, without much money, ill-equipped in French, and completely unconcerned.

Since both his mother and father were going out of the country, young William, now aged twelve, was removed from Dawson's Academy and sent to the Revd William Evans's school at Tavistock. This choice was probably on the recommendation of Godwin, who was acquainted with Evans, a schoolmaster of some reputation, and in any case it brought the boy near to his grandmother and his aunt. Evans was in town: he called on Godwin on 19 July;[96] and we may assume that shortly thereafter, the school year being about to begin, Hazlitt saw the lad off from the Bell and Crown on his long journey in the Exeter coach, accompanied by his new master. On 28 August Hazlitt paid a farewell visit to Godwin's and there met Mary Shelley. His review of her edition of her dead husband's poems had appeared in the *Edinburgh* a few days since and had not pleased her, but she was so startled by Hazlitt's careworn look (she had not seen him since 1818, before *Blackwood's*, before the collapse of his

---

[93] F. Martin, *The Life of John Clare* (1964), 181.

[94] [12] July 1824, Sotheby's cat., 2 Apr. 1973, quoted by Miss Mitford in a letter to Talfourd (see Baker, 440).

[95] Unpub. letter of 4 Nov. 1824, Henry Leigh Hunt to Hazlitt, Buffalo.

[96] Godwin Diary.

marriage, before the useless divorce) that resentment vanished. 'I never was so shocked in my life — gau[nt] and thin, his hair scattered, his cheek-bones projecting — but for his voice and smile I shd not have known him — his smile brought tears into my eyes, it was like a sun-beam illuminating the most melancholy of ruins — lightning that assured you in a dark night of the identity of a friend's ruined and deserted abode.'[97]

[97] *Letters of Mary Shelley* (n. 81 above), i. 452.

16
On the Continent

Hath Britain all the sun that shines?
Shakespeare

Since Hazlitt has already copiously narrated in *Notes of a Journey through France and Italy* his long-anticipated trip to the South, it would be supererogatory to go over the same ground and expose one's lame prose by inviting a comparison with his lively step. Who of those who have traversed the dizzying slopes from Mont Cenis to Susa would attempt to rival his magnificent page describing that plunging descent, in which his admiration for the dramatic scenery is equalled only by his adulation of the hero who made the road that carried him down? I will therefore add only a few names and a few dates omitted in the account, to supplement what he has said. His own dates and names are here more plentiful than anywhere else in his work. In fact the period from September 1824 to September 1825 is better documented by Hazlitt himself than any other period of his life is by any of his biographers.

They crossed the Channel in the *Rapid* on 1 September, a delightful, calm, moonlit, night-passage. Their most notable fellow-traveller was John Jones, MP for Carmarthen (one of the few members praised by the *Examiner*), glad to escape, Hazlitt could not help remarking, from 'late hours and bad company'.[1] Hazlitt too was glad to escape from England, but he could not shake off ill health. This was why, when they left Rouen on 7 September, they left separately. He had need of fresh air, and made part of the journey on foot, while Isabella went on alone in the stage-coach and, arriving in Paris, engaged rooms at the Hôtel des Étrangers, rue Vivienne. It says much for her independence and self-reliance that she took charge of their arrangements. His ex-wife seized the chance of their presence in Paris to dun him for some money. If Isabella was startled by the visit Sarah paid them in Down Street, she must have been even more put out by her intempestive invasion of their rooms at the hotel. She must also have been aware of his inability to scrape together the full instalment of his son's maintenance.[2]

---

[1] Archives nationales, Paris, Enregistrement de Passeports, F7* 2560, fo. 1490; *Examiner*, 22 July 1821, 462–3.

[2] Le Gallienne, 351–2; Bonner, 261.

There seems to be no doubt that his new wife kept him on a short rein. Her Scottish practicality and prudence, the example of her spendthrift brother, the straits in which her widowed sisters had been left and from which others of her sisters may never have emerged, all would incline her to view with a wary eye her second husband's cavalier attitude to money. He remarked to his scandalized Edinburgh landlady that he never checked bills presented to him: it was 'one of the troubles [he got] rid of'. Well, it was one he could not now shirk, with his equally scandalized Inverness wife, who held the purse-strings, peering over his shoulder. The orphan who had lived in London at the age of eighteen, who had soon after travelled by herself from the north of Scotland to the Caribbean, and had had the management of her own affairs since the death of her husband, was unlikely to hand over the entire control of her money to her new husband. She may have made him some kind of allowance, over and above the money required to settle bills for accommodation and travel, but despite his confident advice to Bewick to 'marry some girl with a small independence', if he now had to depend on money doled out piecemeal, his situation, for a man so impatient of restraint, must have been galling. It may even be that it was she who took charge of the bills, both abroad and when they returned home. It is pretty certain that no part of the annuity chargeable upon Bridgwater's estate, and paid, probably quarterly, to Isabella, can have come directly into his hands. We cannot doubt that he was eager to earn what he could by his writings in order to be as seldom as possible obliged to apply to his wife.

Colburn had declined to finance his trip, Taylor & Hessey likewise,[3] and I think it likely that the Hunts, with the moderate income that had been theirs since the circulation of the *Examiner* had fallen, although they held out some hope of publishing his travel-book when he returned, found it impossible to provide him with a purse for the expense of the journey. When he left England he had reached only a tentative understanding with the *Morning Chronicle* to publish a serial account of his journey, and several instalments had already appeared before he came to a formal agreement with Black. Perhaps it was a distrust of the mails, perhaps of businessmen, but whenever he was absent from London and had an important affair in view, it was his habit, in addition to writing to the principal, to write to ask a friend, an intermediary, to call and back up his application. Thus we have a letter from Henry Leigh Hunt dated 4 November.

Dear Sir,
    You must not think we have forgotten you. Immediately on receipt of your

---

[3] Le Gallienne, 350; Bonner 260; *Letters*, 212.

letter I went to Mr Black but was agreeably surprised to find that he had by the previous night's post arranged all to your wishes . . . My father would like much to publish your volume himself and would endeavour to comply with the condition. He will thank you to say what sum (in all) you expect for the copyright and he will then write to you finally on the subject. There is a little budget of letters accumulated at our office since you left England which my cousin William [Reynell, who goes to Paris shortly] will take charge of for you. Should you have left Paris when he arrives he will . . . forward them to you. My father and mother desire their best respects to Mrs. H. and yourself.[4]

Uppermost in his mind when he arrived in Paris on 8 September was the Louvre. There he went on the first morning, and his first article for the *Morning Chronicle* after he got to Paris does not deal with the city, but with what in his memory was its spiritual centre. It was not until after he had again seen his cherished pictures that he embarked on the serious business of his stay, which was to arrange with Messrs Galignani, to whom he must have written from England, for the publication of the Paris edition of *Table-Talk*. Their shop was only a few doors along the street from his hotel, but he evidently found it better to conduct his business in writing. He began negotiations on 27 October, sending them the two volumes of the English edition (and — probably in sheets — a further two volumes already in print, which were eventually to become *The Plain Speaker*) with the essays marked which he wished to be included.[5] The Galignanis were deplorably dilatory. In December he complained of their delays, saying he could not stay to superintend the publication unless they moved faster. The principal objective of his journey had always been Italy. He had confidently told Sarah on 17 September that he was going to leave for Italy in three weeks. He had to postpone his departure, and postpone it again, until eventually he left at the worst time of year, in the dead of winter. On 28 December he made a last attempt to get his publishers to move, declaring that he was leaving the following day. In the end, he set off on 14 January 1825, very dissatisfied with his treatment. It was not as if he were unknown in Paris: the *Revue encyclopédique* had given a very favourable notice of his *Political Essays* in February 1821 and spoken of the celebrity of his *Analyses des pièces de Shakespeare* (a very exact translation of the title); in Paris he had had much to do with the energetic Stendhal,[6] and with Stendhal's friend, Bartholomew Stritch, who appears to have been Galignani's English

---

[4] Buffalo.

[5] These are the 'four volumes of *Table-Talk*, printed in London', referred to in Hazlitt's preface to the Galignani edn., and also in Keynes, 85–6. For an exact and detailed account of the printing history of this edn. see C. E. Robinson, 'William Hazlitt to His Publishers, Friends and Creditors: Twenty-seven New Holograph Letters', *The Keats–Shelley Review*, 2 (August 1987), which also prints all the letters to Galignani here mentioned.

[6] Henri Martineau, *Le Coeur de Stendhal* (Paris, 1952–3), ii. 132–3.

adviser;[7] and a few weeks after his departure *Le Globe* spoke very handsomely of the *Spirit of the Age* ('*Portraits contemporains*') by 'l'aristarque romantique, William Hazlitt'.[8]

We may reasonably suppose that it was in the gloomy, ill-lit, bourgeois salon of his old schoolfellow of thirty years earlier, Dr William Edwards, that he finally met Stendhal. Edwards, a small, delicate-looking, sober man with an even more staid wife, enjoyed a considerable scientific reputation in Paris, where he had long settled. He lived in the rue du Helder, ten minutes' walk from Hazlitt's hotel. Hazlitt speaks of Edwards with the same heartiness as of Kilner Hazlitt, and his words leave no doubt of the cordiality of their reunion: 'It is wonderful how friendship, that has long lain unused, accumulates like money at compound interest.'[9] There must have been a good deal in Edwards for Stendhal, who dreaded boredom, to have frequented his wife's 'Wednesdays' so assiduously. He had contrived his entré in 1822, when Edwards was in the public eye, but with difficulty, through Edwards's embarrassingly bibulous brother, and in spite of his own reputation, which caused him to be greeted with some reserve.[10] These were difficulties which had come within Hazlitt's own experience, and mark another point in the remarkable parallel between the two men.

Hazlitt and Isabella arrived in Florence on 5 February, where Hazlitt met, for the first time in nearly four years, Leigh Hunt, who had recently dwelt so pleasantly in the *Examiner* on his memories of Hazlitt's first visit to him in the Horsemonger Lane Gaol in 1813.[11] He also called on the eccentric Landor, who thought him odd. However curious, watchful, open to impressions, and prepared to stifle his own prejudices Hazlitt might be in his travels, he was still ready to speak out when principle required, and especially in congenial company. 'A funny fellow he was,' said Landor, 'He used to say to me, "Mr Landor, I like you, sir — I like you very much, sir, — you're an honest man, sir; but I don't approve, sir, of a great deal that you have written, sir. You must reform some of your opinions, sir".'[12]

Hazlitt found he could not reconcile modern Rome with the luminous city that had grown in his imagination from such eighteenth-century prints as Claude Lorrain's *Arch of Constantine*, which hung in his room at Winterslow. 'Rome hardly answers my expectations; the ruins do not

[7] Stendhal, *Correspondance* (Pléiade edn. Paris, 1962–8) , ii. 37; Robinson (n.5 above), letter of 11 Mar. [1825].

[8] No. 62, 29 Jan. 1825, p. 303.

[9] 10. 246.

[10] Henri Martineau, *Petit dictionnaire stendhalien* (Paris, 1948), 209–11.

[11] 18 July 1824, subsequently incorporated in Hunt's *Autobiography*.

[12] R. S. Super, *Walter Savage Landor* (New York, 1954), 544 n. 4.

prevail enough over the modern buildings, which are commonplace things.'[13] After a month in the via Gregoriana, in a house where Salvator Rosa had once lived, they set off north again, travelling via Florence and Venice to Milan, where they arrived in the middle of May, to find the unwelcome festivities for the visit of the autocratic Emperor of Austria in full swing.[14] They retreated to Lake Como, and then to the Lago Maggiore, finally making their way via the Simplon to Vevey, the long-cherished goal of his journeyings, haunted by the memory of Julie and Saint-Preux.

They arrived at Vevey at the end of the first week in June.[15] The north shore of Lake Geneva from Lausanne to Chillon was then as popular with the English as Paris or Florence (Kemble retired there in 1817). But Hazlitt was indifferent to fashion and had in view another object than the mild climate and cheap living. The visit to Switzerland was a private matter, and, as its omission in the title of *Notes of a Journey through France and Italy* suggests, was not planned to be included in that volume. What he sought was the country of *La Nouvelle Héloïse*. The thought of Julie and Saint-Preux, as always, brought him back all the joy, hope, and tenderness that had accompanied his first reading of that book thirty years before. At the height of his infatuation with Sarah Walker his greatest wish had been to bring her to these shores, to tell her the story of Julie d'Étange and show her where Saint-Preux, returning to the Valais after long absence, scaled the rocks of Meillerie to gaze longingly across the lake to the Pays de Vaud and Julie's home. It made no odds now that Sarah had forsaken him, and that in his own history the unheroic equivalent of Meillerie was a London doorway in which he skulked at night, in September 1823, to peer across the street at Sarah's darkened window. The feeling was the same. And the associations of *La Nouvelle Héloïse* remained. That feeling, and those associations, compelled him to take the same road despite Sarah's absence.

It made no difference either that, in travelling, he had long penetrated the illusions of the imagination, recognized the impossibility of finding the anticipated ideal in the present reality (since, in any case, the imagination can only attach itself to what is absent), and that he had only recently, in Rome, noted again to his mortification that 'imagination is entirely a *thing imaginary* and has nothing to do with matter of fact, history, or the senses'.[16] Its power over his mind, and its charm, remained; and mortification, as always, receded. But although his Continental journey had brought new scenes, new experiences, now that

---

[13] *Letters*, 338.
[14] *Courier*, 9 June 1825.
[15] They 'remained fifteen weeks' (10. 288) and left on 20 September (ibid. 295).
[16] 10.232.

it was nearly over it merely confirmed him once more in the view that life is lived through the imagination ('all that is worth remembering in life is the poetry of it') and that reality subsists on a lower, if not an inferior plane. Or if not through the imagination, then through the memory, so that Rousseau's exclamation, 'Ah! voilà de la pervenche!', on seeing the flower he had first seen thirty years before with Mme de Warens, seemed to him deeper, truer, more essential than any excitement felt 'at the sight of a palm-tree, or even of Pompey's Pillar'.[17] Now, with a different travelling-companion, he doubtless still thought of Sarah, but with thoughts merging into that mixed and mingled strain he had long recognized as inextricable, as irreducible, in the human heart: regret at her loss, and at the same time the final, wry, wondering incomprehension with which he acknowledged the same truth that rose in the mind of Charles Swann at the end of his love-affair with Odette when he murmured to himself, in the closing lines of *Un Amour de Swann*: 'Dire que j'ai gâché des années de ma vie, que j'ai voulu mourir, que j'ai eu mon plus grand amour, pour une femme qui ne me plaisait pas, qui n'était pas mon genre!'

Almost the first thing Hazlitt and his wife did when they arrived was to visit the local circulating library. They hoped to find news of home in Galignani's latest publishing venture, the weekly *London and Paris Observer*, but it was evidently too recent to have reached Switzerland, so on 12 June Hazlitt wrote to the rue Vivienne himself to order it.[18] Their other desire, easily gratified, was to borrow the latest of the Waverley novels, the *Tales of the Crusades*: they found the local circulating library well stocked with Scott's works, in all the main languages, as in any other Continental town. They had been on the road since the middle of April, had covered nearly eight hundred miles, and were pleased to be settled at last and not obliged to stir beyond the gate of their new demesne. Hazlitt had often longed for a period of uninterrupted leisure to devote to rereading the novels of Scott, for whom his admiration remained as warm as his hatred of his politics was still icy.

The succeeding weeks renewed his delight in *The Fortunes of Nigel*, *Ivanhoe*, *The Heart of Midlothian*, but in the reading he also suffered a renewal of the jolts he had often received from the slovenliness of the style and the bad grammar. He had remarked on this as far back as 1820, in his review of *The Pirate*, in which he had taken the unusual step of providing a handful of specific examples, with page-references. In 1822 he alluded to Scott's 'villainous style' and to his trick of 'ring[ing] the changes unconsciously on the same words in a sentence, like the same rhymes in a couplet'. And in 1824 he made his point more explicitly

---

[17] 20. 156.   [18] See the writer's 'Some New Hazlitt Letters', *NQ*, July-Aug. 1977, 340.

when he ventured 'to hint [his] astonishment at the innumerable and incessant instances of bad and slovenly English in them'.[19]

He was half-annoyed with himself for dwelling on these instances of bad writing. And yet there was no denying that they offended his sense of what was owing to the reader. He himself had always gone to immense lengths to observe such decorum. His own prose style, not only clear, idiomatic, and vigorous far beyond the slack and smudged late-Augustan conventions of his time, but also studiously correct, rested upon a painstaking fastidious practice which obliged him to take account, over and above precision merely, of such elements as balance and stress, rhythm and euphony. The labour he expended on his writing disinclined him from overlooking the slipshod in others. If the picking of faults was a discourtesy so also was their commission. He was not disposed to consider bad writing as anything other than bad manners.

These numerous and persistent blemishes were an unwelcome reminder of the political warfare of the past, and in particular of the summer of 1823 and the *John Bull*. In May 1823 *The Times* had praised *Liber Amoris* as 'written . . . with great power and eloquence', and a week later criticized Scott for a 'sanguinary' toast at a Pitt Club dinner to the memory of Sir Alexander Boswell, killed in the duel with James Stuart, in a speech which was 'a jargon made up of drunken prose and drivelling poetry'.[20] The *John Bull* retorted furiously:

To the extraordinary dissemination of learning, through philanthropic institutions, charity schools, hospitals, and other receptacles of the lower classes, is owing, the power exercised by *The Times*, to deteriorate from the literary merits of SIR WALTER SCOTT, and to laud to the skies the beastly trash of BILLY HAZLITT [. . .] The moment Sir Walter Scott puts a work before the public, he becomes subject to public criticism. All this is quite right — it is one of the few advantages of the boasted liberty of the press; and when the Cocknies can write down SIR WALTER SCOTT, or write up BILLY HAZLITT, we will admit the Cockney lecturer, who is not able to write English, to be a greater man than him, to whom only SHAKESPEARE himself is, of British geniuses superior.[21]

His resentful memories of these attacks, and of Scott's support of the Tory press, grew with the mounting tally of blemishes of style, and eventually annoyed him so much that he found relief in copying out the places he had marked, adding an appropriate epigraph (' ''We desire our readers to take notice that there is not a single sentence of Common English'', The *John Bull*'), and posting the collection off to the *Examiner*, where it appeared in October.[22]

---

[19] 19. 93–4 (1820); 12. 17 (1822); 11. 67 (1824).
[20] 30 May, 7 and 9 June.
[21] 8 June.
[22] See the writer's 'Bad English in the Scotch Novels', *The Library*, 6/3 (1981), 202–16.

His 'bitter persecution' also recurred to his mind when he read in Galignani's *London and Paris Observer* an article, 'Authors and Editors', about contributors indignant at rejection or blue-pencilling, and read at the same time in the April *Edinburgh Review* Jeffrey's patronizing notice of *The Spirit of the Age*. He dashed off some satirical verses, 'The Damned Author's Address to his Reviewers', and sent them to the *London and Paris Observer*, where they appeared on 18 September.[23] He also sent a copy (via Hodgskin) to Black, who did not, however, print them.

When he returned to London in October Haydon came to see him in Down Street and was, as usual, warm and friendly, but a difficulty arose over Haydon's reluctance to allow Isabella and his wife to meet. He was surprisingly strait-laced, and had already similarly alienated Barnes by preventing Mary from visiting Barnes's common-law wife.[24] Although he liked Isabella, and even admired her, he thought the marriage not quite respectable.

At Vevey Hazlitt had been visited by Captain Medwin, who found him reading Sir Walter Scott, and to whom we owe the first hint that he meant to write a Life of Napoleon, when the time was ripe.[25] Although Hazlitt thought it 'too early', his hand was forced by Scott, whose intended Life was announced in the *Examiner* in June, and which was stated in the *Literary Gazette* in December to be so far advanced that it would probably be published in the autumn of 1826.[26] Rather than allow a writer of Scott's political bias a clear field, Hazlitt determined to start work at once on his own Life, and in July 1826, after having brought out *The Plain Speaker* and *Notes of a Journey through France and Italy*, he again crossed the Channel with Isabella and began upon the long arduous task-work he had set himself.

His departure provoked, exactly as in 1824, an outburst of spleen and resentment from Haydon. Its proximate cause was the appearance in the *New Monthly* of the first of Hazlitt's 'Conversations' with Northcote, in which things were said about art and artists that Haydon, smarting from a fresh humiliation by the Royal Academy, took to be aimed at himself and considered as treachery. It was a thought that certainly never entered Hazlitt's head, but it caused Haydon to rip up once again all his old grievances against him, in his diary, in an angry, invented, and improbable report of a conversation between himself and Northcote, whom he detested and never saw, modelled on the first of the 'Boswell

---

[23] For the letter to Galignani (30 Aug. 1825), see Robinson (n. 5 above). For 'Authors and Editors', see *London and Paris Observer*, 24 July 1825, 122–3. The verses, unsigned, were the sole item under section 'Poetry' on 18 Sept.

[24] Pope, iii. 641.

[25] *Fraser's Magazine*, 19 (1839), 282.

[26] 26 June, 408, and 24 Dec., 830, respectively.

Redivivus' series he had just read (a series subsequently incorporated in the *Conversations of Northcote* (1830)).[27] Haydon, in his tortuous way, searching for motives, attributed Hazlitt's putative treachery to Isabella's influence, to the 'malice' she felt because he 'did not sanction her, and would not suffer [his wife] to visit her'.[28]

In Paris the Hazlitts took a house and garden at 58 rue du Mont-Blanc (rue Chaussée d'Antin). If the Faubourg Saint-Germain was the quarter of the old nobility, the Chaussée d'Antin was that of the dashing young aristocrats and the *nouveaux riches*, 'where the newest creations of the fashionable houses were first seen, where the boulevards were thronged with elegance, where the world of pleasure, the opera, and the theatre, were close at hand'.[29] There has been no reason thus far, in the Hazlitts' marriage, to suspect Isabella of harbouring social ambitions, but she was the widow of a Chief Justice, her brother-in-law was a General, and she had independent means. In Hazlitt she must have thought she had captured someone whose fame entitled him to move in circles of distinction. Hardly anyone who had made a name in the past twenty years was unknown to him, and she assumed in him a willingness to multiply relations of this kind. She had not reckoned with the realities of the existence of literary grass-widows, and the long hours of silence and exclusion. But the signs are, that she was determined to make a splash. It would have been more convenient for him if they had stayed in or near the rue Vivienne, close to Galignani's *cabinet de lecture* and the *Bibliothèque royale* in the rue Richelieu. It looks as if she deliberately chose the Chaussée d'Antin for its reputation. But it was expensive. If Hazlitt had hoped for entire peace to meditate his Life of Napoleon, he was disappointed. He had lost the lavish payments of the *Edinburgh Review* and was forced to resume his work for the *New Monthly*, and when remittances were delayed he found himself in difficulty.

In London *The Plain Speaker* and *Notes of a Journey* were not selling well.[30] Trade was stagnant. The great bank failures of late 1825 and Constable's spectacular collapse had caused panic in the City. On 14 July Hazlitt wrote to Galignani to propose a forlorn hope — a volume of selections from sixteenth- and seventeenth-century English drama, but some of the suggested titles would have made even a London publisher stare, and it was turned down.[31] On 7 August he wrote to

---

[27] See the writer's 'Haydon and Northcote on Hazlitt: A Fabrication, RES NS 24 (1973), 165–78. After the publication of this article Professor Pope informed the writer that upon reference to the holograph diary entry he found it to be headed, very faintly, 'Bos. red.'

[28] B. R. Haydon to J. Johns, 13 Dec. 1838, inserted in grangerized copy of *The Hazlitts* in BL.

[29] *Œuvres complètes de Honoré de Balzac* (Paris, 1912–40), iii. 273 (my trans.).

[30] *The Plain Speaker* was still on sale in 1840, reduced from 24*s*. to 11*s*. (Keynes, 98), and *Notes of a Journey* in 1837, remaindered at half-price (BL SC 706, cat. of C. Davis, 48 Coleman St.).

[31] Robinson (n. 5 above).

Patmore to ask him to raise £20 with Colburn on a promise of copy. A
month later he wrote, begging an immediate answer, to Thomas Allsop
to ask the same service. What came of this we do not know, but the letter
introduces us to the amiable Allsop, not hitherto identified as one of
Hazlitt's friends.[32] A silk-mercer in Pall Mall, thirteen years younger
than Hazlitt, he was a friend of Lamb's and a disciple and benefactor of
Coleridge. Poor Allsop was a victim of the recession of 1826, and the
following February Lamb was in great anxiety for his ruined friend.

Isabella cannot have been pleased with her husband's failure to bring
in money, nor with his planning a vast Life which would take him at least
two years, when he was already forestalled by no less a colossus than the
author of *Waverley*, whose biography of Napoleon was so far advanced
that the *Literary Gazette* of 19 May 1827 was able to print a long extract.
As Hazlitt told Patmore in his letter, he '[got] into nothing but rows and
squabbles'.

Unlike Scott, who, thanks to Croker, was given access to Government
archives in London, Hazlitt had to depend almost entirely on printed
sources. Not that Scott benefited much from this facility, any more than
he did in Paris, where he spent twelve days in the autumn of 1826, and
was so fêted that he saw even less source-material. On 30 October he was
in Galignani's reading-room, and Hazlitt, who seems to have been there
at the same time, was delighted with his simple, unassuming demeanour. Hazlitt was less charmed with the pretensions of 'the American Sir
Walter Scott', Fenimore Cooper, who was also in Paris.[33]

Scott was presented to King Charles X; Hazlitt was presented to the
grand old Liberal La Fayette, or at least attended one of his brilliant
soirées, at which he was probably introduced by Dr Edwards, and where
the aged General did the honours with Republican simplicity. The
disdainful Cooper observed him there and noted that his neighbours
were Benjamin Constant, Humboldt, and the artists Ary Scheffer and
David d'Angers.[34] It was probably about this time that he wrote to
Galignani for a copy of *The Last of the Mohicans*.[35]

Nothing is known of his life in the summer of 1827, but at the
beginning of autumn his marriage came to an abrupt end. One
important cause of the parting was money. Isabella had thought that she
was marrying a writer who made a good deal. We do not know how
much his books brought in in 1824, probably not much, but we can

---

[32] *DNB* says the hospitable Mrs Allsop made her husband's house 'a favourite resort with
Lamb, Hazlitt, Barry Cornwall, etc.'.

[33] 11. 276.

[34] *The Letters and Journals of James Fenimore Cooper*, ed. J. F. Beard (Cambridge, Mass., 1960–8),
i. 202, 204.

[35] Robinson (n. 5 above).

calculate that his periodical earnings must have been between £180 and £200.[36] With this, and her £300 (and some addition from his books), they might expect to live comfortably. She found he had other plans, and that she would have to keep him for as long as it took him to write the Life of Napoleon. In 1825 his journalistic earnings dropped to about £80. In the following year to about £40. So the prospect she envisaged was that with each successive year he would spend more time on his Life, and less on his periodical writing. A second reason was that she found herself committed to maintaining not only Hazlitt but his son, now a lad of fifteen or sixteen, who had joined them in Paris at the end of the summer term of 1826, and seemed likely to remain with them indefinitely. Having no children of her own, she was ill-prepared for this adolescent, who was very close to her husband but on very distant terms with herself. I believe, from a number of half-concealed signs, and although there is at present no means of proving the fact,[37] that Hazlitt, when his son arrived in the city of the Louvre, remembered the ambition he had harboured of his son's succeeding where he had failed, as an artist, and proposed another trip to Italy, so that the boy could see the galleries of Florence and Venice. Perhaps Isabella hoped this would temporarily alleviate an increasingly exasperating situation. If so, she was mistaken. The enforced intimacy of overnight stops at small hotels, and of cramped coach-journeys, was bound to exacerbate tension, and when they returned to Paris, Isabella had had enough of young William's resentfulness and irritability, as well as of his father's improvidence. Some chance word caused an 'explosion'[38] which persuaded Hazlitt to get young William out of the way. He took him back to London. But when he wrote to Isabella to say he had got the boy settled, she refused to join him, and announced that their separation was permanent, and that she was going on to Switzerland with her sister.

[36] The *London Magazine* paid 15 gns. a sheet, the *New Monthly* 16 gns. (Mitford, Letters, iv. 434; Lady Morgan, *Memoirs* (1862), ii. 187).

[37] My reasons for so believing are set out in 'Dating Hazlitt's Essay "On Reading New Books": Bibliography and Biography', *EA* 33 (1980), 188–98.

[38] Robinson (n. 5 above), letter of 3 May 1828.

17

The Final Years

spatio brevi spem longam reseces

Horace

Nothing more clearly shows our essential ignorance of Hazlitt's life in his last years than the silence which closes around his second marriage after his wife's defection. Even the word *defection* may be of hazardous use, depending as it does upon a brief statement made forty years later by Hazlitt's grandson.[1] Nor can we be entirely sure that what appears to be Hazlitt's only reference to the separation, in a recently discovered letter, really does refer to it. In the summer of 1828, writing to an obscure Parisian friend, he says, 'I was thrown out of Paris by a bit of an explosion, just as I had quietly settled down there'.[2] If this does mean the autumn of 1827, it is sufficiently uninformative. We are left to guess whether he was upset, or angry, or resigned, or relieved.

A comparable reticence marks the whole of the succeeding period. The obscurity which enshrouds many earlier moments of his life becomes well-nigh impenetrable in his last years. His activity, almost as great as in the *Examiner* years, is evident from the frequency and volume of his publications in the *London Weekly Review*, the *Atlas*, the *New Monthly*, and the *Edinburgh Review*, but he himself is hardly to be discerned. His personal life is hidden from us. We do not know where he was, what he did, or whom he saw. The old friends from whom we formerly learned a good deal, Lamb, Robinson, Haydon, Patmore, are silent. In Godwin's diary, where at one time his visits were recorded almost daily, his name appears only once in four years.[3]

His estrangement from Godwin appears to have come about with the publication of *The Spirit of the Age*, although there is no hint in the diary that there was such a connection. Their only encounter thereafter was an accidental one in Northcote's studio. It is strange that they did not meet there more often, since Godwin had at about the same time taken to calling on the old painter with greater zeal than he had shown for fifteen or more years. This change also coincided with Hazlitt's departure for

---

[1] *Memoirs*, ii. 196, the only place where it is stated that Isabella refused to rejoin him.
[2] Robinson, (ch. 16, n. 5 above), letter of 3 May 1828.
[3] 25 Oct. 1829.

France, but more strikingly with the publication of the first of the Conversations with Northcote (which had so annoyed Haydon) in August 1826. Whether it was that Hazlitt had lauded Northcote's reminiscences of Dr Johnson's circle during his many meetings with Godwin in the early part of the year, or that he dropped a hint that Godwin had been mentioned and would consequently appear in the first number, Godwin, conscious of the uses of publicity, was assiduous in his attendance on Northcote from 9 July onwards. It is equally striking that when the series of Conversations ends, four years later, in the book, *Conversations of James Northcote, Esq., R.A.*, it ends with a conversation about Godwin. The theme of these last pages was one which had haunted Hazlitt from the beginning of his writing career, namely, the inadaptability of literary men to life, the indifference of 'the literary character' to the real world, in its blind preference for words as opposed to things. And the person adduced as proof of this defect of feeling, and in illustration of this deformation, is Godwin. Another criticism of Godwin, graver because more explicit, was made earlier in the volume. Conscious of the enormity of levelling so heavy a charge at the revered author of *Political Justice*, Hazlitt nevertheless declared: 'There is another thing (which it seems harsh and presumptuous to say, but) he appears to me not always to perceive the difference between right and wrong. . . . He is satisfied to make out a plausible case, to give the *pros* and *cons* like a lawyer; but he has no instinctive bias or feeling one way or other . . . Common sense is out of the question: such people despise common sense,' [and then comes the sting in the tail] 'and the quarrel between them is a mutual one.'[4] Northcote must have given Godwin a pre-publication copy of the book, which came out on 26 August 1830: Godwin read it on 27–8 June. At the end of the following month he called on Lamb, who probably then told him that Hazlitt was gravely ill. At the beginning of August he reread *The Spirit of the Age* and must have remembered finding there the same judgements in milder but not less conclusive terms. On 20 August he found a critique of his latest book *Cloudesley* in the *Edinburgh Review* making yet again the same fundamental points. He did not record his reactions in his diary. But neither did he go to see the dying man, his friend of thirty-six years' standing, before he finally set down under the date 18 September the curt valediction 'Hazlitt dies'.

The inadaptability of the 'literary character' to the real world was the theme of the essay 'On the Shyness of Scholars' in December 1827, where Hazlitt ponders for the thousandth time the reasons for his disappointed life, and sees again that it could not have been otherwise,

[4] 11. 262.

that it was in the nature of things that he should have failed in love and
failed in social life. Again, in 'On Personal Identity' in January 1828, he
reflects on the disadvantages of intellectual superiority, and on the
consequences of the capital mistake he had made all his life, which was to
suppose that there was a common language between those who judge of
things from books and those who judge of them from their senses.

These essays he wrote at Winterslow, where he seems to have spent in
these years almost as much time as he did in London. He was principally
occupied, however, with the *Life of Napoleon*, which was giving him
anxiety. On 25 December 1827 he wrote to Galignani proposing a
publication in Paris, and met with a further vexation in that publisher's
chronic delays.[5] He did not hear until the following February, and then
only with a refusal. In the mean time, on 11 February, the first two
volumes had been brought out by Hunt & Clarke, but in the spring came
a further setback when the firm failed, plunging him once more into
financial difficulties. The reviews were not more encouraging. It was
true that the *Morning Chronicle* was helpful, printing nearly five columns
of praise and extracts, in three issues from 11 to 18 September. But the
unremitting hostility of the Tory press again showed itself in its familiar
guise in the *Literary Gazette*, whose reviewer asserted that 'our author's
grammar does not always enable us to see clearly through his meaning,
and . . . his style is, like his sentiments, seldom what can be called
English', and that '[his] rubbish has offended grammar, taste, sense,
feeling, and judgment'.[6]

He may have gone to Paris for another period of work on the
remaining volumes of the Life in the summer of 1828, but this is
uncertain.[7] In January and February 1829 he spent over five weeks at
Winterslow.[8] He was still in difficulties, and on 14 February wrote to the
editor of the *London Weekly Review*, offering a notice of Southey's
Dialogues of Sir Thomas More, Colloquies on the Progress and Prospects of Society:
'I will do the Sir Thomas More . . . and the Somerset House exhibition.
Let the columns be 15*s*. apiece . . . I wish to God you could let me have
£5 more on account.'[9]

---

[5] Robinson, (ch. 16, n. 5 above).   [6] 8 Nov. 1828, p. 709.

[7] For the argument against, see Robinson (ch. 16, n. 5 above), n. 2 to letter of 3 May 1828.
Maclean (548) was not convinced that he went to Paris, but Jules Douady, author of *Vie de William
Hazlitt, l'essayiste* (Paris, 1907), claimed to have discovered evidence (unspecified) of such a visit,
BL Add. MS 38911, fo. 69v, letter to W.C. Hazlitt, 20 Feb. 1906.

[8] A bill in the collection of Mr Michael Foot, MP, presented to Hazlitt by one Sam. Williams,
establishes that he was at Winterslow (at the Hut, or perhaps at the Lion's Head) from 21 Jan. to 11
Feb., and from the amount of a previous bill there mentioned as carried forward we may deduce
that he had been there since the middle of the first week in January. A puzzling reference in
Godwin's diary to having met 'Ollier and Hazlitt' on 4 Feb. may be explained either by Hazlitt's
having made a hasty visit to town or by Godwin's having meant John Hazlitt, but there is no
mistaking the dates on the bill.

[9] See the writer's 'Some New Hazlitt Letters', *NQ*, July-Aug. 1977, 340.

Time was beginning to close in, and soon 'from existence to nonexistence gone he would be by all as none perceived', but there were nevertheless gratifying moments in those last years. In October a few lines at the end of an article in the *Examiner* on the fashionable Sir Thomas Lawrence placing him once more in the company of Lamb and Coleridge must have brought a wry smile to his lips:

In the meantime men of genius must consent to be immortal only in their works, for Sir Thomas cannot do heaped justice to all; or they must sit to Mr Jackson or Mr Pickersgill, for Sir Thomas is a courtier and might hesitate to commit his reputation for grace and elegance by pourtraying the wild wilfulness of genius as it shoots across the face of Mr Hazlitt or gleams in the luminous countenance of the Sibyl Coleridge. Mr Lamb should sit to Mr Jackson — Mr Hazlitt to himself.[10]

In April 1830, when he had already moved into the house where he was to die, he acknowledges, rather stiffly, a letter from an unknown admirer: 'Mr Hazlitt presents his compliments to Mr Pearson and begs leave to acknowledge the receipt of his obliging letter, and to thank him for the favourable opinion expressed in it.'[11] The formality is understandable if the writer was an autograph hunter. Another letter of a little earlier date in this period refers to that race of collectors, but shows also Hazlitt's accessibility, even in his disappointed and lonely later years, his affection, his solicitude even, for the young, to whom he had always been an unpompous and conversable elder brother (as he had been to Keats), and his readiness to do more than was asked of him to help. It is to Rebecca Reynell, elder sister (aged about thirty) of Catherine, and was written in 1829 when he and his son shared lodgings:

> 3 Bouverie Street, Strand
> Saturday. 22 Aug.[1829]

Dear Rebecca,

According to your wish, I send you five autographs. I have also enclosed Leigh Hunts & Lambs as the Gentleman is a collector. Mr Watts isnt it. What night could you get the Eng. Opera order, for I would ask Black, * & then the young lady, heaven knows her name, can go conveniently, with you, for I have some idea that you wish them principally for her, & that would be better than Vauxhall for her, let John say what he likes. Will you take tea with us tomorrow evening? You can call at 36, & see the pictures in your way, for there's no one there now.

> Your's ever

*I called this morning
but Black was gone.

> W.H.[12]

---

[10] 4 Oct. 1829.   [11] 'Some New Hazlitt Letters' (n. 9 above), 339.

[12] Buffalo, wrongly ascribed to William Hazlitt the younger. The writing is Hazlitt's, and quite unlike his son's, who was then 17. This was probably the letter sent by Rebecca Reynell to W. C. Hazlitt early in 1867, but not published by him (BL Add. MS 38899, fo. 251). 'John' was Rebecca's brother (see Le Gallienne, 354).

In the summer of 1830 he fell seriously ill. His old friends, Martin
Burney, Lamb, Hessey, Procter, and, it appears, Patmore, were
concerned and came to see what they could do. He seems to have been
living alone, in Frith Street, where he had moved in the spring, but his
son was in London, and in attendance. It is strange that we hear nothing
of young William's mother during his father's last illness. After Sarah's
visits to Hazlitt in Down Street, upon his return from Melrose, and in
Paris, we would have expected her apparently unchanged regard for him
to prompt her to go to see him when he lay dying. It may be that she was
discouraged, or prevented, from doing so. There is a story, reported by
Robinson from a conversation with J. P. Collier, J. H. Reynolds, and
Sheridan Knowles (the names guarantee that it would have been
contradicted if there was no truth in it) that he was tormented as he lay
helpless in his bed by the thought of being unable to provide for 'a
worthless woman by whom he was at last fascinated'.[13] The only
biographer to allude to this story is Maclean, who, committed to a
romantic yet respectable image of Hazlitt, accuses Robinson of prejudice and distortion, implying that no woman was mentioned, or that if
there was, she was not worthless.[14] Now Robinson was temperamentally
and professionally incapable of invention. Nor was this the first time that
such a thing had happened to Hazlitt. And it does indeed seem that, in
Hazlitt's Frith Street bedroom, something apart from the pain racking
his wasted frame preyed on him, something he would not disclose.
Hessey declared at the time that his mind was quite as ill at ease as his
body, and his own son, in equal ignorance, thought that whatever it was
that worried him was more dangerous than the disease and 'would kill
him if he did not dispel it'.[15]

An involvement with some such woman in his closing years seems to
be confirmed by the following undated entry in the notebooks of John
Mitford, written after Hazlitt's death. 'Collier mentioned Hazlitt's
attachment to one of the girls of the Theatre. She was passée, not
handsome. He owned that these [an illegible word] etc. but was
infatuated. Made her presents of necklaces etc.'[16] It was evidently
Collier who was also Robinson's informant, and the phrase 'worthless
woman' seems to imply that by 'one of the girls of the Theatre' was
meant not an actress but a frequenter of the lobbies and saloons that had
so fascinated him. Here we are on unstable, conjectural ground, but the
story seems plausible. Patmore never knew him when he was not in love,
or imagining himself to be so (just as Mérimée remembered Stendhal),

[13] Morley, 424,
[14] Maclean, 607.
[15] BL Egerton MS 2248, fo. 263; *Memoirs*, ii.238.
[16] BL Add. MS 32566, fo. 171v.

and there is no area of human behaviour in which patterns of expectations, attitudes, reactions, are more inveterate. It was a bias of his, as he recognized, to seek affection in the disregarded, the oppressed, the outcast even — servants, peasants, prostitutes. It was perhaps a legacy of Rousseau, perhaps it was imprinted in his own origins, born as he was on the margin of established society, or was a way of making amends. But the pattern of his successive amours, with the recurring cycle that we see in *Liber Amoris* of exigencies, concessions, surrenders, humiliation, and self-abasement, would be unlikely to cease when Isabella left him. As the years passed, the will to break the mould of behaviour would have become weaker. The habits acquired in his earlier infatuations would determine the form and course of his later. Once such a pattern is established — and in his case it had had thirty years to become entrenched — it does not thereafter greatly vary.

At the last it seems as if his hopes of private happiness on the one hand, and of human liberty and dignity on the other, had both finally succumbed. The first had shrunk to a fascination with a disregarded 'girl of the theatre'. The second, which had occupied his mind and heart for just as many years as the first, in fact for the whole of his life, had been worn down by many successive discouragements, and when young Reynell brought him the news of the July Revolution and the overthrow of the Bourbons, his gratification was cancelled out by his scepticism. 'Ah, I am afraid, Charles', he said, 'that things will go back again.'

Hessey, Edward White (Lamb's colleague), and his son were with him when he died, on Saturday, 18 September. In the following days, however, his dead body, unattended, neglected, as he had been in life, was not even allowed to lie in dignified peace, but was thrust aside, hidden under a piece of furniture, by his landlady, so that she might not lose the chance of showing the room to a prospective new lodger.[17]

Doubt has been cast on the authenticity of his last words, 'Well, I've had a happy life'.[18] Yet it was what he would have said. He *had* had a happy life, and the proof is to be found everywhere in the essays. No one was more aware of his 'glassy essence', or relished more keenly the passing moments of his existence. And no one was more certain that happiness is not separable from sadness, that pain is to be lived no less than joy, and that there is bound to be fear as well as hope.

[17] R. L. Hine, *Charles Lamb and his Hertfordshire* (1949), 66, quoting the contemporary diary of one William Lucas.

[18] *Memoirs*, ii. 238; Maclean, 608.

Epilogue

SARAH HAZLITT spent the last decade of her life in Palace Street, a stone's throw from her old house in York Street, and died there on 2 November 1840.[1] Her old friend Mary Lamb was still alive, in her seventy-sixth year, but in Edmonton, at too great a distance for visits. However, Sarah's endearing gift of fidelity to early attachments is attested by her death certificate, which names as witness Elizabeth Pinney, with whom, according to her grandson, she had then been living, and with whose family — those early hosts of her brother — she had stayed in Wimpole Street thirty-seven years before.[2]

After Isabella and Hazlitt parted we hear no more of her until she crosses the Channel to Calais, alone, in 1836.[3] She may have lost her sister, Mary, six years before, if this was the 'Mary Downie, widow of John Downie' who died 'aged approximately 45' at Boulogne in 1830.[4] She herself died at Tranent, Haddingtonshire on 13 September 1869.[5]

Sarah Walker, who was living in Gloucester Street with John Tomkins in 1841, is next seen, indirectly, on 25 August 1849, when her son Frederick, a solicitor's clerk, married Caroline Jane Scarborough, a messenger's daughter, without, it seems, the approval of his father, who is nevertheless described on the certificate, as a 'solicitor'. Frederick died on 6 October 1852.[6] A few years later, his widow married a 'merchant's

---

[1] Her death certificate gives 'aged 67', but if her elder brother was b. 6 Feb. 1773 this is unlikely to be right.

[2] *Memoirs*, i, p. xxi (W. C. Hazlitt misspells the name as 'Penny').

[3] Archives nationales, Paris, Enregistrement de Passeports, F7*2568, at 22 Sept. 1835. (This document should serve as a caution to biographers: one of Isabella's fellow-passengers was a 'Sara Walker, dame anglaise'.)

[4] Archives communales, Hôtel de Ville, Boulogne, 'décès enregistré le 2 Mars 1830, Mme Mary Downie, veuve de John Downie, âge approximatif 45 ans'. We may also note that 'Mary Downie, Anglaise' landed at Calais on 16 Nov. 1829 (Enregistrement de Passeports F7*2564), along with a 'Jibora Merle' whom I strongly suspect to have been Isabella: the spelling of English names was never the strong suit of the French.

[5] 'Isabella Hazlitts formerly Bridgewater, widow of William Hazlitts, Gentleman. Died 1869, September 13th, 7.30 am., at Tranent (Haddingtonshire) age 70. Father, James Shaw, Gentleman, deceased; Mother, Jane Shaw, deceased, maiden surname McKenzie. Cause of death: old age. No immediate medical attendant. Marion Scott, X her mark, nurse, present. William Hamilton, Registrar, witness.' Her practicality seems to have declined in later life: when she died there was £3 in the house and £82 in the bank; it was left to a surviving sister to reclaim £1,350 owed her by the Bridgwater estate (Record of Commissary Court, Haddington, vols 18, fos. 94, 143; and 19, fo. 239).

[6] His age was given as 26 on marriage, 28 at death. If the first is right he was b. between 26 Aug. 1822 and 25 Aug. 1823; if the second, between 7 Oct. 1823 and 5 Oct. 1824. In view of the 1823 'diary', the year 1824 seems the more likely. See Ch. 14, n. 59 above.

clerk', a widower, Henry Eastwood, but there is no sign here of Sarah, her first mother-in-law: her witnesses were her father and mother, as in the earlier marriage.[7] But Sarah and her daughter-in-law must have been fond of each other: after Sarah's common-law husband John Tomkins died in 1858, she went to live around the corner from the Eastwoods, at Newington. And there, at 65 Penton Place, with Caroline at her bedside, 'Sarah Tomkins, widow of John Tomkins, a solicitor', died of 'old age', at seventy-eight, on 7 September 1878, eleven years after the publication of the *Memoirs* spelling out in detail, for all the world to see, her abortive love-affair of half a century earlier with William Hazlitt, author.

[7] Henry Horatio Neville Eastwood and Caroline Jane Tomkins were married at Holy Trinity Church, Islington, 8 July 1855.

Index

H = William Hazlitt